Graphene Materials Handbook

Graphene Materials Handbook

Edited by
Chris Patterson

STATES
ACADEMIC PRESS
www.statesacademicpress.com

Published by States Academic Press,
109 South 5th Street,
Brooklyn, NY 11249, USA

ISBN: 978-1-63989-246-4

Cataloging-in-Publication Data

Graphene materials handbook / edited by Chris Patterson.
p. cm.
Includes bibliographical references and index.
ISBN 978-1-63989-246-4
1. Graphene. 2. Nanostructured materials. 3. Materials science. I. Patterson, Chris.
TA455.G65 G73 2022
620.115--dc23

For information on all States Academic Press publications
visit our website at www.statesacademicpress.com

STATES
ACADEMIC PRESS

Contents

Preface

Graphene is an allotropic form of carbon which has a single layer of carbon atoms arranged in a two-dimensional honeycomb lattice. Each carbon atom in a graphene sheet is connected to three other atoms by a sigma bond. It exhibits unusual electronic properties and is considered a semimetal. While the two dimensional structure is termed as graphene, the three dimensional structure is referred to as graphite. It is a good conductor of heat and absorbs light of all visible wavelengths. It can be produced through various techniques such as fullerene splitting, nanotube slicing, epitaxy, sodium ethoxide pyrolysis, etc. Graphene finds extensive application in various devices such as solar cells, touch panels, light emitting diodes and sporting equipment. This book includes some of the vital pieces of work being conducted across the world, on various topics related to graphene materials. It includes contributions of experts and scientists which will provide innovative insights into this field.

This book has been the outcome of endless efforts put in by authors and researchers on various issues and topics within the field. The book is a comprehensive collection of significant researches that are addressed in a variety of chapters. It will surely enhance the knowledge of the field among readers across the globe.

It gives us an immense pleasure to thank our researchers and authors for their efforts to submit their piece of writing before the deadlines. Finally in the end, I would like to thank my family and colleagues who have been a great source of inspiration and support.

Editor

1

Reduced Graphene Oxide–Based Microsupercapacitors

Zhi Jiang, Yang Wang and Cheng Yang

Abstract

Recent development in miniaturized electronic devices has been boosting the demand for power sources that are sufficiently thin, flexible/bendable, and even tailorable and can potentially be integrated in a package with other electronic components. Reduced graphene oxide can be a promising electrode material for miniaturized microsupercapacitors due to their excellent electrical conductivity, high surface-to-volume ratio, outstanding intrinsic electrochemical double-layer capacitance, and facile production in large scale and low cost. Therefore, the routes to produce high-quality reduced graphene oxide as electrode materials, along with the typical fabrication techniques for miniaturized electrodes, are deliberately discussed in this chapter. Furthermore, breakthroughs in the area of the advanced packaging technology, deciding the electrochemical performance and stability of these miniaturized microsupercapacitors, are highlighted. Lastly, a summary of the overall electrochemical properties and current development of the reported devices is presented progressively to provide insights into the development of novel miniaturized energy storage technologies.

Keywords: reduced graphene oxide, miniaturized supercapacitor, flexible/bendable supercapacitor, laser, packaging technique

1. Introduction

The recent boom in portable electronic devices has raised great demand for miniaturized, flexible/bendable, and even tailorable energy storage devices, which can be directly integrated with a series of electronic devices, to provide unprecedented high-performance functionalities [1, 2]. Currently, the thickness of commercial miniaturized supercapacitors is still faced with great challenges to be further shrunk with the size to the level of surface mountable resistors and ceramic capacitors, which has hindered their application in integrated electronic devices. Microsupercapacitors (MSCs) with a thickness of 10–100 μm can fulfill the above-mentioned requirements because of their excellent rate capability, high power density, and long lifetime [3–5]. Also, ultrathin MSCs with polymer substrates showed great flexibility, which can make full use of interstitial spaces among components in an integrated system [2, 6].

This chapter reviews the recent progress of fabrication strategies for reduced graphene oxide (r-GO) and exemplifies some very promising r-GO–based MSC technologies; besides, it shows the breakthroughs in packaging techniques for these novel devices. It will provide insights to the development of novel miniaturized energy storage technologies.

2. Typical fabrication methods of high-quality r-GO as the electrode material

Graphite oxide consists of layered graphene oxide (GO) nanosheets that contain many oxygen functional groups, such as hydroxyl, epoxide, carbonyl, and carboxyl groups, which promote the formation of few-layer GO nanosheets [7]. However, these functional groups also destroy the planar sp^2 carbons of graphite by converting them into sp^3 carbons and alter van der Waals interactions between these layers. The structural changes bring about two problems of GO nanosheets for supercapacitors: one is that the electrical conductivity of GO is much lower than that of graphene and the other is the strong hydrophilicity of GO nanosheets. Therefore, GO cannot be used as an electrode material for supercapacitors without further treatments. Notably, the electrical conductivity of r-GO can be increased close to the level of graphene, and the specific capacitance of r-GO has been found to be significantly higher than that of GO. Herein, the reduction routines of GO to r-GO are systematically introduced.

2.1. Chemical reduction of GO

The chemical reduction of GO route involves reducing agents to eliminate most of oxygen-containing functional groups of GO nanosheets and partially restore the π-electron conjugation within the aromatic system of graphite [8]. The reducing agents can be categorized into toxic ones and environment-friendly ones. Though toxic reducing agents, such as hydrazine, hydrazine monohydrate, dimethylhydrazine, sodium borohydride, hydroquinone, and hydroxylamine, have been shown to efficiently reduce GO into r-GO [9–13], they are not suitable for the future large-scale fabrication of electrode materials due to their damages to the environment. Therefore, environmental friendly reducing agents, such as amino acids and ammonia, have attracted more attention. Pham et al. developed a green and facile approach to produce r-GO by using an environmental friendly reagent, namely L-glutathione as a reducing agent, under mild condition in aqueous solution [14]. The L-glutathione-r-GO could be dispersed in both distilled water and some polar aprotic solvents, including Dimethylformamide (DMF), Dimethyl sulfoxide (DMSO), and Tetrahydrofuran (THF) with simple ultrasonication, because the oxidized product of L-glutathione could play an important role as a capping agent in the stabilization of r-GO. As this chemical reduction process of GO could produce r-GO in a bulk quantity in a relatively short time and mild experimental conditions without any environmentally hazardous chemicals, it offers excellent flexibility for various potential applications not only in electronic devices but also in biocompatible materials. Chen et al. prepared r-GO through the reduction of GO employing L-cysteine as the reductant under a mild reaction condition [15]. During this chemical reduction process, the thiol group of L-cysteine released protons, commonly owning high-binding affinity to the oxygen-containing groups of GO, such as hydroxyl and epoxide groups, to form water molecules, which was similar to L-glutathione. Furthermore, the electrical conductivity of r-GO produced in this way increased by about 10^6 times in comparison to that of GO.

Jiang et al. reported an ammonia-assisted hydrothermal method to obtain N-doped r-GO by simultaneous N-doping and reduction of GO [16]. The effects of hydrothermal temperature on the surface chemistry and the structure of N-doped rGO were investigated. Their results showed that N doping was accompanied by the reduction of GO with a decrease in oxygen levels from 34.8% in GO down to 8.5% in N-doped r-GO. Meanwhile, electrochemical measurements demonstrated that the N-doped product showed a higher capacitive performance than that of pure graphene, and a maximum specific capacitance of 144.6 F/g could be obtained by the pseudocapacitive effect from the N-doping. Li et al. obtained bulk quantities of N-doped r-GO through thermal annealing of GO in ammonia [17]. X-ray photoelectron spectroscopy (XPS) study of GO annealed at various reaction temperatures revealed that N-doping occurred at a temperature as low as 300°C while the highest doping level of about 5% N was achieved at 500°C. Oxygen groups in GO were found responsible for reactions with NH_3 and C-N bond formation, and r-GO obtained by the annealing of GO in NH_3 exhibited higher electrical conductivity than that of r-GO by the annealing GO in H_2.

R-GO obtained by the chemical reduction method showed good electrical conductivity, high specific capacitance value, and excellent dispersibility in different organic solvents. However, the totally removal of oxygen functionalities from the surface of GO nanosheets is impossible by using chemical reducing agents. Therefore, other reduction methods of GO for the large-scale production of high-quality r-GO have been proposed.

2.2. Thermal reduction of GO

Thermal reduction of GO is one of the most well-known methods for the removal of oxygen-containing functional groups [18]. In this type of reduction process, most of oxygen functionalities were transformed into water, carbon dioxide, and carbon monoxide. Usually, the GO sample is placed in a specific furnace filled with inert gas environment and then heated to a relatively high temperature. During the reduction process, reductive gases are generated to reduce GO and to expand their layers at the same time.

Schniepp et al. obtained r-GO by annealing GO nanosheets in a closed quartz tube at 1050°C under the protection of argon atmosphere for 10 min, and the carbon content in the final r-GO product increased with the rise of the annealing temperature [19]. They fabricated polymer nanocomposites with stacks of these r-GO nanosheets as the filler, and the as-fabricated polymer showed outstanding thermal, mechanical, and electrical properties. Such functionalized r-GO nanosheets can provide even further benefits, not only for structural property enhancements at lower filler contents, but also for the development of new materials such as electrically conducting polymers, conductive inks, and supercapacitors. Vallés et al. reported a considerable recovery of the sp^2 network structure of r-GO by a relatively simple thermal treatment at 700°C under argon or hydrogen atmosphere [20]. The as-fabricated r-GO paper exhibited an excellent electrical conductivity as high as 8100 S/m, which was five orders of magnitude higher than that of initial GO paper, and also much higher than that of r-GO paper by a chemical reduction method. Moreover, this gentle and direct thermal reduction method allowed r-GO paper to remain the good structural integrity and mechanical flexibility of the initial GO paper, overcoming problems of brittleness typically encountered in the annealing process. Mattevi et al. reported that the effect of annealing GO at 700°C or above was equivalent to the chemical reduction of GO via hydrazine monohydrate at 80°C followed by heating at 200°C [21]. For fully reduced GO nanosheets, they found the oxygen content was about 8% with the ratio of C to O equaling to 12.5:1 and the sp^2 concentration being 80%, and they demonstrated that the presence of residual oxygen with only 8% content significantly hampered the carrier transport among the graphitic domains.

The thermal reduction of GO is a direct, cost-effective, and environmentally friendly route for the fabrication of high-quality r-GO. However, it usually needs a high temperature to obtain r-GO with a high electrical conductivity, and the experimental set-up is very complicated, all of which have hindered its further application.

2.3. Electrochemical reduction of GO

The routine of electrochemical reduction of GO nanosheets was newly proposed for the fabrication of high-quality r-GO whose defect concentration is almost negligible. Electrochemically reduced GO (Er-GO) nanosheets are contaminant-free and of high electrical conductivity, which makes them suitable for applications in the areas of supercapacitor devices, touch screens, and flexible electronics.

Li et al. firstly reported a fiber-shaped solid electrochemical capacitor based on Er-GO [22]. The typical fiber-shaped electrode can be prepared by electrochemically reducing GO on an Au wire with a diameter of 100 mm in 3-mg/mL GO aqueous suspension containing 0.1-M lithium perchlorate ($LiClO_4$) at an applied potential of about 1.2 V (vs. saturated calomel electrode, SCE) for 10 s, and it exhibited high specific capacitance and rate capability, long cycling life, and also attractive mechanical flexibility. Yang et al. demonstrated the application of Er-GO as supercapacitor electrodes by reducing GO in a convenient, cost-effective, and eco-friendly electrochemical way [23]. During the electrochemical reduction process, p-conjugated structure in graphene was almost completely repaired by removing electrochemically unstable oxygen groups, yielding improved electron transfer kinetics and capacitive properties of Er-GO. At the same time, residual oxygen groups on Er-GO nanosheets were very stable and reversible in capacitive performance providing additional faradic pseudocapacitance without damaging their electronic properties. As a result, the Er-GO exhibited a specific capacitance of as high as 223.6 F/g at 5 mV/s, showing a promising potential as an electrode material for supercapacitors.

Chen et al. reported a general method for the fabrication of three-dimensional (3D) porous Er-GO–based composite materials [24]. First, 3D porous Er-GO was prepared electrochemically by reducing a concentrated GO aqueous solution containing 0.15-M $LiClO_4$ as shown in **Figure 1a**. The cyclic voltammogram of a copper foil electrode in the GO and $LiClO_4$ electrolyte at the scan rate of 50 mV/s was shown in **Figure 1b**, and the redox peak appeared at about −1.15 V. The X-Ray diffraction (XRD) pattern of Er-GO film obtained by reducing 3-mg/mL GO solution at a potential of 1.15 V for 300 s, confirmed that most of electrochemically unstable oxygen groups were removed from initial GO film. The as-fabricated Er-GO–based paper electrode, prepared by electrochemically reducing 3-mg/mL GO at a potential of 1.15 V for 600 s, was shown in **Figure 1d**, and the Er-GO–based paper electrode exhibited a 3D porous structure, which was good for ion and electron transportation during the charge/discharge process. Furthermore, by the growth of polyaniline layer on 3D Er-GO paper electrodes, a high specific capacitance of 716 F/g at 0.47 A/g was achieved. Sheng et al. reported an electrochemical double-layer capacitor based on 3D interpenetrating r-GO electrodes fabricated by the electrochemical reduction of GO [25]. At 120 Hz, the Er-GO–based MSCs exhibited a phase angle of −84°, a specific capacitance of 283 mF/cm^2, and a resistor-capacitor (RC) time constant of 1.35 ms, making it capable of replacing activated carbon-based MSCs for AC-line filtering.

The electrochemical reduction of GO process is very fast and avoids the use of any harmful chemical reducing agents, which means that there is no need for further purification of Er-GO [26].

2.4. Laser reduction of GO

Laser reduction of GO routine uses laser as the major energy source to obtain porous r-GO nanosheets without the need for any additional chemical agents; and it leaves abundant defects in r-GO nanosheets, which can highly improve the electrochemical characteristic of r-GO [27–29]. Except from the reduction function, laser with a higher energy level can selectively remove GO/r-GO nanosheets without any damage to substrates due to the instant heating and cooling feature, which can make it possible for graphical electrodes [27, 30].

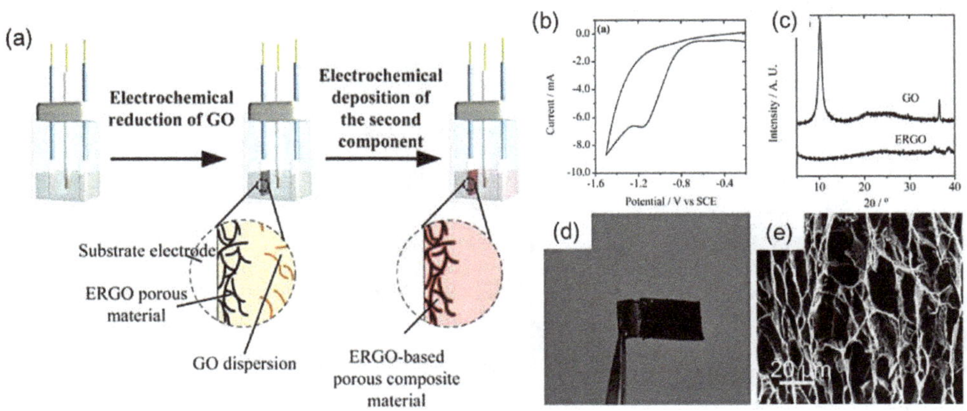

Figure 1. (a) Schematic illustration of the preparation method for Er-GO–based composite materials, (b) cyclic voltammogram of a copper foil electrode in 3-mg/mL GO aqueous solution containing 0.15-M $LiClO_4$ at a scanning rate of 50 mV/s, (c) XRD patterns of GO film and as-prepared wet Er-GO porous material by reducing 3-mg/mL GO solution at a potential of 1.15 V for 300 s, (d) optical image of Er-GO paper electrode prepared by electrochemically reducing 3 mg/mL GO at a potential of 1.15 V for 600 s. (e) SEM image of lyophilized Er-GO paper (Reproduced with permission from Ref. [24]).

Abdelsayed et al. reported a facile laser reduction method for the preparation of r-GO from GO [31]. They obtained few-layer r-GO nanosheets in water under ambient conditions without the use of any chemical reducing agent indicated by changes in the color of GO solutions, as shown in **Figure 2a**. Furthermore, XRD results of the products after different irradiation time confirmed that only 10-min irradiation of the 532-nm laser could totally reduce GO into r-GO. By this mild reduction treatment in water, the few-layer character of GO nanosheets was perfectly preserved in the final products with a thickness of 0.99 nm as shown in **Figure 2b** and **c**.

Sokolov et al. obtained r-GO via continuous-wave (532 nm) and pulsed (532 and 355 nm) laser irradiation of GO [32]. They investigated the graphene degree of r-GO with different laser power and background gas and fabricated microsized defects in r-GO film using laser with a higher energy, indicating laser could be an effective technique for reducing r-GO with high spatial resolution and patterning feature. Many attentions have been paid to the mechanisms of laser reduction and ablation of GO, and most of them focused on the interaction among electron, photo, exciton, and ions [32–34]. When GO nanosheets are exposed to a laser beam, deoxygenation reaction is triggered instantaneously via both photochemical and photothermal reduction pathways. Though electron-lattice temperature equilibration occurs on the picosecond time scale, strain associated with the oxygen in GO nanosheets can lead to exciton self-trapping, hole localization, and subsequent material removal, as indicated by the release of CO, CO_2,

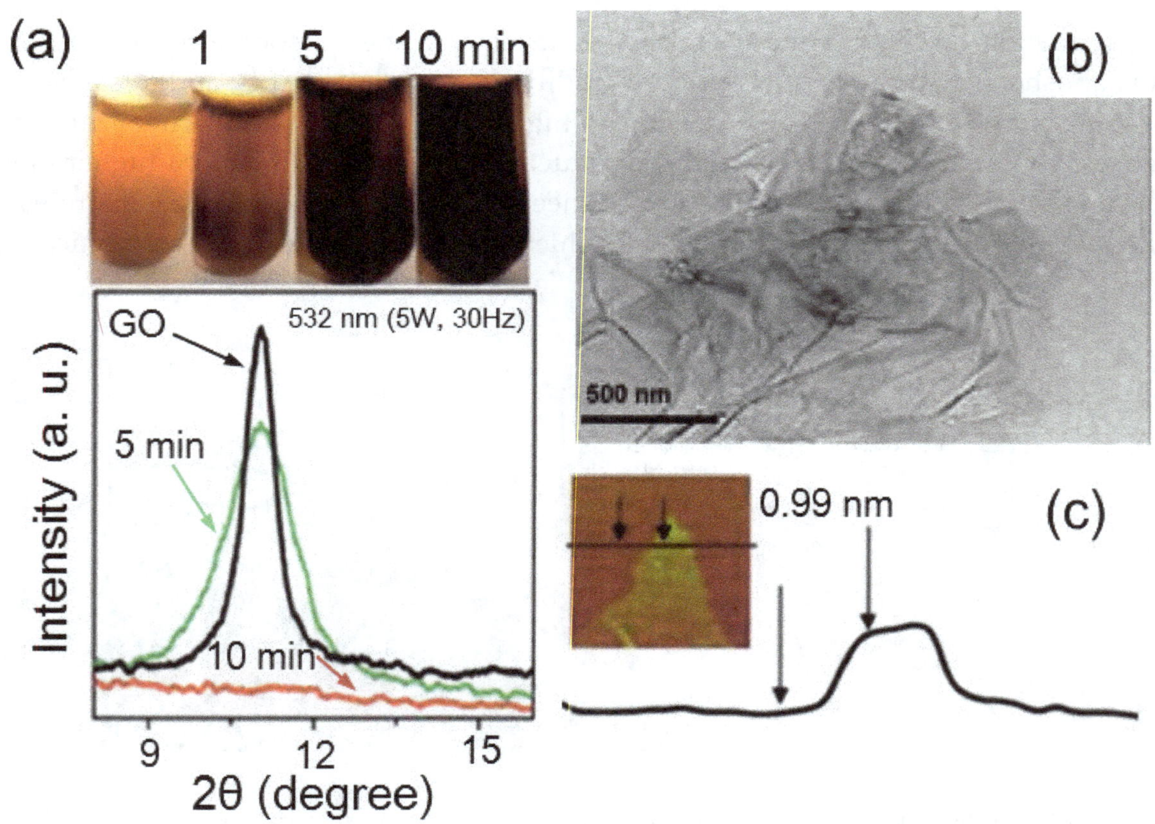

Figure 2. (a) Digital photographs displaying the changes in the color of the original GO solution (0 min) after irradiated with the 532-nm laser for 1, 5, and 10 min and the corresponding XRD results, (b) TEM images of r-GO nanosheets prepared by the 532-nm laser irradiation to the aqueous GO solution for 10 min, (c) AFM image and cross-sectional analysis of the r-GO nanosheets (Reproduced with permission from Ref. [31]).

and H_2O from GO nanosheets by laser irradiation [33]. During the ablation process of GO/r-GO nanosheets by laser, the release of high mass carbon cluster ions had been observed during photon- and low-energy electron-stimulated desorption, which can be deposited on the surface of r-GO nanosheets and served as seeds for growth of larger graphene particles or nanosheets [31].

Gao et al. demonstrated the scalable fabrication of an all-carbon monolithic supercapacitor by laser reduction and patterning of GO films [35]. They patterned both in-plane and sandwiched structured electrodes consisting of r-GO nanosheets with micrometer resolution, between which GO served as a solid-state electrolyte. The substantial amounts of trapped water in GO nanosheets made them simultaneously an excellent ionic conductor and an electrical insulator, allowing them to serve as both an electrolyte and an electrode separator with the ionic transport characteristics. The resulting microsupercapacitor devices showed an excellent cyclic stability and energy storage capacities as compared to the existing thin-film supercapacitors. El-Kady et al. used a standard LightScribe DVD optical drive to directly reduce GO films into r-GO by laser, and the as-fabricated r-GO–based films were mechanically robust and showed high electrical conductivity (1738 S/cm) and specific surface area (1520 m^2/g) [36]. The r-GO–based supercapacitor exhibited a high energy density of as high as 1.36 mWh/cm^3, which was approximately two times higher than that of the commercial activated carbon-based electrochemical capacitors (AC-EC). Moreover, a power density of about 20 W/cm^3 can be achieved, which was 20 times higher than that of the AC-EC and three orders of magnitude higher than that of the 500-mAh thin-film lithium battery.

Though laser reduction of GO cannot achieve mass production of r-GO, combined with film technology, r-GO fabricated via laser-reducing GO nanosheets is a promising electrode material for flexible, ultrathin, and binder-free microsupercapacitors and other electronic devices, such as sensors, displays, and batteries.

3. Recent progress of r-GO–based miniaturized microsupercapacitors

Rapid developments in semiconductor devices, low-power integrated circuits, and packaging technology are accelerating the reduction in the size and volume of current electronic devices, especially portable mobile communication equipment, implantable medical devices, and active radio frequency identification (RFID) tags. It makes rechargeable high-performance miniaturized energy storage devices become an urgent need [37]. R-GO–based miniaturized MSCs exhibit excellent rate capability, high power density, long lifetime, and good 3D micro-integration capability with other components, showing a promising perspective for the use as miniaturized energy storage devices. Here, some of recent representative works in the area of r-GO–based miniaturized MSCs are progressively introduced.

El-Kady et al. demonstrated scalable fabrication of high-power graphene microsupercapacitors for on-chip energy storage by direct laser writing on graphite oxide films using a standard LightScribe DVD burner [38]. **Figure 3a** shows the features of r-GO–based MSCs on a chip, and **Figure 3b** shows the morphological change of GO films after direct laser irradiation. The as-fabricated r-GO–based MSCs showed an excellent electrochemical performance and good flexibility on a flexible polymer substrate. Furthermore, the adaptability of r-GO–based MSCs for serial/parallel combinations was demonstrated by connecting four devices together both in series and in parallel configurations, as shown in **Figure 3c** and **d**. The tandem MSCs exhibited a very good control over the operating voltage window and capacity, thus enabling them to be considered for practical applications. **Figure 3e** shows a Ragone plot comparing the performance characteristics of r-GO–based MSCs with different energy-storage devices designed for high power microelectronics. Remarkably, compared with activated carbon-based MSCs, r-GO–based MSCs exhibited three times higher energy and about 200 times higher power density. Furthermore, r-GO–based MSCs demonstrated power densities comparable to those of the aluminum electrolytic capacitor while providing more than three orders of magnitude higher energy density. This superior energy and power performances of the r-GO–based MSCs should enable them to compete with microbatteries and electrolytic capacitors in a variety of applications.

Xie et al. developed device-level tailorable gel-based MSCs with symmetric electrodes prepared by electrochemically reducing and depositing GO on a nickel nanocone array (NCA) current collector using a unique packaging method [39]. **Figure 4a** shows the fabrication process of the tailorable Er-GO–based MSCs, which connected in series can be used to drive small gadgets, such as a light-emitting diode (LED) and a mini-motor propeller. **Figure 4b** and **c** displays the morphology and structure of the Er-GO electrode, which exhibited a high areal specific capacitance (57.1 mF/cm^2) and an outstanding cycle stability (20,000 cycles with only 3.5% capacitance loss). Compared with conventional chemical reduction, the porous Er-GO layer can be homogeneously deposited on a conductive substrate and be used as the electrode material without any binder; besides, the production process of Er-GO can be conveniently scaled up, all of which made the electrochemical reduction routine more promising. CV curves of the MSC under different scan rates from 50 to 5000 mV/s in **Figure 4d** demonstrated the excellent rate capability of Er-GO–based MSCs, which can be attributed to the excellent electrical conductivity and porous structure of Er-GO itself. Furthermore, Er-GO–based MSCs exhibited excellent mechanical reliability, indicated by the little capacitance loss under pressure or bending, showing great potential in flexible, bendable, wearable, and even tailorable electronics.

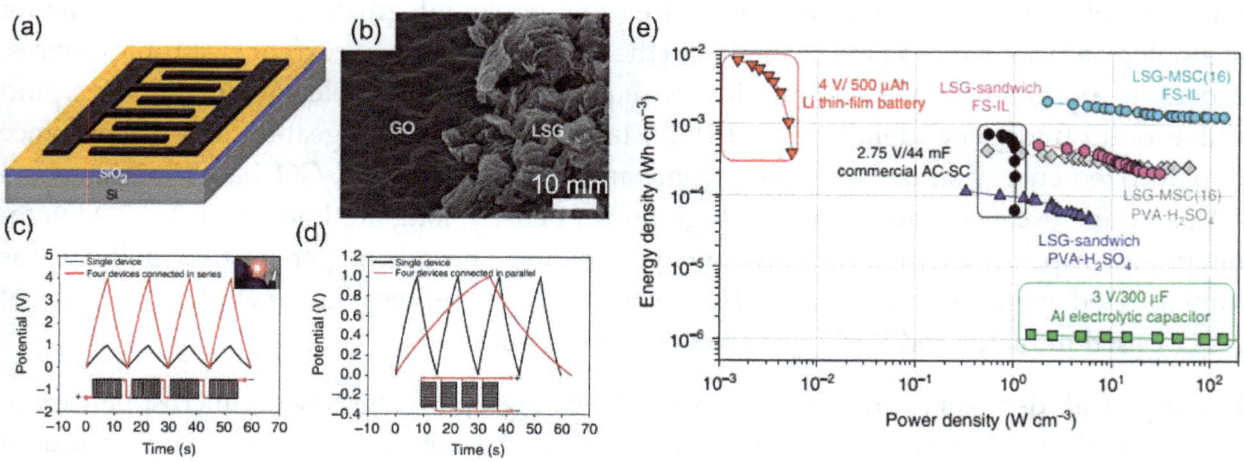

Figure 3. (a) Schematic illustration of r-GO–based MSCs on a chip, (b) SEM image showing the direct reduction and expansion of the GO film after exposure to the laser beam (scale bar is 10 mm), galvanostatic charge/discharge curves for four tandem microsupercapacitors connected (c) in series, and (d) in parallel, (e) energy and power densities of r-GO–based MSCs compared with commercially available energy-storage systems (Reproduced with permission from Ref. [38]).

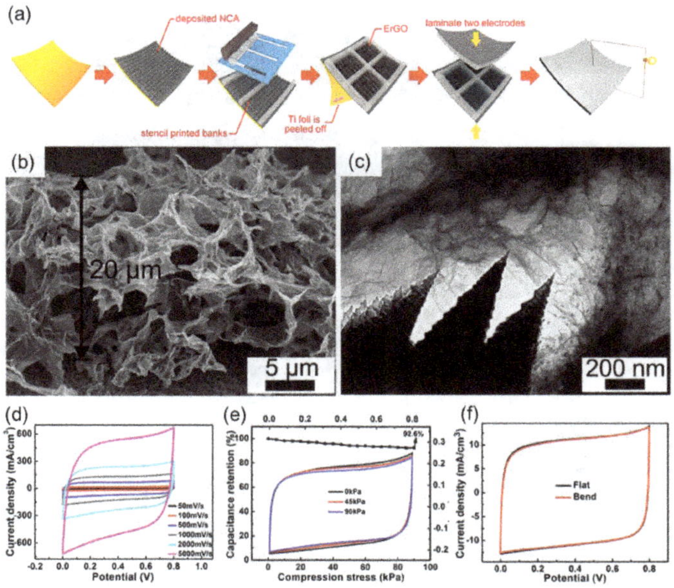

Figure 4. (a) Schematic illustration of the shape-tailorable r-GO–based ultra-high-rate supercapacitor (Regular nickel nanocone arrays (NCAs) deposited on a Ti substrate were used as current collector, and perpendicularly stencil-printed EVA glue was used as cofferdams, and Er-GO electrochemically deposited onto the NCAs was electrode material, and Na_2SO_4/PVA gel was used as electrolyte), (b, c) SEM and TEM images of NCAs and Er-GO, (d) CV curves of the MSC under different scan rates from 50 to 5000 mV/s, (e) capacitance retention performance of the MSC under different compression strength, (f) CV curves of the MSC tested at 100 mV/s with and without bending (curvature radius: 1.7 cm) (Reproduced with permission from Ref. [39]).

Xie et al. in the same group also carried out a benchmark study of the state-of-the-art well-packaged ultrathin microsupercapacitor compared with commercial microsupercapacitor and aluminum electrolyte capacitor [40]. **Figure 5a** shows the fabrication process of the ultrathin and flexible MSCs using laser to complete the reduction and ablation of GO film to obtain patterned r-GO–based electrodes. **Figure 5b** gives the optical image of as-fabricated r-GO–based MSCs arrays, and **Figure 5c** demonstrates the porous structure of laser-processed r-GO film with a thickness of about 4 μm, providing more surface area and being beneficial to a high mobility of ions, both of which helped deliver high energy and power densities. The as-fabricated r-GO–based MSCs demonstrated great flexibility and bendability indicated by the almost same CV curves under different bending angles around a cylinder with a diameter of about 3 cm in **Figure 5d**. The Ragone plot in **Figure 5e** suggests that the energy density of this r-GO–based MSC with LiCl-PVA gel electrolyte achieved 0.98 mWh/cm^3 at the power density of 300 mW/cm^3, which was almost three orders of magnitude higher than that of a commercial activated carbon supercapacitor. With the power density raised from 300 mW/cm^3 to 2000 mW/cm^3, the energy density of this r-GO–based MSC had no significant degradation in capacitance, showing excellent rate performance. Furthermore, using the ionic liquid electrolyte, r-GO–based MSCs with a thickness of 18 μm can be fabricated with a maximum energy density up to 5.7 mWh/cm^3 at 830 mW/cm^3, which can be comparable with lithium thin-film battery but with more than two orders of magnitude higher power density. However, their energy density was still not large enough for high-performance MSCs compared with those of Refs. [1, 38, 41]. The ultrathin thickness and foldability of as fabricated r-GO–based MSCs made it possible for them to be folded to fit in restricted spaces integrated with other

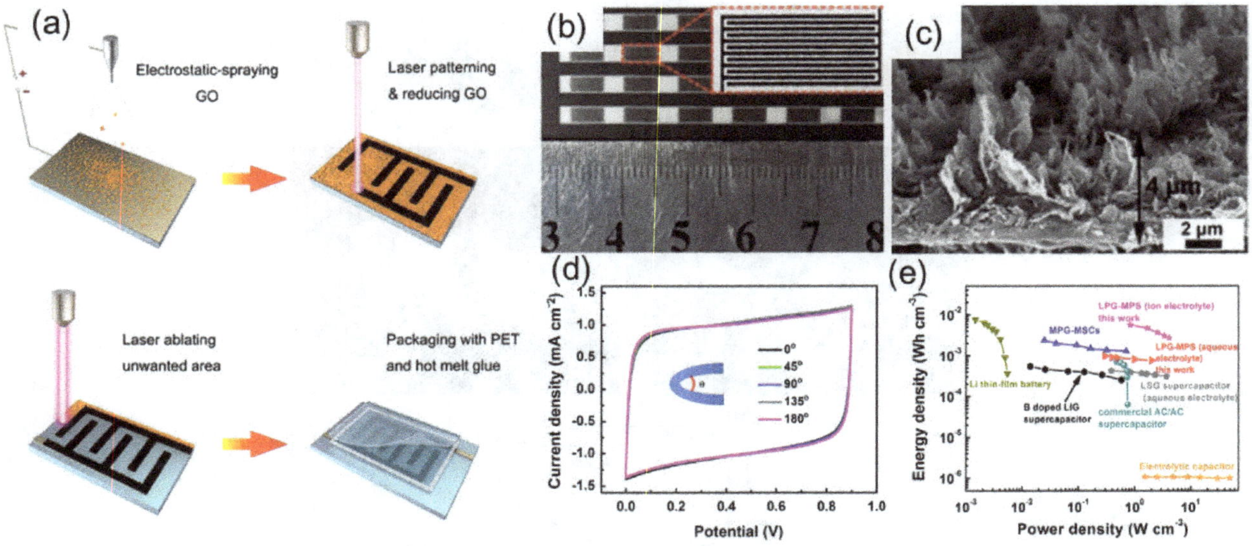

Figure 5. (a) Schematic illustration of ultrathin laser-processed r-GO–based MSCs (the PET substrate with a thickness of 6 μm sputtered with Ni layer (500 nm) was homogenously deposited with GO nanosheets by electrostatic spray deposition technique, and the reduction and ablation of GO film were completed by different power of laser treatment. Then, electrolyte was added and a thin PET film printed with cross-linkable PU cofferdams was used as the cover layer.), (b) photographic image of r-GO–based MSCs arrays (Inset is a single unit), (c) cross-view SEM image of laser processed r-GO with a height of about 4 μm, (d) CV curves of r-GO–based MSCs observed in different bending state, (e) Ragone plot of r-GO–based MSCs and some reported supercapacitors (Reproduced with permission from Ref. [40]).

components. Similarly, Meng et al. reported an integrated, flexible, and all-solid-state MSC technology with 3D microintegration capability [37]. They firstly embedded the entire electrical routing path and contact pads within the effective device area, utilizing the novel fabrication of through-via bottom electrodes. The entire device was successfully packaged with a solid-state electrolyte at the microscale (720 μm × 720 μm × 50 μm) and has surface-mounted integration capability.

Wu et al. demonstrated r-GO–based in-plane interdigital MSCs with high power and energy densities on arbitrary substrates [1]. **Figure 6a** shows the fabrication process of the device. The resulting r-GO film had an electrical conductivity of 297 S/cm, and the as-fabricated r-GO–based MSCs showed the features of large-area uniformity, good transparency, and mechanical flexibility. The plot of the discharge current as a function of scan rate in **Figure 6b** demonstrated an enhanced electrochemical performance. A linear increase of discharge current can be observed up to 200 V/s, suggesting the ultrahigh power ability of r-GO–based MSCs. The as-fabricated r-GO–based MSCs can be operated at an ultrahigh rate up to 1000 V/s and showed an excellent capacitance. For example, an areal capacitance of 78.9 mF/cm^2 and a volumetric capacitance of 17.5 F/cm^3 were obtained for the device at 10 mV/s in **Figure 6c**. The Ragone plot in **Figure 5e** suggests that the r-GO–based MSCs delivered a volumetric energy density of 2.5 mWh/cm^3, which was an order of magnitude higher than that of the typical supercapacitors of activated carbon (≤1 mWh/cm^3) and well comparable to that of lithium thin film batteries (10^{-3}–10^{-2} Wh/cm^3). Furthermore, the r-GO–based MSC manifested an ultrahigh power density of 495 W/cm^3, which was 50 times higher than that of conventional

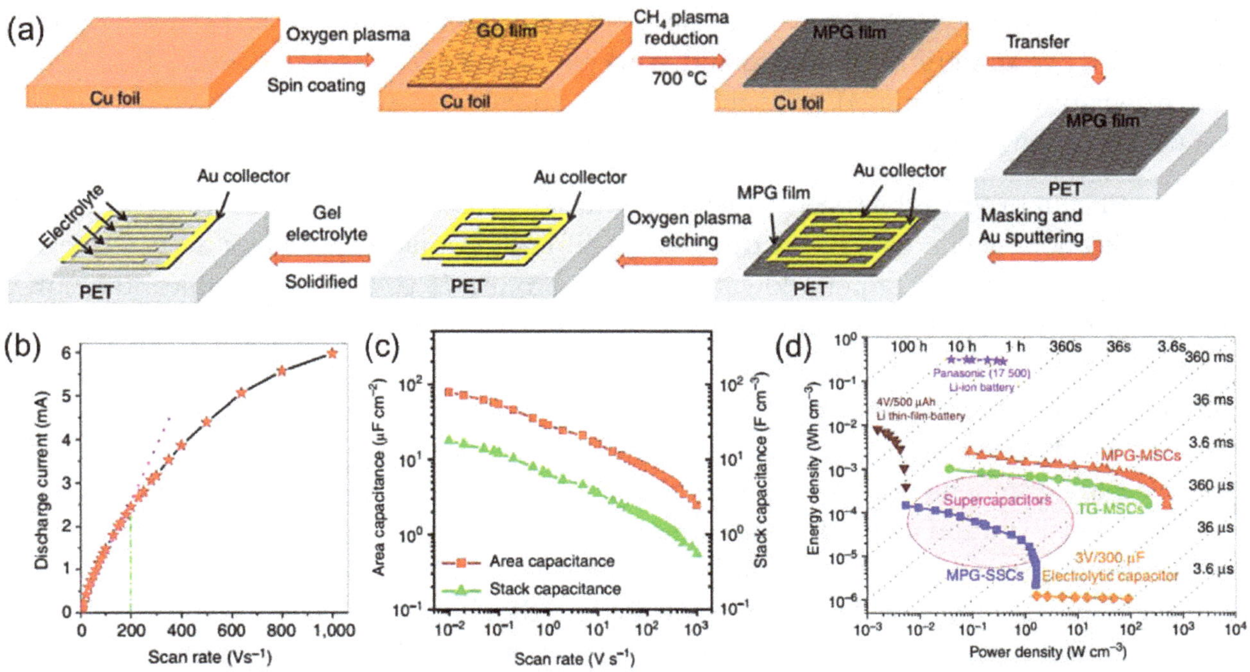

Figure 6. (a) Schematic illustration of the fabrication of flexible r-GO–based flexible MSCs (The fabrication process included a sequence of spin-coating of GO solution on Cu foil, CH_4 plasma reduction of GO film, transfer of r-GO film from the Cu foil to PET substrate, masking pattern and deposition of gold current collector, oxidative etching, drop casting of H_2SO_4/PVA gel electrolyte and solidification of gel electrolyte.), (b) a plot of the discharge current as a function of the scan rate (red star line) (linear dependence (magenta dot line) was observed up to at least 200 V/s (green dash dot line), suggesting the ultrahigh power ability of r-GO–based MSCs.), (c) areal capacitance and stack capacitance of r-GO–based MSCs, (d) the comparison of energy and power density of as fabricated MSGs with commercially applied electrolytic capacitors, lithium thin-film batteries, Panasonic Li-ion battery, and conventional supercapacitors (indicated by the pink region), demonstrating that the as-fabricated MSCs exhibited exceptional electrochemical energy storage with simultaneous ultrahigh energy density and power density (Reproduced with permission from Ref. [1]).

supercapacitors (even at a high energy density of 0.14 mWh/cm^3) when discharged within an extremely short time (1 ms) and much higher than those in Refs. [38–40].

Wu et al. in the same group also described the development of large-area, highly uniform, ultrathin nitrogen and boron co-doped graphene (BNG) films for high-performance MSCs [41]. The BNG film was prepared using a layer-by-layer (LBL) assembly technique of anionic GO nanosheets and cationic poly-L-Lysine (PLL) as a nitrogen-containing precursor, followed by intercalation of H_3BO_3 within the layers and annealing treatment. Remarkably, BNG-MSCs allowed for operation at an ultrahigh scan rate of up to 2000 V/s, which was three orders of magnitude higher than that of conventional supercapacitors and represented superior value for high-power MSCs. The volumetric capacitance of BNG film for MSCs recorded at 10 mV/s was ~488 F/cm^3, which can be attributed to the co-doping with dual heteroatoms. On one hand, the co-doping strategy created new electrochemically active moieties (e.g., B-N-C) with a synergistic effect that provides additional pseudocapacitance contribution, and on the other hand, the new binding environment of neighboring B and N atoms directly incorporated into the graphene lattice was favorable for improving the interface wettability of the electrode with the electrolyte, resulting in a thickened electrochemical double layer. Furthermore, based on the volumetric capacitance and working voltage (1.0 V), the energy density and power density were calculated to be about 16.9 mWh/cm^3 and 910 W/cm^3, which was among best performances of the-state-of-art carbon-based MSCs.

Even significant progress has been achieved in the field of carbon-based MSCs in recent years, their application as energy storage devices is still limited due to their low energy density, which results in a high cost per kilowatt-hour, i.e., around 20,000$/kWh, significantly higher than those of well-established storage technologies (such as batteries) [42]. One direct way to lower their cost is to find suitable electrode materials with high energy storage capability without any damage to their power handling ability, for example, the MSCs based on graphene and MnO_2 hybrid materials demonstrated a two times higher energy density than that of initial pure graphene due to the Faradaic pseudocapacitance effect of MnO_2 [43]. Another way is to use fabrication methods compatible with efficient large-scale production throughout the fabrication process of MSCs, including preparation of electrode materials, fabrication of electrodes, and assembly of electrodes, electrolyte, and/or separator as well as their packaging process, for instance, combining with low-cost printing technology during electrode fabrication process can greatly lower the total cost [44].

4. Conclusions

In summary, this chapter reviews the recent research advances of r-GO as a promising electrode material for miniaturized MSCs. R-GO exhibited many advantages such as excellent rate capability, high power density, long lifetime, facile production in large scale and low costs. However, further enhancement in performance characteristics and functionality of r-GO–based miniaturized MSCs need more attentions before this promising technology may be adopted by industry. Combined with novel film-fabrication, patterning, and packaging techniques, miniaturized r-GO–based MSCs with superior capability to be integrated with other components on chip and system levels can be developed due to their ultrathin thickness, flexibility, bendability, foldability, and even tailorability. The examples given in this chapter also discussed about their merits and limitations, we expect that they may give elicitations for developing advanced techniques for the fabrication of high-performance and multifunction miniaturized r-GO–based MSCs in the near future.

Acknowledgements

The authors acknowledge the financial support from National Nature Science Foundation of China Project Nos. 51578310 & 51607102; China Postdoctoral Science Foundation No. 2016M601017, Guangdong Province Science and Technology Department Project Nos. 2014B090915002 & 2015A030306010, and Nanshan District "Rising Stars" Project No. KC2014JSQN0010A.

Author details

Zhi Jiang, Yang Wang and Cheng Yang*

*Address all correspondence to: yang.cheng@sz.tsinghua.edu.cn

Division of Energy & Environment, Graduate School at Shenzhen, Tsinghua University, Shenzhen City, Guangdong Province, China

References

[1] Wu Z S, Parvez K, Feng X, Mullen K: Graphene-based in-plane micro-supercapacitors with high power and energy densities. Nature Communnications. 2013;4:2487. DOI: 10.1038/ncomms3487

[2] Wu Z S, Feng X, Cheng H M: Recent advances in graphene-based planar micro-supercapacitors for on-chip energy storage. National Science Review. 2013;1:277–292. DOI: 10.1093/nsr/nwt003

[3] Wen F, Hao C, Xiang J, Wang L, Hou H, Su Z, Hu W, Liu Z: Enhanced laser scribed flexible graphene-based micro-supercapacitor performance with reduction of carbon nanotubes diameter. Carbon. 2014;75:236–243. DOI: 10.1016/j.carbon.2014.03.058

[4] Zhu J, Yang D, Yin Z, Yan Q, Zhang H: Graphene and graphene-based materials for energy storage applications. Small. 2014;10:3480–3498. DOI: 10.1002/smll.201303202

[5] Zhang Q, Scrafford K, Li M, Cao Z, Xia Z, Ajayan P M, Wei B: Anomalous capacitive behaviors of graphene oxide based solid-state supercapacitors. Nano Letter. 2014;14:1938–1943. DOI: 10.1021/nl4047784

[6] Kim B C, Hong J-Y, Wallace G G, Park H S: Recent progress in flexible electrochemical capacitors: electrode materials, device configuration, and functions. Advanced Energy Materials. 2015;5:1500959. DOI: 10.1002/aenm.201500959

[7] Kuila T, Mishra A K, Khanra P, Kim N H, Lee J H: Recent advances in the efficient reduction of graphene oxide and its application as energy storage electrode materials. Nanoscale. 2013;5:52–71. DOI: 10.1039/c2nr32703a

[8] Singh R K, Kumar R, Singh D P: Graphene oxide: strategies for synthesis, reduction and frontier applications. RSC Advances. 2016;6:64993–65011. DOI: 10.1039/c6ra07626b

[9] Gao X, Joonkyung J, Shigeru N: Hydrazine and thermal reduction of graphene oxide reaction mechanisms, product structures, and reaction design. Journal of Physics Chemistry C. 2010;114:832–842. DOI: 10.1021/jp909284g

[10] Ren P G, Yan D X, Ji X, Chen T, Li Z M: Temperature dependence of graphene oxide reduced by hydrazine hydrate. Nanotechnology. 2011;22:055705. DOI: 10.1088/0957-4484/22/5/055705

[11] Pei S, Cheng H M: The reduction of graphene oxide. Carbon. 2012;50:3210–3228. DOI: 10.1016/j.carbon.2011.11.010

[12] Zhou X, Zhang J, Wu H, Yang H, Zhang J, Guo S: Reducing graphene oxide via hydroxylamine: a simple and efficient route to graphene. The Journal of Physical Chemistry C. 2011;115:11957–11961. DOI: 10.1021/jp202575j

[13] Stankovich S, Dikin D A, Dommett G H, Kohlhaas K M, Zimney E J, Stach E A, Piner R D, Nguyen S T, Ruoff R S: Graphene-based composite materials. Nature. 2006;442:282–286. DOI: 10.1038/nature04969

[14] Pham T A, Kim J S, Kim J S, Jeong Y T: One-step reduction of graphene oxide with l-glutathione. Colloids and Surfaces A: Physicochemical and Engineering Aspects. 2011;384:543–548. DOI: 10.1016/j.colsurfa.2011.05.019

[15] Chen D, Li L, Guo L: An environment-friendly preparation of reduced graphene oxide nanosheets via amino acid. Nanotechnology. 2011;22:325601. DOI: 10.1088/0957-4484/22/32/325601

[16] Jiang B, Tian C, Wang L, Sun L, Chen C, Nong X, Qiao Y, Fu H: Highly concentrated, stable nitrogen-doped graphene for supercapacitors: simultaneous doping and reduction. Applied Surface Science. 2012;258:3438–3443. DOI: 10.1016/j.apsusc.2011.11.091

[17] Li X, Wang H, Robinson J T, Sanchez H, Diankov G, Dai H: Simultaneous nitrogen doping and reduction of graphene oxide. Journal of American Chemistry Society. 2009;131:15939–15944. DOI: 10.1021/ja907098f

[18] Thakur S, Karak N: Alternative methods and nature-based reagents for the reduction of graphene oxide: a review. Carbon. 2015;94:224–242. DOI: 10.1016/j.carbon.2015.06.030

[19] Schniepp H C, Li J L, McAllister M J, Sai H, Herrera-Alonso M, Adamson D H, Prud'homme R K, Car R, Saville D A, Aksay I A: Functionalized single graphene sheets derived from splitting graphite oxide. The Journal of Physical Chemistry B. 2006;110:8535–8539. DOI: 10.1021/jp060936f

[20] Vallés C, David Núñez J, Benito A M, Maser W K: Flexible conductive graphene paper obtained by direct and gentle annealing of graphene oxide paper. Carbon. 2012;50:835–844. DOI: 10.1016/j.carbon.2011.09.042

[21] Mattevi C, Eda G, Agnoli S, Miller S, Mkhoyan K A, Celik O, Mastrogiovanni D, Granozzi G, Garfunkel E, Chhowalla M: Evolution of electrical, chemical, and structural properties of transparent and conducting chemically derived graphene thin films. Advanced Functional Materials. 2009;19:2577–2583. DOI: 10.1002/adfm.200900166

[22] Li Y, Sheng K, Yuan W, Shi G: A high-performance flexible fibre-shaped electrochemical capacitor based on electrochemically reduced graphene oxide. Chemical Communnications. 2013;49:291–293. DOI: 10.1039/c2cc37396c

[23] Yang J, Gunasekaran S: Electrochemically reduced graphene oxide sheets for use in high performance supercapacitors. Carbon. 2013;51:36–44. DOI: 10.1016/j.carbon.2012.08.003

[24] Chen K, Chen L, Chen Y, Bai H, Li L: Three-dimensional porous graphene-based composite materials: electrochemical synthesis and application. Journal of Materials Chemistry. 2012;22:20968. DOI: 10.1039/c2jm34816k

[25] Sheng K, Sun Y, Li C, Yuan W, Shi G: Ultrahigh-rate supercapacitors based on eletrochemically reduced graphene oxide for ac line-filtering. Sicentific Reports. 2012;2:247. DOI: 10.1038/srep00247

[26] Shao Y, Wang J, Engelhard M, Wang C, Lin Y: Facile and controllable electrochemical reduction of graphene oxide and its applications. Journal of Materials Chemistry. 2010;20:743–748. DOI: 10.1039/b917975e

[27] Sokolov D A, Rouleau C M, Geohegan D B, Orlando T M: Excimer laser reduction and patterning of graphite oxide. Carbon. 2013;53:81–89. DOI: 10.1016/j.carbon.2012.10.034

[28] Huang L, Liu Y, Ji L C, Xie Y Q, Wang T, Shi W-Z: Pulsed laser assisted reduction of graphene oxide. Carbon. 2011;49:2431–2436. DOI: 10.1016/j.carbon.2011.01.067

[29] Moussa S, Atkinson G, SamyEl-Shall M, Shehata A, AbouZeid K M, Mohamed M B: Laser assisted photocatalytic reduction of metal ions by graphene oxide. Journal of Materials Chemistry. 2011;21:9608. DOI: 10.1039/c1jm11228g

[30] Zhang Y, Guo L, Wei S, He Y, Xia H, Chen Q, Sun H-B, Xiao F-S: Direct imprinting of microcircuits on graphene oxides film by femtosecond laser reduction. Nano Today. 2010;5:15–20. DOI: 10.1016/j.nantod.2009.12.009

[31] Abdelsayed V, Moussa S, Hassan H M, Aluri H S, Collinson M M, El-Shall M S: Photothermal deoxygenation of graphite oxide with laser excitation in solution and graphene-aided increase in water temperature. The Journal of Physical Chemistry Letters, 2010;1:2804-2809. DOI: 10.1021/jz1011143

[32] Kaplan A, Lenner M, Palmer R E: Emission of ions and charged clusters due to impulsive Coulomb explosion in ultrafast laser ablation of graphite. Physical Review B. 2007;76:073401. DOI: 10.1103/PhysRevB.76.073401

[33] Lenner M, Kaplan A, Huchon C, Palmer R E: Ultrafast laser ablation of graphite. Physical Review B. 2009;79:184105. DOI: 10.1103/PhysRevB.79.184105

[34] Carbone F, Baum P, Rudolf P, Zewail A H: Structural preablation dynamics of graphite observed by ultrafast electron crystallography. Physical Review Letters. 2008;100:035501. DOI: 10.1103/PhysRevLett.100.035501

[35] Gao W, Singh N, Song L, Liu Z, Reddy A L, Ci L, Vajtai R, Zhang Q, Wei B, Ajayan P M: Direct laser writing of micro-supercapacitors on hydrated graphite oxide films. Nature Nanotechnology. 2011;6:496–500. DOI: 10.1038/nnano.2011.110

[36] El-Kady M F, Strong V, Dubin S, Kaner R B. Laser scribing of high-performance and flexible graphene-based electrochemical capacitors. Science. 2012;335:1226–1230. DOI: 10.1126/science.1216744

[37] Meng C, Maeng J, John S W M, Irazoqui P P: Ultrasmall integrated 3D micro-supercapacitors solve energy storage for miniature devices. Advanced Energy Materials. 2014;4:1301269. DOI: 10.1002/aenm.201301269

[38] El-Kady M F, Kaner R B: Scalable fabrication of high-power graphene micro-supercapacitors for flexible and on-chip energy storage. Nature Communnications. 2013;4:1475. DOI: 10.1038/ncomms2446

[39] Xie B, Yang C, Zhang Z, Zou P, Lin Z, Shi G, Yang Q, Kang F, Wong C-P: Shape-tailorable graphene-based ultrahigh rate supercapacitor for wearable electronics. ACS Nano. 2015;9:5636–5645. DOI: 10.1021/acsnano.5b00899

[40] Xie B, Wang Y, Lai W, Lin W, Lin Z, Zhang Z, Zou P, Xu Y, Zhou S, Yang C, Kang F, Wong C-P: Laser-processed graphene based micro-supercapacitors for ultrathin, rollable, compact and designable energy storage components. Nano Energy. 2016;26:276–285. DOI: 10.1016/j.nanoen.2016.04.045

[41] Wu Z S, Parvez K, Winter A, Vieker H, Liu X, Han S, Turchanin A, Feng X, Mullen K: Layer-by-layer assembled heteroatom-doped graphene films with ultrahigh volumetric capacitance and rate capability for micro-supercapacitors. Advanced Materials. 2014;26:4552–4558. DOI: 10.1002/adma.201401228

[42] Hadjipaschalis I, Poullikkas A, Efthimiou V: Overview of current and future energy storage technologies for electric power applications. Renewable and Sustainable Energy Reviews. 2009;13:1513–1522. DOI: 10.1016/j.rser.2008.09.028

[43] Liu W W, Feng Y Q, Yan X B, Chen J T, Xue Q J: Superior micro-supercapacitors based on graphene quantum dots. Advanced Functional Materials. 2013;23:4111–4122. DOI: 10.1002/adfm.201203771

[44] Fukuda K, Minamiki T, Minami T, Watanabe M, Fukuda T, Kumaki D, Tokito S: Printed Organic Transistors with Uniform Electrical Performance and Their Application to Amplifiers in Biosensors. Advanced Electronic Materials. 2015;1:1400052. DOI: 10.1002/aelm.201400052

2

Chemical, Thermal, and Light-Driven Reduction of Graphene Oxide: Approach to Obtain Graphene and its Functional Hybrids

Mohammad Razaul Karim and Shinya Hayami

Abstract

The alternative synthetic route of graphene (G) is presented. At first, graphite is oxidized to graphite oxide, which is dispersed in water to form graphene oxide (GO). GO can be reduced to rGO or G. GO being negatively charged can be used to obtain organic functionalized GO or hybrid of GO, reduction of which finally result in the formation of G-based hybrids. The reduction of GO or GO hybrid can be accomplished by light irradiation, thermal annealing, or by treating with reducing agents. The chemical changed from graphite to GO and GO to rGO can be monitored by surface analysis, microscopic investigation, and various spectroscopic methods.

Keywords: graphene oxide, chemical reduction, photo reduction, graphene hybrids, solution processable graphene, multifunctionality

1. Introduction

Apart from chemical vapor deposition (CVD) or micromechanical exfoliation process, thermal, chemical, or light-driven reduction of graphene oxide (GO) is the simplest route for synthesis of graphene (G) in large scale [1–8]. The light-driven reduction is accomplished through free radical reduction process of GO associated with the decomposition of carboxylic, epoxy, and hydroxyl groups. The chemical reduction process includes reaction of GO with various reducing agents. Depending on the temperature and strength of reducing agent, the extent of reduction of GO or extent of formation of graphene can be varied [2, 3].

G being a hydrophobic material is difficult to be processed in polar solvents especially in water. But for considering chemical reactions or processing with other, ingredient G necessarily needs to be dispersed in solvents [2]. As an alternative route, reduction of GO in the presence of the water soluble polymers such as polyvinylpyrrolidone results in linking between graphene nanosheet and water molecules, making it solution processable [9]. Graphene being unreactive itself can be functionalized by reacting GO with other reagents with subsequent reduction. The synthesis of GO from graphite is usually accomplished by three classical graphite oxidation techniques, including Hummers', Staudenmaier, and Brodie's method [10–12]. The synthesized GO is reduced by light irradiation, high-temperature refluxing, or reacting with reducing agents including hydrazine and its derivatives, ammonia, sodium borohydride, and so on to obtain graphene. To obtain graphene hybrids, GO suspension in water is treated with metallic salts or organic bulk molecules containing amino or sulfonic acid terminals. The formation of chemical bonds or electrostatic bindings stabilizes the resultant GO hybrid, while subsequent reduction results in the formation of G hybrids. The conversions of graphite to GO and GO to G are confirmed from X-ray photo electron spectroscopy (XPS) analysis, Raman spectra, infrared spectra (IR) spectroscopy, and powder X-ray diffraction (PXRD) analysis [2, 13]. The intercalation of materials within graphite results in significant change in chemical and physical properties [14]. The ionic and electronic conductivity can be justified as a function of oxygen content or sp^2 carbon sites. The hybrids usually display multifunctional property including the conductivity of G and additional functionality attributed by the metal ions, semiconductor ingredients, or organic molecules.

The overall process includes three major steps. Firstly, graphite is converted into graphene oxide (GO) by surface oxidation and exfoliation. Secondly, GO is reduced to graphene by reduction. Thirdly, composite of graphene is formed by reaction of GO with metal ions, organic molecules, or semiconductor nanoparticles and subsequent reduction. Multifunctionality can be achieved in the G hybrids formed by this way.

2. Strategy for indirect approach of graphene synthesis

After finding the extraordinary physical, chemical, and mechanical properties, surface area, and optoelectronic characteristics of graphene (G) [13–20], the necessity for chemical and physical modification of G with formation of hybrid materials became a vital issue. Scaffolding of various functional ingredient on G matrices became a facile route for material devising and fabrication. But, due to the hydrophobic nature and chemical inertness, G has been found to be a poor candidate for the as-mentioned purposes. Regarding these facts, GO, the oxidized form of G, became a more potential candidate due to its hydrophilic nature, processability in polar solution and ease for chemical functionalization [21]. Hence, both the theoretical and experimental researches reveal the utilization of G/GO as some applied materials. Though pristine G possesses enormous possibility for various applications, the great opportunity of G or GO primarily is associated with their extent of functionalization and formation of hybrids with other organic and inorganic materials. There exist a number of reports in favor of such strategy. To obtain G through an indirect route, at first GO is obtained from G. The GO thus synthesized is reduced to from G. For obtaining G-based hybrid, at first GO-based hybrids are obtained from chemical reaction or physical interaction between GO and a second ingredient. The as-prepared GO hybrid is then reduced to form rGO hybrid or the G-based hybrid materials [2, 21]. **Figure 1** represents the scheme for both the approaches.

There exist a number of routes to obtain G. Numerous modified version for synthesis of G, including micromechanical exfoliation, chemical vapor deposition, reduction of GO, or most recently developed electrochemical exfoliation of graphite, are most commonly used [1, 3–6, 22]. Usually, bulk graphite is intercalated by a chemical agent to separate the layers of graphitic stake. Though this process, results contaminated graphitic mass due to the intervened atoms or molecules functioning as impurity, the processes are widely employed for the cheap, nontoxic, and easily affordable ways. In epitaxial growth method, single and multilayer G can be grown using the process for chemical vapor deposition of hydrocarbons on the surface of some metal substrate. Typical thermal decomposition of Si also can be utilized. The micromechanical cleavage is the most facile techniques to isolate pure G from graphite. All these methods are suitable for separation of single-layer G, which is chemically very stable. The problem with this very pure form of G is its limited chemical affinity for a second chemical entity, metal ions, semiconductor ingredient, metal complex, or bulk organic molecules. However, the purpose of the addition of such an entity is to decorate a second functionality in the hybrid in addition to the conductivity and stability of G. G is composed of the sp^2 hybridized carbon atoms. The perpendicular electronic currents that circulate above and below the planar network of graphite layers are very stable and difficult to be destroyed. Also, it is not facile to generate defect sites in G, which can aid chemical bond formation with other ingredient.

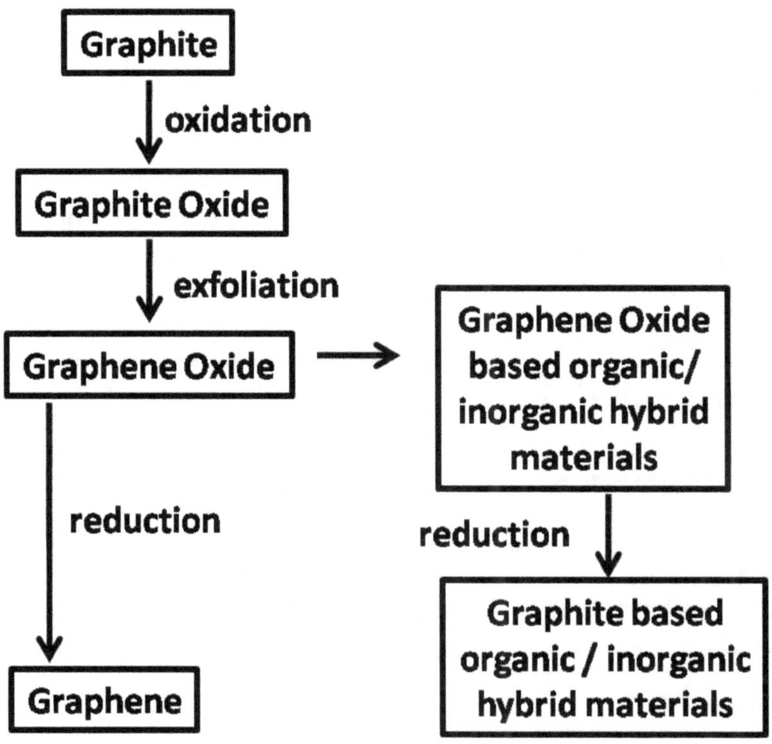

Figure 1. Scheme for the indirect synthesis of G and G-based organic/inorganic hybrid materials.

These intrinsic limitations can be overcome by a three-step process. Firstly, graphite materials are chemically converted into GO. Then, the GO is used for making chemical bond or physical attachments with active functional ingredient. Finally, the hybrids are reduced to form G-based complex structures. Initially for graphite → GO conversion, the conductivity of G is destroyed significantly due to the destruction of sp^2 conductive π networks. The sites where oxygen atoms become attached to G take the form of sp^3 hybridized carbon [23, 24]. But due to the attachment of oxygenated sites at G basement, the chemical affinity is increased. Thus, the presence of oxygenated sites in terms of epoxy, hydroxyl, and carboxyl groups increases the possibility of chemical reaction or physical affinity of GO toward other functional ingredients. Hence, the formation of GO-based nonconductive inorganic hybrid materials (GO-M) takes place in the second stage. In the final stage, GO-M is reduced to form rGO-M or G-M type hybrid. During reduction, the oxygenated sites scaffolding M remain unaffected, whereas the rest of the bare functional sites is reduced to recover some of graphene's conductivity and other functionality. The G-M hybrid thus possesses the functionality of both G and M [21].

3. Oxidation of graphite to GO with subsequent reduction to graphene

Graphite usually can be oxidized by Hummers', Staudenmaier, and Brodie's method (**Figure 2**). The typical synthetic routes involved are as following:

3.1. Hummers method

In Hummers' method, 1.0 g graphite powder, 1.0 g finely meshed $NaNO_3$, and 50 mL 97% H_2SO_4 need to be cooled to 0°C by stirring in an ice bath for 15 min [10]; 3.0 g finely meshed $KMnO_4$ powder needs to be added slowly with vigorous stirring with maintaining the temperature of the mixture below 20°C. After 30 min, the mixture needs to be warmed to 35°C for 30 min. Then, 200 mL water should be added slowly. The temperature will rise gradually in this stage and it should be maintained around 95°C for another 30 min. Then, 400 mL water and 12 mL 30% H_2O_2 must be added with stirring. The mixture

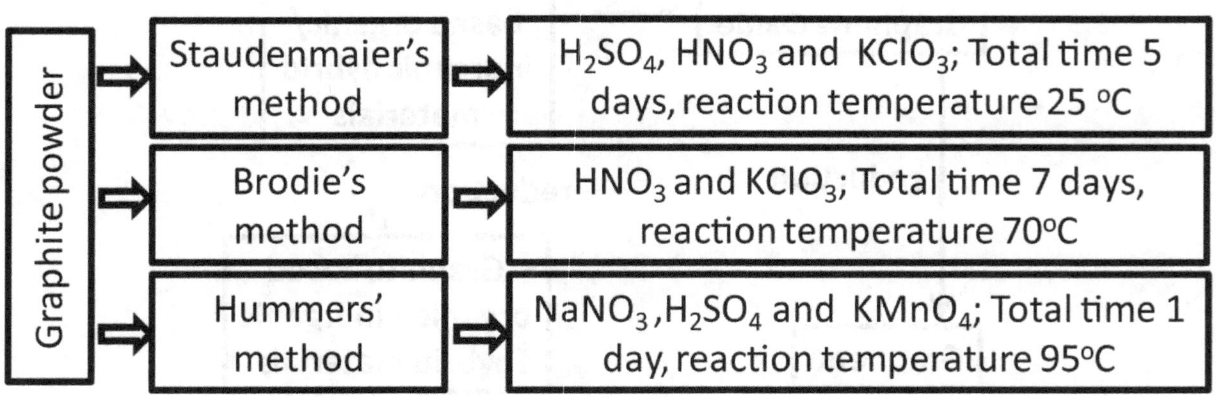

Figure 2. Scheme for the various graphite oxidation techniques through slight modification to reported classical methods.

needs to be centrifuged at 3000 rpm for 10 min. The precipitate should be washed one time with 5% HCl solution and three times with water. The precipitate finally needs to be dried at 70°C for 24 hours to obtain H-GO.

3.2. Staudenmaier method

In Staudenmaier process, a mixture of 35 mL H_2SO_4 (95–98%) and 18 mL fuming HNO_3 in a round bottom flask need to be cool down to 0°C by stirring in an ice bath [11]. Then, 2 g graphite will be added to the mixture under vigorous stirring to obtain a homogeneous dispersion. While keeping the reaction flask in the ice bath, 22 g $KClO_3$ must be added slowly (over 15 min) to the mixture in order to avoid any sudden rise in temperature and formation of explosive chlorine dioxide (ClO_2) gas. After complete dissolution of $KClO_3$, the reaction flask will be removed from the ice bath with capping loosely to allow evolution of gases. The mixture must be stirred vigorously for 5 days at room temperature. After completion of the reaction, the mixture should be poured into 2-L deionized water and filtered after keeping overnight for sedimentation. The precipitate thus obtained should be dispersed and stirred repeatedly by HCl (5%) solutions to remove sulfate ions. The mass obtained will be washed several times with deionized water until neutral pH of the filtrate is obtained. The suspension obtained should be dried in a vacuum oven at 60°C for 48 h to obtain S-GO.

3.3. Brodie's method

In Brodie's method, 6 g graphite powder should be meshed finely with 21 g $KClO_3$ in a retort. The powdered mixture might be poured into a round bottom flask fitted with magnetic stirrer [12], and 75 ml fuming HNO_3 needs to be added to the mixture with stirring continuously at 70°C for 1 day until the disappearance of yellow vapor. The mixture obtained thus should be washed repeatedly with water. The final product will be obtained from drying. The whole procedure was repeated three times by Brodie using the final products of the former operation as the starting materials for the consecutive operation.

4. Reduction of GO into rGO or graphene

The oxygenated functional groups of GO disappear due to the reduction. For mild reduction, there exist some residual oxygenated sites. This form is usually presented by rGO. In fact, during treatment with strong reducing agent, rGO sometimes contains insignificant amount of oxygenated sites and reveal similar characteristics of G. The as-prepared G displays some electronic and chemical properties, which are comparable with pristine G. In general, oxygen presents on GO is available mainly in the forms of epoxy (the bridge site oxygen) and hydroxyl (—OH groups) and carboxylic acid (—COOH) groups. The relative composition in terms of C:O:H ratio varies according to the synthetic route, degree of oxidation, and variations in the synthesis conditions. In average, the percent of oxygen in GO remains around 30% by weight. To remove these oxygenous sites, GO can be reduced through several ways including: (1) thermal reduction, (2) photo reduction, and (3) chemical reduction.

4.1. Thermal reduction

Thermal reduction of GO is a simple heating process accomplished by gradual heating of solid GO mass or its dispersion in solvents under inert atmosphere. The absence of oxygen and reactive chemical specious needs to be ensure as it might result in the loss of carbon content due to the formation of gases or volatile substances. Gradual decrease in weight loss in thermogravimetric analysis indicates the decomposition of oxygenated functional groups and conversion of GO into rGO or G. Complete reduction of GO by heat treatment is possible, and the process is defined as thermal annealing reduction. Fast heating of GO at a rate above 2000°C min^{-1} results in the exfoliation of graphite oxide into G. Such exfoliation of the staked GO mass is aided by the swift expansion of CO or CO_2 gases evolved at the interlayer space of G sheets due to the fast heating of GO. At high temperature, the oxygenous groups decompose, and the gases make very high pressure within the layers. Calculations show that at 300 and 1000°C, a pressure of 40 and 130 MPa, respectively, is created on the surface of G. However, a pressure of only 2.5 MPa is enough to separate two stacked GO platelets. Thus, the rapid heating process not only exfoliates GO but also reduces the functionalized graphene sheets by decomposing oxygen-containing groups at high temperature. Due to these multiple effect, thermal reduction of GO is preferred over other methods to generate G. In spite, this procedure reveals some drawback, as it can produce small sized and wrinkled graphene sheets only. This is mainly because the decomposition of oxygen-containing groups is associated with the removal of some carbon atoms from the G skeleton. However, ultimately chopping of the large-sized G sheets into tiny pieces also results in the distortion in morphology of carbon skeleton (**Figure 3**). The thermal annealing also generates defects and damages at the structure of G due to the evolution of CO_2 gas. Around 30% weight loss takes place during the conversion of GO into G. Although the G obtained is very pure, the electronic property of rGO-based G decreases significantly compared with G obtained from CVD method. Such incident takes place due to the decrease in ballistic transport path length and introduction of scattering canters. The electrical conductivity of rGO-based G is around 10–23 S cm^{-1}, which is significantly lower than that of perfect G. The thermal reduction of GO at liquid phase results in larger sized G [25]. In this method, the thermal treatment is applied after fabricating macroscopic films, powders or thin pellets.

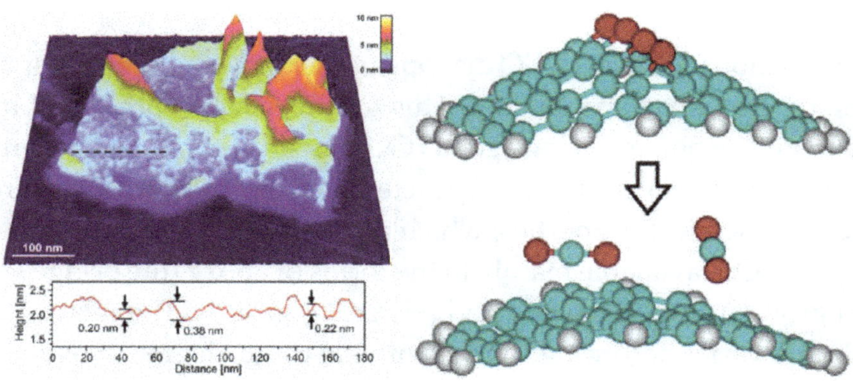

Figure 3. Three dimensional representation of a 600 nm × 600 nm AFM scan of a G nanosheet obtained from heat treatment of GO. It shows the wrinkled and rough structure of the surface.

4.2. Photo reduction

The photo-aided reduction is usually performed in solution state. Exfoliated GO water solution in a vial is needed to be irradiated by 365 nm light beam flashed from a 500 W high-pressure Hg lamp. Typically, 99.99% H_2 or N_2 at 1 atm pressure needs to be flown through a quartz cell–filled GO sample at a rate of about 50 mL/min. The light intensity can be about 67 mW/cm^2, and the light source can be positioned at a distance of about 40 cm from the sample surface. The degree and extent of GO → rGO/G conversion can be controlled by tuning the exposure time. Photoreduction can also be resulted in by employing a UV cutoff filter (<420 nm) to generate visible light irradiation. But, in such condition, the degree of reduction becomes low. In contrast, UV irradiation obtained from a visible-light cutoff filter (>390 nm) can cause in photoreduction through the same extent of reduction taking place in absence of any filter. Thus, UV irradiation has been found to be a very effective way to reduce GO. Light irradiating reduction can also be accomplished using an intense light source such as laser or photocatalyst such as TiO_2 or ZnO.

Light-driven reduction takes place through a mechanism associated with free radical cleavage method. Epoxy groups locating at the basal planes of GO are relatively unstable due to the bond strain and angular distortion associated with the bending of regular sp_3 hybrid orbitals. In the reduction process, π–π* electronic excitations within the sp_2 domains on the GO surface take place as the initial step of UV irradiation. The electron–hole pairs generated this way, then migrates to the epoxy groups and aids C–O–C bond breaking process. Associated release of O_2 and formation of larger sp_2 domains take place simultaneously. It was found that some ions including Ag^+ can promote the destruction of the C–O bond and escaping of O_2 during photoreduction. In this case, the reduction of GO takes place along with the formation of Ag nanoparticles from Ag^+ under the presence of ethylene glycol and $NaBH_4$. Here, some strong interaction happens between Ag nanoparticles and the remaining surface hydroxyl O atoms. The reduction of GO can initially be justified by change in color of GO dispersion and respected UV–vis spectra (**Figure 4**). The epoxy bond cleavage study of X ray photo electron spectra suggests that the extent of reduction of oxygenated sites follows the trend as epoxy > ketonic > carboxylic > hydroxyl groups.

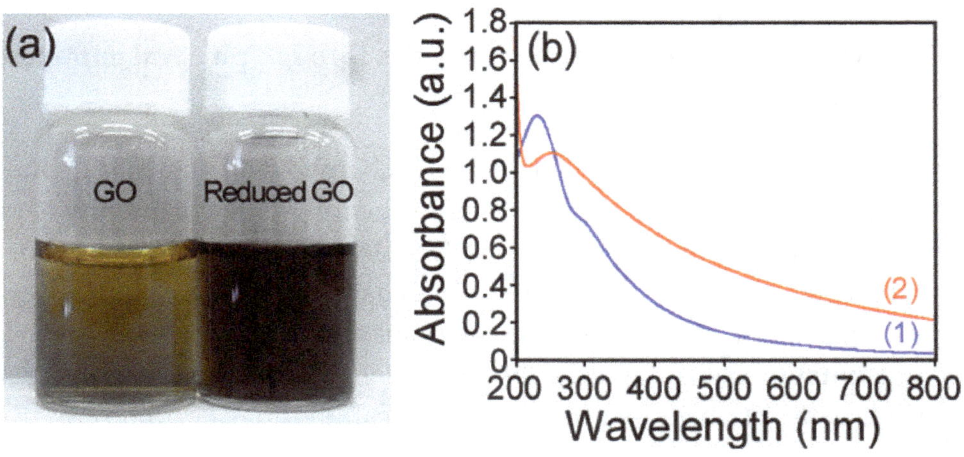

Figure 4. (a) Change in color and (b) UV-vis spectra of dispersion of GO (1) and rGO (2) generated from photoreduction of GO in H_2 for 2 h. The color turns to black after photoreduction.

4.3. Chemical reduction

A list of reducing agent can be used for the reduction of GO chemically. Typical strong reducing agent includes hydrazine, dimethylhydrazine and other related derivatives, sodium borohydrate ($NaBH_4$), ammonia, hydrogen plasma and so on. Depending on the strength of chemical agent, the reduction can be accomplished from room temperature to elevated temperature. The commonest way of chemical reduction is treating GO with hydrazine hydrate at 100°C for 24 hours [26]. Instead of the solution phase, GO film can be exposed to hydrazine vapor from long to short time to obtain desired extent of reduction. In average, the conductivity of rGO films obtained from hydrazine reduction is 99.6 S cm^{-1} with a C/O ratio of around 12.5 [27]. Strong metal hydride reducing agent, including sodium hydride, sodium borohydride ($NaBH_4$), and lithium aluminum hydride, cannot be applied through facile way as these reducing agents exhibit slight to very strong reactivity toward water. As water is the major dispersion phase for GO, the presence of water cannot be avoided.

$NaBH_4$ hydrolyzes in water through a slow rate. As a result, compared with hydrazine solution, freshly prepared $NaBH_4$ solution serves as a more effective reducing agent for reducing C=O groups in GO [28]. However, the efficiency of $NaBH_4$ is low to moderate for reduction of epoxy, carboxylic acids, and alcohol groups. To solve this problem after reduction with $NaBH_4$, dehydration with concentrated sulfuric acid (98%) at 180°C can be applied to reduce the oxygenated sites. Thus, reaching to a C/O ration of 8.6 with conductivity value as 16.6 S cm^{-1} is possible [29]. Ascorbic acid (Vitamin C) as a comparatively water stable and nontoxic materials can reduce GO with a C/O ration of 12.5 and conductivity as 77 S cm^{-1}. Glucose solution at 100°C can also be applied for the reduction of GO. Exposure to hydrogen plasma for several seconds is the shortest time reduction route. Heating GO with urea as a simple and affordable method. Reduction of GO by HI is possible in various physical condition such as GO colloid, powder, or film [30]. Some other mild reducing agent for GO includes hydroquinone, pyrogallol, hot strong alkaline solutions (KOH, NaOH), hydroxylamine, and thiourea.

5. Formation of GO hybrid

5.1. Scaffolding metal/metal oxide or bulk molecules through physical entrapments to form GO-M hybrid

The columbic interaction between the negatively charged GO nanosheet and positively charged chemical ingredients including metal ions function as the major driving force behind the formation of this type of hybrid (**Figure 5**). Initially, some primary metal ions bound to some of the partially charged oxygenated sites of GO nanosheets and prohibit these sites from further reaction with other metal ions. The GO nanosheet thus losses its charge, becomes heavier, and precipitates as sediments in the form of GO-M hybrid. The GO-M hybrid then can be converted to G-M hybrid by means of chemical, thermal, or photo reduction. When GO-M hybrid is reduced to G-M, along with the change of GO → G, the growth of metal precursor takes place by gradual combination with oxygen atoms, metal oxide molecules (produced during reduction process), and more free metal ions. As a result, in the final stage, irregular-shaped metal oxide nanoparticles become anchored to the rGO matrices, which in fact function as hybrid of G with metal/ metal oxide.

5.2. Functionalization of GO by organic compounds to form GO-Org hybrids

Numerous G-based hybrid materials can be generated from organic functionalization of GO (GO-Org) and its successive reduction process. In the form of GO-Org, the materials show excellent solubility and solution processability in both the water and organic solvents. As G-Org form is inferior with respect to these properties, prior to the formation of G-Org, a common stand includes obtaining GO-Org as the earlier step. At the second step, the GO-Org is reduced to G-Org through a suitable reduction process discussed above. Most of the outstanding properties of G and the organic molecules sustain in G-Org without being perturb. The durability, stability, and extent of functionality have been found to be excellent. Both the noncovalent and covalent functionalization of GO with organic compounds thus became the subject of intensive research for developing G-based hybrid nanocomposites with new functions and applications. From noncovalent aspect, due to the tendency of GO for forming π-π and hydrophobic interaction, strong adsorption between organic aromatic compounds or bulk molecules and GO nanosheets takes place. Thus, the formation of numerous GO-Org hybrids becomes possible. For covalent modification, the oxygenated sites in GO serve as the active site toward various chemical agent. Such reaction needs some pre-requisite condition including the perfect diffusion of the organic precursors toward the reactive moiety of GO nanosheet with successful approach for undergoing surface reactions. A wide range of organic compounds including polyanilines, phenolic derivatives, porphyrins, aromatic dyes, alkylamines, doxorubicin, hydrochloride, ionic liquids, pyrene, perylenediimide, cyclodextrin, tetracyanoquinodimethane, and aryl diazonium compounds can react

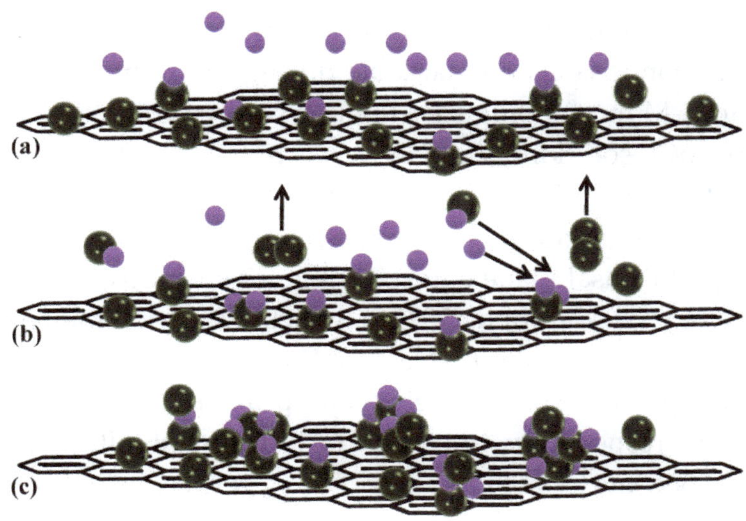

Figure 5. Scheme for the formation of rGO–metal oxide nanoparticle hybrids. (a) In the first step, negatively charged oxygenated site (deep green) of GO electrostatically attracts some primary metal ions (violet) through columbic force. (b) In the second step, reduction takes place. During reduction, some oxygen escapes from the system, whereas metal-masked oxygenated sites remain unreduced and attract metal oxide precursors, oxygen, and further metal ions. (c) Gradual deposition finally results irregular-shaped metal oxide nanoparticles anchored on rGO matrices [21].

noncovalently or covalently for being attached on GO surface. Prior to the conversion of G-Org hybrid, additional facilities obtained from the formation of GO-Org hybrid include the increase in the dispersibility due to the existence of residual polar groups (after some being engaged for chemical bonding with the organic molecules) in GO-Org.

6. Monitoring the graphite → GO → G conversion and characterization of the products

The black-colored graphite powder changes to stake layered mass in graphite oxide. Color of GO dispersion in water is usually brown, though this color varies according to the route of oxidation. GO dispersion in water turns to black-colored coagulated nanosheet due to the reduction (**Figure 6**). Although these changes in morphology are visible by bare eyes, there exist numerous well characterized physical and chemical methods to monitor the changes. The surface morphology can be characterized by scanning electron microscopy (SEM), atomic force microscopy (AFM), and transmission electron microscopy (TEM). In case of hybrid formation with metallic ingredients, TEM in combination with EDX becomes an excellent characterization tool. The granular pattern and changes in interlayer distances with reaction can be monitored by powder X-ray diffraction (PXRD) study. The incorporation of oxygenated groups can be monitored by Raman spectra. The extent of oxygenation and nature of functional groups can be justified by infrared spectra (IR), and finally, the presence of various functional groups can be confirmed from X-ray photo electron spectroscopy (XPS).

When graphite powder is oxidized into GO, the intercalation of oxygenated groups separate layers of graphite. But, the hydrophilic functional groups at the graphitic surface make the sheet coagulated with each other. The graphite oxide thus seems to be some massive pellets. When graphite oxide is ultrasonicated in water, the nanosheets become separated and sustain in water as colloid. The colors of the GO solutions vary depending on the synthetic route (**Figure 6**). The color change is also remarkable during the formation of GO to GO-M and finally from GO-M to G-M form. In **Figure 6** (h–j), the brown-colored GO starts to precipitate due to the mixing of $CoCl_2$ solution. When GO-Co is reduced by hydrazine, the final color of G-Co becomes dark. Change in PXRD pattern (**Figure 6k**) clearly depicts that due to oxidation, the interlayer distance of graphite increases significantly. The extension of oxygenated groups from the graphitic basal plane is responsible for such observation. Graphite usually show characteristic sharp peak around 27° (2θ), which represents the reflection plane (0 0 1) of graphite and indicates its amorphous structure [31]. Due to oxidation, the graphitic sharp peak disappears and appearance of relatively wider peak around $2\theta = 9.13$, 11.4, and 11.6 for H-GO, S-GO, and B-GO, respectively, represent the lowering of reflection planes. In this case, the 'd-spacing' values become as 9.68, 7.76, and 7.62 Å, respectively. When GO is reduced to G, the wide peak disappears and a broad hollow rises around 20° (2θ).

Raman spectroscopy is commonly employed to monitor change and doping on graphitic materials and GO. Raman spectra of carbon materials show two bands at ~1580 cm^{-1} (G band) and ~1350 cm^{-1} (D band; **Figure 7a**). These G and D bands are responsible for E_{2g} photon

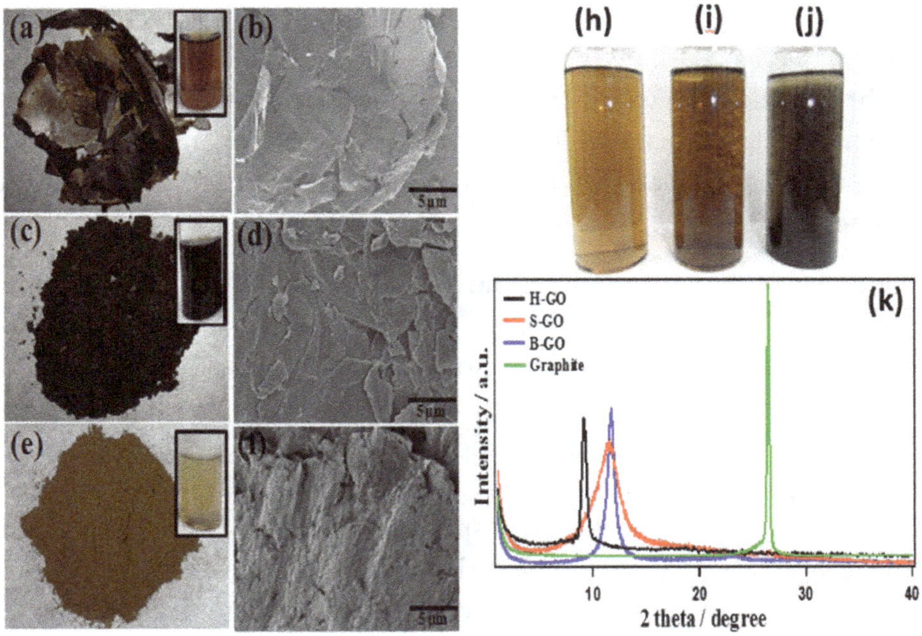

Figure 6. Physical properties and morphologies of the samples. Photographs of GO prepared from Hummers' method (H-GO) (a), Staudenmaier method (S-GO) (c), and Brodie's method (B-GO) (e). The inset images show the respective solution (1 mg/mL) in water after ultrasonication for 2 h. Respective SEM images of H-GO (b), S-GO (d), and B-GO (f) display their surface morphologies. (h)–(j) The change in color and physical state of GO solution, GO-M hybrid and G-M hybrid (M represents the Co^{2+} ion). (k) The powder X-ray diffraction patterns of graphite powder, H-GO, S-GO, and B-GO [35].

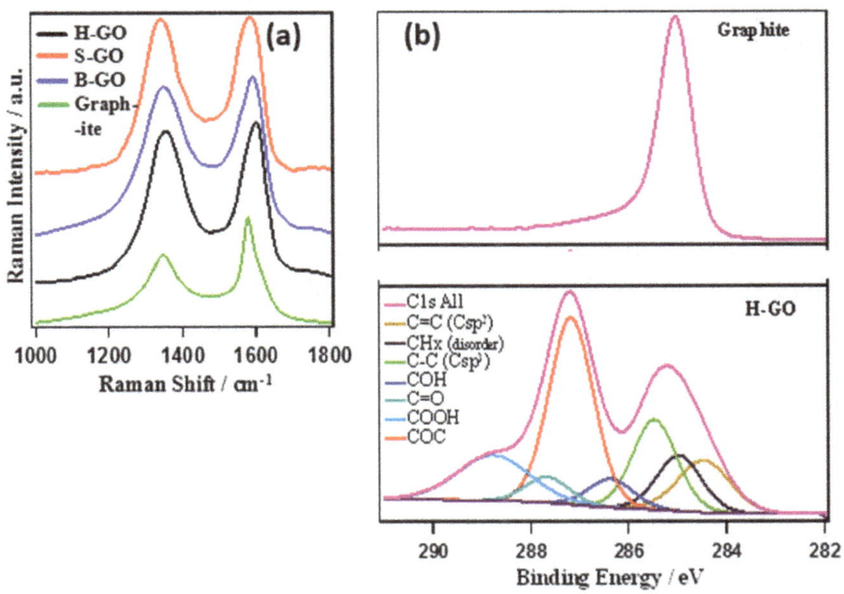

Figure 7. (a) Raman spectra and (b) XPS spectra of graphite and GO synthesized by various routes. GO shows characteristic oxygenous functional groups.

corresponding to sp_2 atoms and κ-point photon of A_{1g} symmetry of breathing mode resulting from local defects and disorder [32, 33]. During chemical conversion, the hole and electron doping in carbon materials results in the shifting of G-band, as well the intensity ratio of D and G-band varies significantly with doping. Peak ratio (I_D/I_G) gradually increases for G → GO; however, decreases along the conversion: GO → GO/GO-metal ion hybrids → G/G-metal oxide hybrids. The I_D/I_G value changes proportionally with the extent of sp_2 domain [34, 35]. Thus, the gradual decrease in I_D/I_G value implies expansion of this domain. The expansion of sp_2 domain indicates conversion of GO to G, whereas shrinking in this value signifies conversion of G to GO.

XPS is the most powerful tool for marking the oxidation of graphite into GO and reduction of GO into G. Pure graphite shows a sharp peak arising from the sp_2 hybridized carbon atoms. After oxidation, all the samples exhibit two individual peaks near 285 and 287 eV in their XPS spectra [36–38]. The deconvolution of spectra shows that GO exhibits characteristic peaks comprising several particular oxygenated functional groups. The existence of epoxide (—O—) and hydroxyl (—OH) groups at 286.8–287.0 eV, and carbonyl (—C=O) and carboxyl (—COOH) groups at 287.8–288.0 and 289.0–289.3 eV, respectively, confirms the presence of oxygenated sites in GO and graphite → GO conversion (**Figure 7b**). When GO is reduced into rGO or G, the peak intensity gradually reduces depending on the extent of reduction. For complete reduction, the spectra take the pattern of graphite almost, except revealing the existence of some residual oxygenated sites [39–41].

7. Conclusions

Indirect method for synthesis of graphene includes conversion of graphite into graphite oxide and GO dispersion by various oxidation method including Hummers', Staudenmaier, and Brodie's techniques with successive reduction into rGO. The graphite oxide synthesized can be dispersed into water by ultrasonication to obtain GO solution. Finally, the GO treating for reduction by means of thermal, photo irradiation, or chemical method can transform into rGO or G. In addition, GO having higher chemical affinity than G can easily be used for the formation of various functional GO hybrids, successive reduction of which can result in the formation of G-based hybrid materials. The series of chemical changes and progress of the reactions can be monitored by various surface analysis techniques and spectroscopic investigations.

Author details

Mohammad Razaul Karim[1] and Shinya Hayami[2*]

*Address all correspondence to: hayami@sci.kumamoto-u.ac.jp

1 Department of Chemistry, Shahjalal University of Science and Technology, Sylhet, Bangladesh

2 Department of Chemistry, Graduate School of Science and Technology, Kumamoto University, Kumamoto, Japan

References

[1] Eda G., Fanchini G., Chhowalla M. Large-area ultrathin films of reduced graphene oxide as a transparent and flexible electronic material. Nat. Nanotechnol. 2008;3:270–274.

[2] Zhu Y., Murali S., Cai W., Li X., Suk J.W., Potts J.R., Ruoff R.S. Graphene and graphene oxide: Synthesis, properties, and applications. Adv. Mater. 2010;22:3906–3924.

[3] Sutter P.W., Flege J.I., Sutter E.A. Epitaxial graphene on ruthenium. Nat. Mater. 2008;7: 406–411.

[4] Lee Y., Bae S., Jang H., Jang S., Zhu S.E., Sim S.H., Song Y.I., Hong B.H., Ahn J.H. Wafer-scale synthesis and transfer of graphene films. Nano Lett. 2010;10:490–493.

[5] Hernandez Y., Nicolosi V., Lotya M., Blighe F.M., Sun Z.Y., De S., McGovern I.T., Holland B., Byrne M., Gun'ko Y.K., Boland J.J., Niraj P., Duesberg G., Krishnamurthy S., Goodhue R., Hutchison J., Scardaci V., Ferrari A.C., Coleman J.N. High-yield production of graphene by liquid-phase exfoliation of graphite. Nat. Nanotechnol. 2008;3:563–568.

[6] Wang J.Z., Manga K.K., Bao Q.L., Loh K.P. High-yield synthesis of few-layer graphene flakes through electrochemical expansion of graphite in propylene carbonate electrolyte. J. Am. Chem. Soc. 2011;133:8888–8891.

[7] Shih C.J., Vijayaraghavan A., Krishnan R., Sharma R., Han J.H., Ham M.H., Jin Z., Lin S.C., Paulus G.L.C., Reuel N.F., Wang Q.H., Blankschtein D., Strano M.S. Bi-and trilayer graphene solutions. Nat. Nanotechnol. 2011;6:439–445.

[8] Liu N., Luo F., Wu H.X., Liu Y.H., Zhang C., Chen J. One-step ionic-liquid-assisted electrochemical synthesis of ionic-liquid-functionalized graphene sheets directly from graphite. Adv. Funct. Mater. 2008;18:1518–1525.

[9] Zhang Y., Hu W., Li B., Peng C., Fan C., Huang Q. Synthesis of polymer-protected graphene by solvent-assisted thermal reduction process. Nanotechnology 2011;22:34.

[10] Hummers W.S., Offeman R.E. Preparation of graphitic oxide. J. Am. Chem. Soc. 1958;80: 1339–1339.

[11] Staudenmaier L. Verfahren zur darstellung der graphitsäure. Ber. Dtsch. Chem. Ges. 1898;31:1481–1487.

[12] Brodie B.C. On the atomic weight of graphite. Philos. Trans. R. Soc. Lond. 1859;149:249–259.

[13] Land, T.A., Michely T., Behm R.J., Hemminger J.C., Comsa G. STM investigation of single layer graphite structures produced on Pt(111) by hydrocarbon decomposition. Surf. Sci. 1992;264:261–270.

[14] Dresselhaus M.S., Dresselhaus G. Intercalation compounds of graphite. Adv. Phys. 2002; 51:1–186.

[15] Nagashima A., Nuka K., Itoh H., Ichinokawa T., Oshima C., Otani S. Electronic states of monolayer graphite formed on TiC(111) surface. Surf. Sci. 1993;291:93–98.

[16] Van Bommel A.J., Crombeen J.E., van Tooren A. LEED and Auger electron observations of the SiC(0001) surface. Surf. Sci. 1975;48:463–472.

[17] Forbeaux I., Themlin J.-M., Debever J.M. Heteroepitaxial graphite on 6H-SiC(0001): Interface formation through conduction-band electronic structure. Phys. Rev. B 1998;58:16396–16406.

[18] Berger, C., Song Z., Li T., Li X., Ogbazghi A.Y., Feng R., Dai Z., Marchenkov A.N., Conrad E.H., First P.N., De Heer W.A. Ultrathin epitaxial graphite: 2D electron gas properties and a route toward graphene-based nanoelectronics. J. Phys. Chem. B 2004;108:19912–19916.

[19] Berger C., Song Z., Li X., Wu X., Brown N., Naud C., Mayou D., Li T., Hass J., Marchenkov A.N., Conrad E.H., First P.N., de Heer W.A. Electronic confinement and coherence in patterned epitaxial graphene. Science 2006;312:1191–1196.

[20] Ohta T., Bostwick A., Seyller T., Horn K., Rotenberg E. Controlling the electronic structure of bilayer graphene. Science 2006;313:951–954.

[21] Karim M.R., Shinoda H., Nakai M., Hatakeyama K., Kamihata H., Matsui T., Taniguchi T., Koinuma M., Kuroiwa K., Kurmoo M. et al Electrical conductivity and ferromagnetism in a reduced graphene-metal oxide hybrid. Adv. Funct. Mater. 2013;23:323–332.

[22] Ambrosi A., Pumera M. Electrochemically exfoliated graphene and graphene oxide for energy storage and electrochemistry applications. Chem. Eur. J. 2016;22:153–159.

[23] Kudin K.N., Ozbas B., Schniepp H.C., Prud'homme R.K., Aksay I.A., Car R. Raman spectra of graphite oxide and functionalized graphene sheets. Nano Lett. 2007;8:36–41.

[24] Das A., Pisana S., Chakraborty B., Piscanec S., Saha S.K., Waghmare U.V., Novoselov K.S., Krishnamurthy H.R., Geim A.K., Ferrari A.C., Sood A.K. Nat. Nanotechnol. 2008;3:210.

[25] Zhao J., Pei S., Ren W., Gao L., Cheng H.-M. Efficient preparation of large-area graphene oxide sheets for transparent conductive films. ACS Nano 2010;4:5245–5252.

[26] Kovtyukhova N.I., Wang Y., Berkdemir A., Cruz-Silva R., Terrones M., Crespi V.H., Mallouk T.E. Non-oxidative intercalation and exfoliation of graphite by Brønsted acids. Nat. Chem. 2014;6:957–963.

[27] Fernandez-Merino M.J., Guardia L., Paredes J.I., Villar-Rodil S., Solis-Fernandez P., Martinez-Alonso A., et al Vitamin C is an ideal substitute for hydrazine in the reduction of graphene oxide suspensions. J. Phys. Chem. C 2010;114:6426–6432.

[28] Shin H.-J., Kim K.K., Benayad A., Yoon S.-M., Park H.K., Jung I.-S., et al Efficient reduction of graphite oxide by sodium borohydride and its effect on electrical conductance. Adv. Funct. Mater. 2009;19:1987–1992.

[29] Gao W., Alemany L.B., Ci L., Ajayan P.M. New insights into the structure and reduction of graphite oxide. Nat. Chem. 2009;1:403–408.

[30] Moon K., Lee J., Ruoff R.S., Lee H. Reduced graphene oxide by chemical graphitization. Nat. Commun. 2010;1:73–78.

[31] Cai, D., Song M. Preparation of fully exfoliated graphite oxide nanoplatelets in organic solvents. J. Mater. Chem. 2007;17:3678–3680.

[32] Tuinstra F., Koenig J.L. Raman spectrum of graphite. J. Chem. Phys. 1970;53:1126.

[33] Ferrari A. C., Robertson J., Interpretation of raman spectra of disordered and amorphous carbon. Phys. Rev. B 2000;61:14095.

[34] Hatakeyama K., Karim M.R., Ogata C., Tateishi H., Funatsu A., Taniguchi T., Koinuma M., Hayami S. and Matsumoto Y. Proton conductivities of graphene oxide nanosheets: single, multilayer, and modified nanosheets. Angew. Chem., Int. Ed. 2014;53:6997–7000.

[35] Karim M.R., Islam Md.S., Hatakeyama K., Nakamura M., Ohtani R., Koinuma M. and Hayami S. Effect of interlayer distance and oxygen content on proton conductivity of graphite oxide. J. Phys. Chem. C 2016;120:21976–21982.

[36] Ikeda Y., Karim M.R., Takehira H., Matsui T., Taniguchi T., Koinuma, Matsumoto Y. and Hayami S. Proton conductivity of graphene oxide hybrids with covalently functionalized alkylamines. Chem. Lett. 2013;42:1412–1414.

[37] Ikeda Y., Karim M.R., Takehira H., Hatakeyama K., Taniguchi T., Koinuma M., Matsumoto Y. and Hayami S. Impaired proton conductivity of metal-doped graphene oxide. Bull. Chem. Soc. Jap. 2014;87:639–641.

[38] Hatakeyama K., Karim M.R., Ogata C., Tateishi H., Taniguchi T., Koinuma M., Hayami S. and Matsumoto Y. Optimization of proton conductivity in graphene oxide by filling sulfate ions. Chem. Commun. 2014;50:14527–14530.

[39] Wakata K., Karim M.R., Islam Md.S., Ohtani R., Nakamura M., Koinuma M. and Hayam S. Superionic Conductivity in Hybrid of 3-hydroxypropanesulfonic acid and graphene oxide. Chem. Asian J. 2017; 12:194–197.

[40] Hatakeyama K., Tateishi H., Taniguchi T., Koinuma M., Kida T., Hayami S., Yokoi H. and Matsumoto Y. Tunable graphene oxide proton/electron mixed conductor that functions at room temperature. Chem. Mater. 2014;26:5598–5604.

[41] Hatakeyama K., Islam Md.S., Koinuma M., Ogata C., Taniguchi T., Funatsu A., Kida T., Hayami S. and Matsumoto Y. Super proton / electron mixed conduction in graphene oxide hybrids by intercalating sulfate ions. J. Mater. Chem. A 2015;3:20892–20895.

3

Degradation of Toxic Organic Contaminants by Graphene Cathode in an Electro-Fenton System

Liu Shuan, Zhao Xia, Jiang Xin, Zhao Xiaorong, Pu Jibin, Wang Liping and Zhou Kaihe

Abstract

A novel composite electrode was constructed by pressing graphene and CuO, using a cathode in an electro-Fenton (EF) system. Cyclic voltammetry, charge/discharge curve and electrochemical impedance spectroscopy (EIS) were used to characterize the composite electrode. The degradation of a toxic organic contaminant, Terramycin, by EF system was studied in an undivided electrolysis cell. The possible degradation products of Terramycin were studied by a Fourier transform-infrared spectrum, and the findings showed that the structure of Terramycin was damaged. The variations of hydrogen peroxide and the relative content of hydroxyl radical (·OH) during the degradation process were traced by enzyme catalysis method and fluorescence spectrometry. The results showed that the electro-catalytic degradation of Terramycin occurred by an ·OH radical mechanism. More importantly, this as-prepared cathode was very stable and could be reused without any catalytic activity decrease, suggesting its potential application in the wastewater treatment.

Keywords: Terramycin, CuO-graphene electrode, electro-Fenton system, hydrogen peroxide, hydroxyl radicals introduction

1. Introduction

Persistent organic pollutants (POPs) resist conventional chemical and biological treatments and accumulate in the aquatic environment after discharge. They tend to be liposoluble and undergo food chain amplification and, if toxic, threaten human health [1, 2].

Electro-Fenton (EF) reaction is one of the most widely used advanced oxidation processes (AOPs) for the destruction of a wide range of recalcitrant organic contaminants such as antibiotics that cannot be eliminated biologically in wastewater [3–5]. In the EF systems, they have the ability to in situ generate highly potent chemical oxidants such as ·OH, which is strong enough to non-selectively oxidize most organic as well as some inorganic compounds [6]. H_2O_2 is simultaneously produced in an aqueous medium at the cathode by the $2e^-$ reduction of the dissolved molecular oxygen (Eq. (1)), and ·OH is formed from the Fenton's reagent according to the following reaction (Eqs. (2, 3)) [7].

$$O_2 + 2H^+ + 2e \rightarrow H_2O_2\,(\text{acidic}) \tag{1a}$$

$$O_2 + 2H_2O + 2e \rightarrow H_2O_2 + 2OH^-\,(\text{basic}) \tag{1b}$$

$$H_2O_2 + 2e \rightarrow 2{\cdot}OH \tag{2}$$

$$H_2O_2 + Fe^{2+} \rightarrow Fe^{3+} + OH^- + {\cdot}OH \tag{3}$$

Therefore, cathode materials determine the efficiency of the EF system. Usually, carbon materials with excellent characteristics of conductivity, gas diffusion, adsorption and catalysis are chosen as cathode materials. Graphene, an excellent functional material, has been widely used in super capacitors [8], transistors [9, 10], fuel cells [11], etc., but as electrode materials in electro-catalytic degradation of organic matter, very few have been reported. In this research, a new CuO-graphene electrode was constructed by pressing nano-CuO and pure graphene together. The electro-catalytic characteristics of the CuO-graphene electrode were studied using cyclic voltammetry, time power method and electrochemical impedance spectroscopy, respectively. In the EF system, the CuO-graphene electrode acted as the cathode and the pt net as the anode to degrade Terramycin. The concentration of Terramycin in the oxidative degradation was traced by HPLC. The variations of hydrogen peroxide and the relative content of hydroxyl radical during the degradation process were traced by enzyme catalysis method and fluorescence spectrometry, and the oxidation mechanism of Terramycin by EF system was studied.

2. Preparation and characterization of CuO-graphene cathode

In this section, we will describe the preparation and the properties of CuO-graphene cathode and discuss its electro-catalytic characterization. Graphene (layer of 8–10, diameter of 20–40 μm, thickness of 6–15 nm, purity 99.5%) was purchased from Ningbo Morsh Technology Company in China. Nano-CuO was prepared by the hydrothermal method. The CuO-graphene composite cathode was prepared by loading CuO onto graphene powder. CuO and graphene were mixed in an ethanol/PTFE emulsion (60% by volume), sonicated for 5 min to disperse the mixture completely, and then dried at 50°C to form a dough-like paste. Two pieces of the paste were fixed on a stainless steel mesh current collector and pressed at 30 MPa for 2 min. The electrode was refluxed in acetone for 24 h to remove ethanol and surface PTFE. The dimensions of the composite cathode were 1.0 cm^2 × 5.0 mm (see **Figure 1**; the composite cathode was firmly compressed with the edge sealed with epoxy coating to yield an exposed area of 1 cm^2).

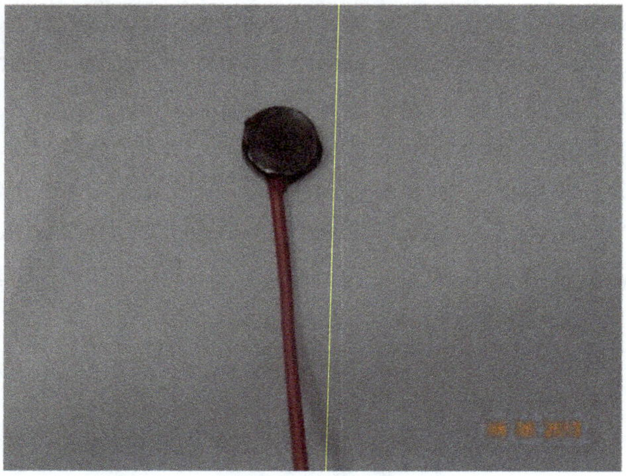

Figure 1. The photograph of the CuO-graphene electrode.

CHI-660E electrochemical station was employed to study the electro-catalytic characteristics of the CuO-graphene cathode at room temperature. A Pt net (99.99% purity, Tianjin Aida Technology Co., Ltd.) of 2.0 cm^2 in area was used as the counter electrode and a saturated calomel electrode (SCE) equipped with a Luggin capillary as the reference electrode. **Figure 2** displayed the cyclic voltammetry curves of the CuO-graphene electrode under neutral pH conditions after being scanned eight times. The electrode showed good electro-catalytic properties under the working voltage. There were two pairs of redox peaks in the system medium. The current peak at 0.246 V vs. SCE is produced by the cathodic reduction of O_2 to H_2O_2 and that at −0.231 V vs. SCE is produced by the cathodic reduction of H^+ to H_2 [12]. Thus, there are two degradation pathways that may occur in the EF system. One is the anodic electrochemical oxidation process on the Pt anode [13] and the other is the cathodic ·OH oxidation process in which the CuO-graphene cathode catalyzes the two-electron reduction of O_2 to H_2O_2 under certain potential, and the H_2O_2 may be converted to ·OH [14].

$$H_2O_2 + e \rightarrow OH^- + \cdot OH \quad (4)$$

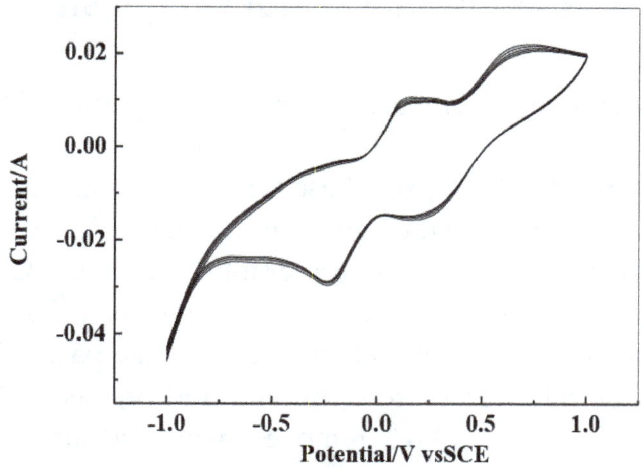

Figure 2. The CV curves of CuO-graphene cathode in a 10 g/L Na_2SO_4 solution, with a scan rate of 1 mV/s.

The specific capacitance is the ratio of capacity and mass of the electrode, which could represent the reversibility and stability of the as-prepared electrode while charging/discharging. **Figure 3** showed the charge/discharge curve of CuO-graphene cathode in a 10 g/L Na_2SO_4 solution with a scan rate of 1 mV/s, a current of 10 mA and a potential range of −0.8 to +0.8 V (vs. SCE). After nine repeated cycles, the charging and discharging curve still showed good symmetry, the potential was essentially linear with time and the charge-discharge curve slope dV/dt was still constant, suggesting that the CuO-graphene cathode material had a good reversibility [15].

EIS is a rapid, in situ and non-destructive technique for investigating the electrochemical properties of the electrode surface and affording information on the impedance changes of the electrode surface during the modification process [16, 17]. **Figure 4** illustrated the Nyquist diagrams and Bode plots of the CuO-graphene cathode under different reduction potentials in 10 g/L Na_2SO_4 solution. It can be seen that all the impedance spectra exhibited a capacitive semicircle at the medium-high frequency and a bigger capacitive semicircle at the medium-low frequency. Two-time constants were observed in the Bode plots under different reduction potential. R_s is the solution resistance and the medium-high frequency circuit Q_m–R_m corresponds to the capacitance and the ohmic process within the porous structure; the medium-low frequency circuit Q_{dl}–R_{ct} is the double-layer capacitance and the kinetic process of electrochemical reaction (the O_2 reduction into H_2O_2) [18, 19]. For the capacitive loops, the coefficients n_1 and n_2 represent a depressed feature in the Nyquist diagram. The EIS data was simulated by ZSimpWin software through **Figure 5** [20, 21]. It was seen that the equivalent circuit fitted the experimental data well in most of the frequency ranges (**Figure 4**), suggesting that the equivalent circuit (**Figure 5**) was suitable.

According to the fitted results in **Table 1**, the values of R_s remained constant as the reduction potential increased. When the reduction potential changed from 0 to −250 mV, the R_{ct} values

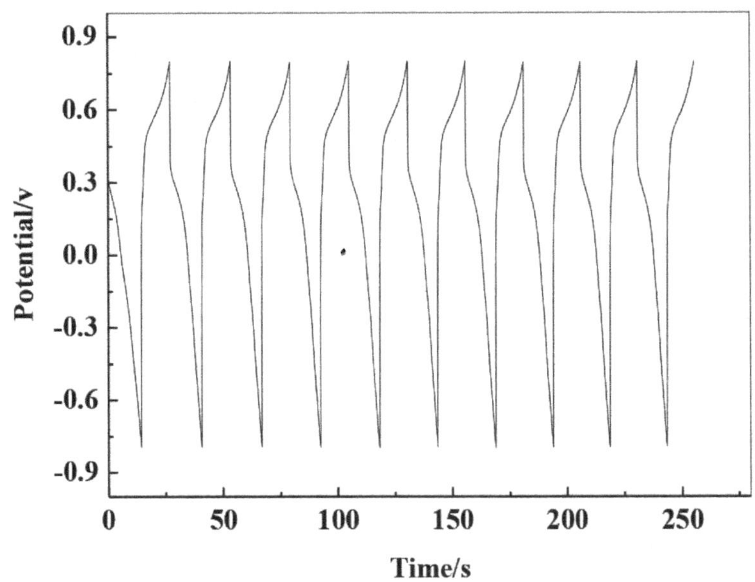

Figure 3. The charge-discharge of the CuO-graphene cathode in a 10 g/L Na_2SO_4 solution, at a scan rate of 1 mV/s and current of 10 mA.

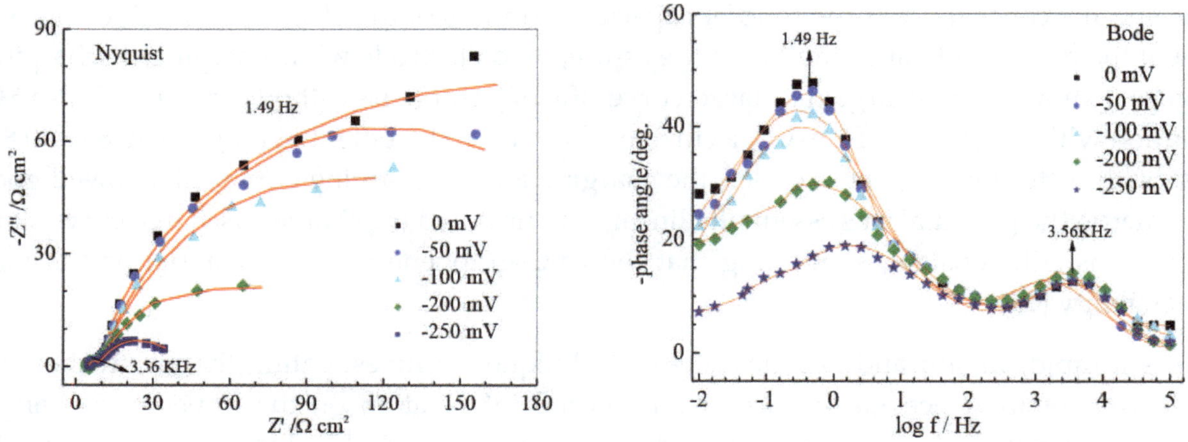

Figure 4. The experimental (dots with different symbols) and the fitted (solid lines) EIS curves of the CuO-graphene cathode according to the presented equivalent circuit immersed in a 10 g/L Na_2SO_4 solution under different reduction potentials.

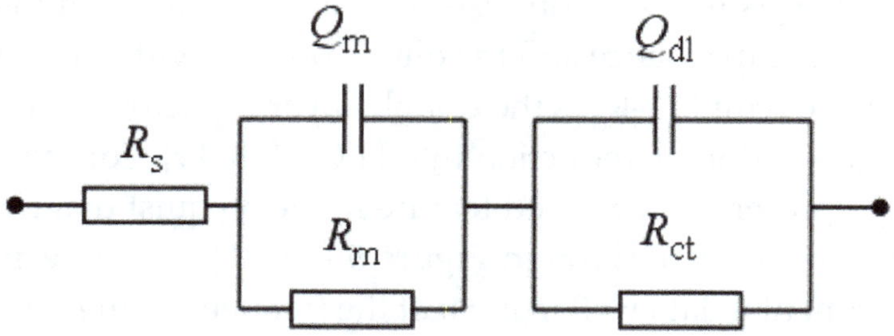

Figure 5. The equivalent circuit models used to fit the experiment impedance data of the CuO-graphene cathode in a 10 g/L Na_2SO_4 solution.

decreased from 420.5 to 30.12 Ω cm^2, and R_m values decreased from 9.51 to 3.45 Ω cm^2, indicating that the reduction rate of O_2 into H_2O_2 on CuO-graphene cathode was very rapid under higher reduction potential, that is to say, the degradation rate based on Cuo-graphene cathode of EF system increased with reduction potential. It should be noted that higher reduction potential might cause the electrolysis of water [22].

Cathodic potential	R_s Ω cm²	Q_m (F cm⁻²)	n_1	R_m (Ω cm²)	Q_2 (F cm⁻²)	n_2	R_{ct} (Ω cm²)
0 mV	3.724	0.0092	0.80	9.51	0.00071	0.89	402.5
-50 mV	4.37	0.00031	0.77	6.06	0.015	0.66	223.9
−100 mV	3.91	0.000054	0.81	4.70	0.016	0.65	175.9
−200 mV	4.40	0.000025	0.99	2.83	0.021	0.54	108.4
−250 mV	4.04	0.000064	0.86	3.45	0.019	0.59	30.12

Table 1. EIS fitting parameters of the CuO-graphene cathode under different reduction potentials (0, −50, −100, −200 and −250 mV) in 10 g/L Na_2SO_4 solution.

3. Degradation mechanism of Terramycin in EF system

3.1. The degradation process of Terramycin in EF system

The EF system degradation of Terramycin was performed in an undivided cell (50 mL) with a two-electrode system at room temperature. A Pt net (99.99% purity, Tianjin Aida Technology Co., Ltd) of 1.0 cm^2 in area was used as the anode. The as-prepared CuO-Graphene electrode was employed as the cathode. The initial concentration of Terramycin was 20 ug/L. Na_2SO_4 aqueous solution of 10 g/L was used as the electrolyte to increase the conductivity. The initial pH of the Terramycin solution was neutral at 7.0. The solution was magnetically stirred at room temperature during the whole reaction period. Before degradation experiments, the system was stirred for 40 min to establish adsorption/desorption equilibrium between the solution and electrodes, and then different voltages were applied and 1 mL samples were collected and analyzed with HPLC spectra to examine the decomposition of Terramycin.

Analysis of HPLC: The mobile phases were pump-mixed dynamically from oxalic buffers (including 1 mmol/L of EDTA solution) and acetonitrile at specified compositions. The flow rate was 1.0 mL/min and the injection volume was 20 mL. The UV detection was at 254 nm, and the column temperature was maintained at 35°C [23].

Figure 6 showed the changes of HPLC chromatograms of Terramycin in the EF system under 5 V in 10 g/L of Na_2SO_4 aqueous solution. The mobile phase was V(acetonitrile): V(oxalic buffers) = 30: 70 of mixture, adding a little EDTA in the mobile phase in order to chelate with Cu^{2+}/Cu^+ which were corroded from the CuO-Graphene electrode. It was observed that the peak of Terramycin at the retention time of 3.06 min gradually decreased with time, and disappeared thoroughly after 180 min, which means that the tetracycline had degraded completely.

3.2. Effect of applied potential on degradation of Terramycin in EF system

Figure 7 depicted the degradation rate of Terramycin in the EF system under different applied potentials. As we know, the generation rate of H_2O_2 on the cathode depended on the applied potential, and the degradation efficiency of the EF system is proportional to the quantity of ·OH produced by Fenton's reaction [24, 25]. When no power was applied, no H_2O_2 was

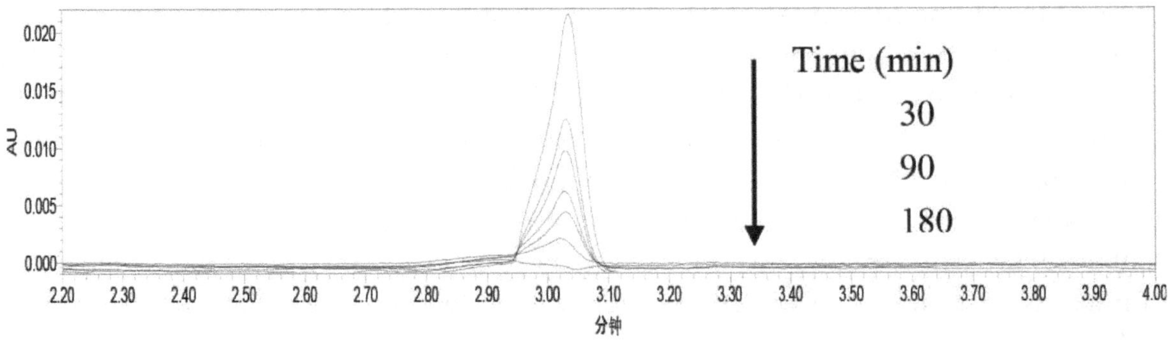

Figure 6. HPLC spectra of Terramycin during degradation in the EF system with the CuO-Graphene cathode: initial concentration of Terramycin, 20 ug/L; electrolyte, 10 g/L of Na_2SO_4; applied potential, 5 V.

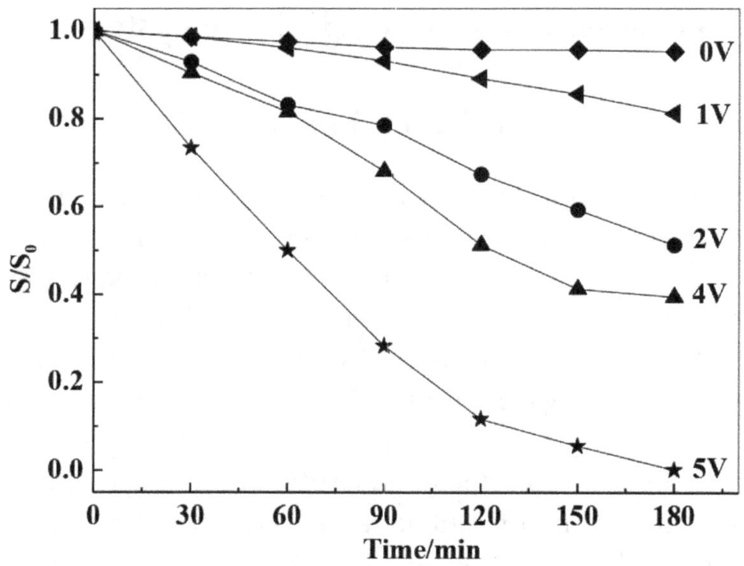

Figure 7. Effects of applied potentials on Terramycin degradation in the EF system with the CuO-Graphene cathode. S: HPLC peak area and S_0: the initial peak area. Initial tetracycline concentration, 20 ug/L; initial pH = 7.0; electrolyte, 10 g/L Na_2SO_4.

generated in EF system and there was no Terramycin degradation in the blank solution [26, 27]. The degradation efficiency of Terramycin increased with the addition of applied potentials; only 60.6% of Terramycin decomposed in 180 min at 4 V, and when the potential reached 5 V, Terramycin decomposed 100% in 180 min. This phenomenon can be ascribed to the fact that a higher potential leads to higher H_2O_2 formation rate, resulting in an increased rate of ·OH formation from Fenton's reaction. When the applied potential exceeded 5 V, the electrolysis of water increased.

3.3. Concentration of oxidizing species during degradation of Terramycin

The effectiveness of the Fenton reaction depends on the catalytic generation of H_2O_2, which yields the highly reactive species, ·OH radical, that oxidizes organic material non-selectively. In the EF system, the electrogenerated levels of H_2O_2 and ·OH determine the rate and efficiency of Terramycin degradation. **Figure 9** displayed the concentration changes of H_2O_2 and relative content of ·OH as the degradation reaction proceeded.

Analysis of H_2O_2: The concentration of H_2O_2 was determined by the oxidation of peroxidase catalytic method. The reaction was carried out in a colorimetric tube (10 mL). 30 μL of 1.0 mg/mL horseradish peroxidase (POD), 1 mL of NaH_2PO_4-NaOH buffer solution (pH = 6.80), 150 μL of 10.0 mg/mL N,N-diethyl-p-phenylenediamine (DPD) and 1 mL sample were added into the colorimetric tube in turns. After oxidation for 15 min, the concentration of H_2O_2 was analyzed spectroscopically at the wavelength of 510 nm [28]. Analysis of ·OH: The relative content of ·OH was determined by fluorescence spectrometry. 1 mL of 0.01 mol/L benzoic acid and 1 mL sample from Fenton's reaction system was added into a 10 mL colorimetric tube and the volume was kept constant at 5mL; after a reaction time of 10 min, the content of ·OH was analyzed with an excitation wavelength of 309 nm and an emission wavelength of 410 nm, and the slit width of excitation and emission was 5 nm [29].

Little amounts of H_2O_2 and ·OH were detected in the blank solution (the applied potential was 0 V). With the electrolysis performed, the amount of H_2O_2 concentration accumulated in the EF system was calculated. As 5 V was applied, the concentration of H_2O_2, generated by cathodic reduction, increased with time. After 90 min, the H_2O_2 concentration reached 2.3 × 10^{-5} mol/L and fluctuated over time. This indicates that the H_2O_2 production increased in the beginning and then was consumed in the degradation of Terramycin. These results were in good agreement with the previous experimental studies by Wang [14], who conducted diethyl phthalate degradation using a self-made Pd/C gas-diffusion cathode and a Ti/IrO_2/RuO_2 anode. By comparing **Figure 8a** and **b**, it can be seen that the electrogenerated H_2O_2 in the EF system was quickly converted to ·OH. The content of ·OH increased by prolonging the electrolysis time and then reached a steady state after about 120 min [30, 31]. The degradation of Terramycin in the EF system was in accordance with the characteristic of the Fenton reaction, indicating that the degradation process involves ·OH [32, 33].

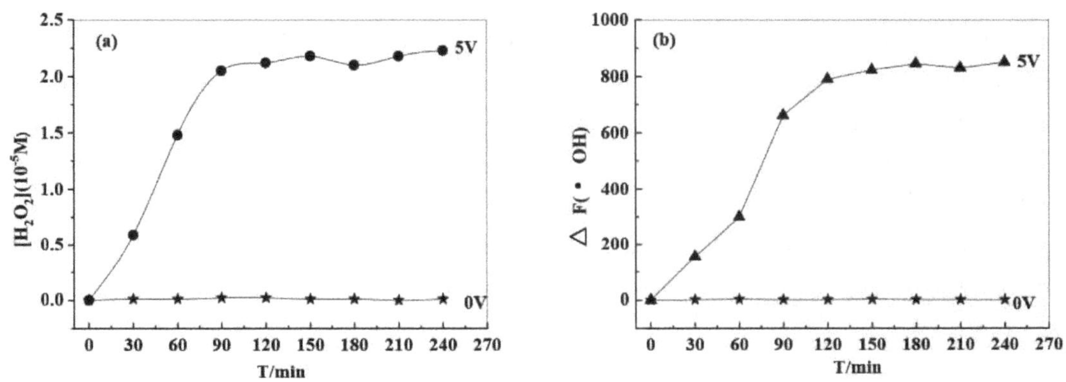

Figure 8. Evolution of the concentration of H_2O_2 (a) and relative content of ·OH (b) during the degradation of Terramycin in the EF system, initial Terramycin concentration, 20 μg/L; initial pH = 7.0; electrolyte, 10 g/L Na_2SO_4.

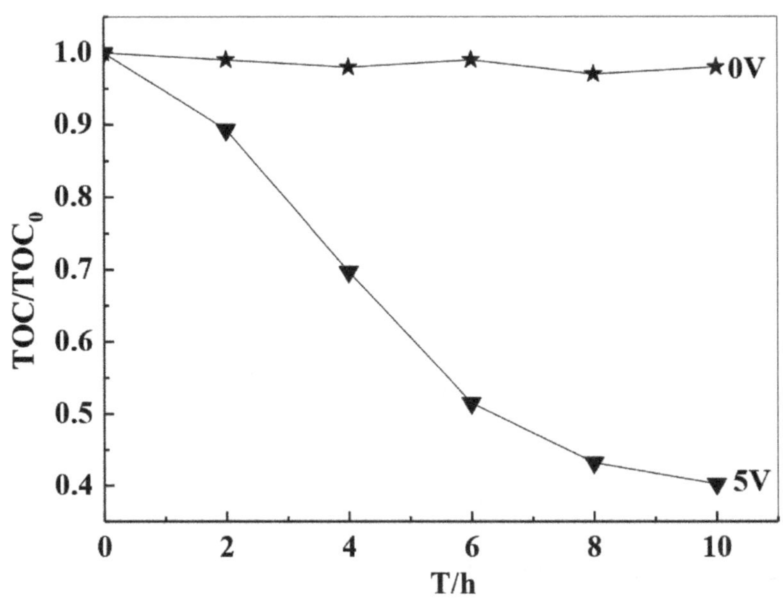

Figure 9. Evolution of TOC for the degradation of Terramycin, initial tetracycline concentration, 60 ug/L; initial pH = 7.0; electrolyte, 10 g/L Na_2SO_4.

3.4. The deep oxidation of Terramycin in EF system

In order to measure the extent of Terramycin mineralization during the degradation process, the TOC change of solution was tracked during degradation, and the results were displayed in **Figure 9**. The TOC did not change when no power was applied to the electrode, but as a voltage of 5 V was applied, the mineralization rate of Terramycin was 59.8% in 10 h. These results indicated that the EF system not only degraded the Terramycin but also oxidized over half of the Terramycin into CO_2 (mineralization) [34, 35].

3.5. IR spectra of the Terramycin in EF system during degradation

IR spectra were used to identify non-mineralized products as the degradation proceeded and the results are displayed in **Figure 10** [36–38]. Initially, the bands 1450 and 930 and 750–800 cm^{-1} corresponded to benzene ring vibration and C—H in the benzene ring plane bending vibration, respectively. The in-plane bending vibration absorption peak at 1620–1650 cm^{-1} indicated the existence of O=C—NH_2, and the bands at 1310 and 1180–1240 cm^{-1} indicated the existence of $(N{-}C{<}^{N}_{N})$. With the photoelectron-catalytic reaction proceeding, bands of $(N{-}C{<}^{N}_{N})$ vibrations disappeared after 10 h, and at the same time, a new band at 1110 cm^{-1}emerged, which indicated that the tertiary amine was wiped off from the skeleton molecule of Terramycin and degraded into small amine molecules. Meanwhile, the 750–800 and 930 cm^{-1} bands of characteristic vibrations in the benzene ring bond disappeared. These results showed that the Terramycin aromatic structure was damaged.

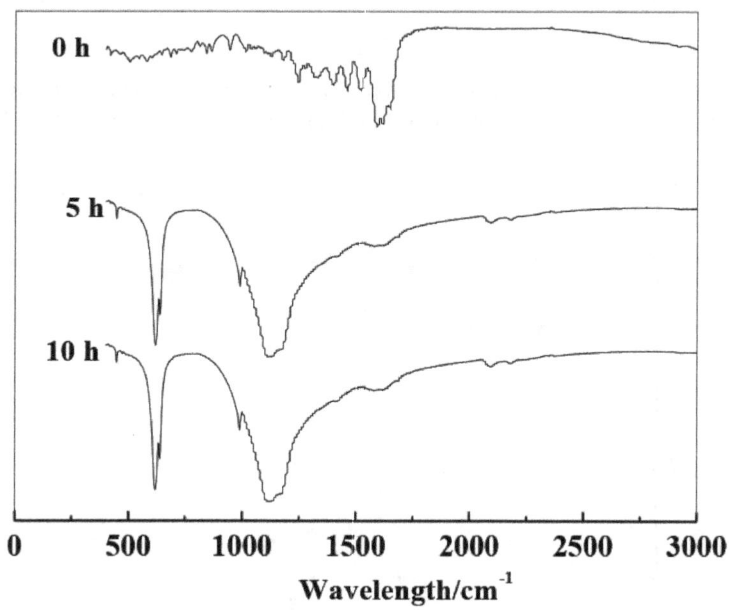

Figure 10. The IR spectra of the Terramycin in the EF system after different times.

3.6. The stability of CuO-graphene electrode in EF system

The stability of the electrode is very important for its application in environmental technology [39, 40]. In order to study the stability of CuO-graphene electrode, the degradation of Terramycin with the same CuO-graphene electrode under the same conditions is also evaluated. **Figure 11** showed the degradation stability of the CuO-graphene cathode over five cycles in the EF system. After being reused five times, the degradation rates of Terramycin for cycle one, two, three, four and five were 97.84, 95.67, 95.21, 96.54 and 93.88%, respectively, indicating that the CuO-graphene electrode was stable and reusable.

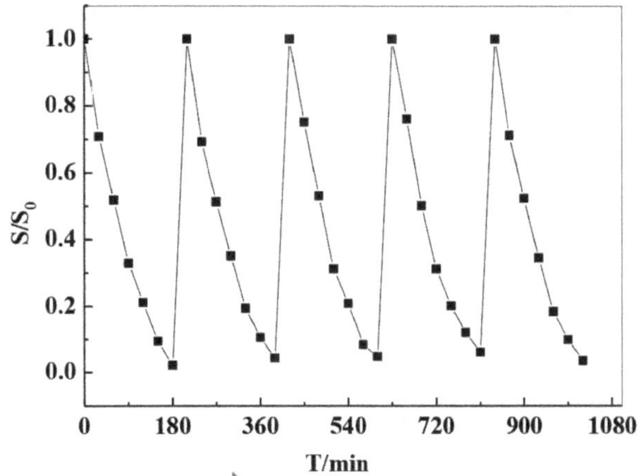

Figure 11. The degradation stability of the CuO-graphene cathode over five cycles in the EF system, initial Terramycin concentration, 20 μg/L; initial pH = 7.0; applied potential, 5 V; electrolyte, 10 g/L of Na_2SO_4.

4. Conclusion

A new cathode-CuO-graphene electrode was prepared by pressing nano-CuO and graphene. The degradation of Terramycin was investigated in an undivided electrolysis cell with a CuO-graphene cathode and Pt net anode. The electro-catalytic performance of the CuO-graphene cathode was studied by cyclic voltammetry, charge/discharge curve and electrochemical impedance spectroscopy. The primary conclusions were listed as follows:

(1) The CuO-graphene cathode was stable and exhibited high catalytic activity in neutral Na_2SO_4 solution.

(2) Under the optimal conditions of using 10 g/L of supporting electrolyte Na_2SO_4 at 5 V with a pH of 7.0 and reaction time of 180 minutes, respectively, up to 100% of 20 ug/L of Terramycin degraded.

(3) Terramycin can be mineralized to carbon dioxide and small molecules; this new EF system may soon be developed for other toxic organic contaminants degradation.

Acknowledgements

The authors wish to acknowledge the financial support of the National Science Foundation of China (No. 41506098), Zhejiang Province Preferential Postdoctoral Science Foundation

(No. BSW1502160) and Open Fund Project of the Key Laboratory of Marine Materials and Related Technologies (LMMT-KFKT-2014-008) in the Chinese Academy of Sciences.

Author details

Liu Shuan[1], Zhao Xia[1], Jiang Xin[1], Zhao Xiaorong[2*], Pu Jibin[1*], Wang Liping[1] and Zhou Kaihe[3]

*Address all correspondence to: rongrong315@ctgu.edu.cn; pujibin@nimte.ac.cn.

1 Key Laboratory of Marine Materials and Related Technologies, Zhejiang Key Laboratory of Marine Materials and Protective Technologies, Ningbo Institute of Materials Technologies and Engineering, Chinese Academy of Sciences, Ningbo, China

2 College of Hydraulic and Environmental Engineering, China Three Gorges University, Yichang, China

3 State Grid Zhejiang Electric Power Corporation Ningbo Power Supply Company, Ningbo, China

References

[1] AM Díez, O Iglesias, E Rosales, et al. Optimization of two-chamber photo electro Fenton reactor for the treatment of winery wastewater. Process Safety and Environmental Protection, 101 (2016) 72–79.

[2] RS Rocha, FL Silva, RB Valim, et al. Effect of Fe^{2+} on the degradation of the pesticide profenofos by electrogenerated H_2O_2. Journal of Electroanalytical Chemistry, 783 (2016) 100–105.

[3] J Meijide, J Gómez, M Pazos, et al. Degradation of thiamethoxam by the synergetic effect between anodic oxidation and Fenton reactions. Journal of Hazardous Materials, 319 (2016) 43–50.

[4] HC Lan, JF Li, M Sun, et al. Efficient conversion of dimethylarsinate into arsenic and its simultaneous adsorption removal over FeCx/N-doped carbon fiber composite in an Electro-Fenton process. Water Research, 100 (2016) 57–64.

[5] H Nakagawa, S Takagi, J Maekawa et al. Fered-Fenton process for the degradation of 1,4-dioxane with an activated carbon electrode: a kinetic model including active radicals. Chemical Engineering Journal, 296 (2016) 398–405.

[6] PY Liang, M Rivallin, S Cerneaux, et al. Coupling cathodic Electro-Fenton reaction to membrane filtration for AO7 dye degradation: a successful feasibility study. Journal of Membrane Science, 510 (2016) 182–190.

[7] JH Shen, YF Li, YH Zhu, et al. Aerosol synthesis of Graphene-Fe_3O_4 hollow hybrid microspheres for heterogeneous Fenton and Electro-Fenton reaction. Journal of Environmental Chemical Engineering, 4 (2016) 2469–2476.

[8] NL Chen, YP Ren, PP Kong, et al. In situ one-pot preparation of reduced graphene oxide/polyaniline composite for high-performance electrochemical capacitors. Applied Surface Science, 392 (2017) 71–79.

[9] JP Cheng, L Liu, KY Ma, et al. Hybrid nanomaterial of α-Co(OH)$_2$ nanosheets and few-layer graphene as an enhanced electrode material for supercapacitors. Journal of Colloid and Interface Science, 486 (2017) 344–350.

[10] K Deshmukh, MB Ahamed, KK Sadasivuni, et al. Graphene oxide reinforced poly (4-styrenesulfonic acid)/polyvinyl alcohol blend composites with enhanced dielectric properties for portable and flexible electronics. Materials Chemistry and Physics, 186 (2017) 188–201.

[11] K Kim, J Bae, MY Lim, et al. Enhanced physical stability and chemical durability of sulfonated poly(arylene ether sulfone) composite membranes having antioxidant grafted graphene oxide for polymer electrolyte membrane fuel cell applications. Journal of Membrane Science, 525 (2017) 125–134.

[12] XR Zhao, S Liu, YP Huang. Removing organic contaminants by an Electro-Fenton system constructed with graphene cathode. Toxicological & Environmental Chemistry, 98 (2015) 530–539.

[13] P Canizares, JA Domingues, MA Rodrigo, et al. Effect of the current intensity in the electrochemical oxidation of aqueous phenol wastes at an activated carbon and steel anode. Industrial and Engineering Chemistry Research, 38 (1999) 3779–3785.

[14] H Wang, DZ Sun, ZY Bian. Degradation mechanism of diethyl phthalate with electrogenerated hydroxyl radical on a Pd/C gas-diffusion electrode. Journal of Hazardous Materials, 180 (2010) 710–715.

[15] S Liu, XR Zhao, HY Sun, et al. The degradation of tetracycline in a photoelectro-Fenton system. Chemical Engineering Journal, 231 (2013) 441–448.

[16] L Gu, S Liu, HC Zhao, et al. Facile preparation of water-dispersible graphene sheets stabilized by carboxylated oligoanilines and their anticorrosion coatings. ACS Applied Materials and Interfaces, 7 (2015) 17641–17648.

[17] S Liu, L Gu, HC Zhao, et al. Corrosion resistance of graphene-reinforced waterborne epoxy coatings. Journal of Materials Science & Technology, 32 (2016) 425–431.

[18] S Liu, HY Sun, HJ Fan et al. Effects of pH and Cl^- concentration on corrosion behavior of the galvanized steel in simulated rust layer solution. Corrosion Science, 65 (2012) 520–527.

[19] H Wei, D Ding, S Wei, et al. Anticorrosive conductive polyurethane multiwalled carbon nanotube nanocomposites. Journal of Materials Chemistry A, 1 (2013) 10805–10813.

[20] HY Sun, S Liu, LJ Sun. A comparative study on the corrosion of galvanized steel under simulated rust layer solution with and without 3.5 wt% NaCl. International Journal of Electrochemical Science, 8 (2013) 3494–3509.

[21] MM Wind, HJW Lenderink. A capacitance study of pseudo-fickian diffusion in glassy polymer coatings. Progress in Organic Coatings, 28 (1996) 239–250.

[22] Y Fan, ZH Ai, LZ Zhang. Design of an Electro-Fenton system with a novel sandwich film cathode for wastewater treatment. Journal of Hazard Materials, 176 (2010) 678–684.

[23] RP Li, Y Zhang, YP Huang, et al. Development and validation of a hydrophilic interaction liquid chromatographic method for determination of aromatic amines in environmental water. Journal of Chromatography A, 1217 (2010) 1799–1805.

[24] M Zarei, A Niaei, D Salari, et al. Removal of four dyes from aqueous medium by the peroxi-coagulation method using carbon nanotube-PTFE cathode and neural network modeling. Journal of Electroanalytical Chemistry, 639 (2010) 167–174.

[25] D Salari, A Niaei, A Khataee, et al. Electrochemical treatment of dye solution containing C.I. Basic Yellow 2 by the peroxi-coagulation method and modeling of experimental results by artificial neural networks. Journal of Electroanalytical Chemistry, 629 (2009) 117–125.

[26] L Fu, S J You, GQ Zhang, et al. Degradation of azo dyes using insitu Fenton reaction incorporated into H_2O_2-producing microbial fuel cell. Chemical Engineering Journal, 160 (2010) 164–169.

[27] GQ Zhang, FL Yang, MM Gao, et al. Electro-Fenton degradation of azo dye using polypyrrole/anthraquinonedisulphonate composite film modified graphite cathode in acidic aqueous solutions. Electrochimica Acta, 53 (2008) 5155–5161.

[28] H Bader, V Sturzenegger, J Hoigne. Photometric method for the determination of low concentrations of hydrogen peroxide by the peroxidase catalyzed oxidation of *N,N*-p-phenylene-diamine. Water Research, 22 (1988) 1109–1115.

[29] S Liu, Y Gu, SL Wang, et al. Degradation of organic pollutants by a Co_3O_4-graphite composite electrode in an Electro-Fenton-like system. Chinese Science Bulletin, 58 (2013) 2340–2346.

[30] HY Zhao, Y Chen, QS Peng, et al. Catalytic activity of MOF(2Fe/Co)/carbon aerogel for improving H_2O_2 and ·OH generation in solar photo–electro–Fenton process. Applied Catalysis B: Environmental, 203 (2017) 127–137.

[31] N Flores, I Sirés, JA Garrido, et al. Degradation of trans-ferulic acid in acidic aqueous medium by anodic oxidation, Electro-Fenton and photoelectro-Fenton. Journal of Hazardous Materials, 319 (2016) 3–12.

[32] ZH Ai, LR Lu, JP Li, et al. Fe@Fe_2O_3 core–shellnanowires as iron reagent. 1. Efficient degradation of Rhodamine B by a novel sono-Fenton process. Journal of Physical Chemistry C, 111 (2007) 4087–4093.

[33] ZH Ai, LR Lu, JP Li, et al. Fe@Fe_2O_3 core–shell nanowires as iron reagent. 2. An efficient and reusable sono-Fenton system working at neutral pH. Journal of Physical Chemistry C, 111 (2007) 7430–7436.

[34] RS Rocha, FL Silva, RB Valim, et al. Effect of Fe^{2+} on the degradation of the pesticide profenofos by electrogenerated H_2O_2. Journal of Electroanalytical Chemistry, 783 (2016) 100–105.

[35] PY Liang, M Rivallin, S Cerneaux, et al. Coupling cathodic Electro-Fenton reaction to membrane filtration for AO7 dye degradation: a successful feasibility study. Journal of Membrane Science, 510 (2016) 182–190.

[36] F Sopaj, N Oturan, J Pinson, et al. Effect of the anode materials on the efficiency of the Electro-Fenton process for the mineralization of the antibiotic sulfamethazine. Applied Catalysis B: Environmental, 199 (2016) 331–341.

[37] C Espinoza, J Romero, L Villegas, et al. Mineralization of the textile dye acid yellow 42 by solar photoelectro-Fenton in a lab-pilot plant. Journal of Hazardous Materials, 319 (2016) 24–33.

[38] Y Zhang, MM Gao, XH Wang, et al. Enhancement of oxygen diffusion process on a rotating disk electrode for the Electro-Fenton degradation of tetracycline. Electrochimica Acta, 182 (2015) 73–80.

[39] OG Rodríguez, JA Bañuelos, AE Ghenymy, et al. Use of a carbon felt-iron oxide air-diffusion cathode for the mineralization of Malachite Green dye by heterogeneous Electro-Fenton and UVA photoelectro-Fenton processes. Journal of Electroanalytical Chemistry, 767 (2016) 40–48.

[40] W Chen, XL Yang, JF Huang, et al. Iron oxide containing graphene/carbon nanotube based carbon aerogel as an efficient E-Fenton cathode for the degradation of methyl blue. Electrochimica Acta, 200 (2016) 75–83.

4

Graphene Oxide–Antimony Nanocomposite Sensor for Analysis of Platinum Group Metals in Roadside Soil Samples

Bongiwe Silwana, Charlton van der Horst,
Emmanuel Iwuoha and Vernon Somerset

Abstract

The present study introduced a very sensitive and low-cost analytical procedure based on voltammetry to study platinum group metals in road dust and roadside soil matrices. Cathodic stripping voltammetry in conjunction with a reduced graphene oxide-antimony nanocomposite sensor and ICP-MS analysis were used to analyse roadside soil and dust samples. The results were processed to evaluate possible pollution in order to map the distribution of the PGMs along specific roads in the Stellenbosch area, outside Cape Town. The results revealed that within each site under investigation, Pd was more abundant than Pt and Rh using both voltammetric and spectroscopic methods. The AdDPCSV results obtained showed concentrations for Pd(II) ranging between 0.92 – 4.0 ng kg^{-1}. For Pt (II), the concentrations ranged between 0.84 – 0.99 ng kg^{-1}. For Rh(III), concentrations ranged between 0.42 – 1.21 ng kg^{-1}. The ICP-MS results showed Pd concentrations ranging between 0.01 – 0.34 μg kg^{-1}. For Pt the concentrations ranged between 0.004 – 0.07 μg kg^{-1}. For Rh, concentrations ranged between 0.002 – 0.26 μg kg^{-1}. The analysis showed significant levels of all PGMs in soil and dust samples analysed. Metal concentration in dust and soil followed the trend Pd > Pt > Rh using both voltammetric and spectroscopic techniques.

Keywords: adsorptive stripping voltammetry, roadside dust and soil samples, ICP-MS analysis, PGMs, Sensitivity, speciation

1. Introduction

The pollutants present in soil not only affect the micro-flora and fauna of soil ecosystems but also contaminate fruits and vegetables in close vicinity. They result in formation of leachates that ultimately contaminate the ground water passing through the soil profiles or by directly entering the surface water system through run off. The consumption of the contaminated food and water causes severe health problems in human beings [1].

Research has shown that platinum group metals (PGMs) could become available to plants and especially to humans and livestock, who consume plants growing on a contaminated soil [2]. In general the Pt(II), Pt(V) and Pd(II) toxic effects on the cellular level are comparable to the toxic effects of Cd(II) and Cr(VI), sometimes even exceeding the damage induced by the above species of Cd and Cr [3].

Numerous studies have concentrated on the heavy metal (e.g. Pb or Cd) concentrations in the environment, while relatively few studies have focussed on PGMs in the environment [4–7]. PGMs released from the catalytic converters are primarily bound to aluminium oxide particles [8–9]. Until recently, they were regarded as inert elements, but studies have shown that they may be soluble and quite reactive. High levels of PGMs in human body may cause asthma, nausea, hair loss, abortion, dermatitis and other health problems [10].

PGMs are emitted both in nanometre and micrometre particle sizes on the surfaces of road dusts, roadside soils and plants during vehicle operation as evidenced by atmospheric particle analysis, but their presence is increasing as a consequence of the use of catalytic converters introduced to reduce gaseous automobile emissions [11–14].

Studies conducted by Barbante et al. [15] showed that the spread of PGMs in the environment is rather a global process, despite some works asserting the low transportability of these pollutants. Therefore, it is important to study how these PGMs species may become chemically/biochemically active and mobile in interactions with various environmental matrices under the changing weather conditions, in order to give a prediction on the possible hazards of PGMs for human health.

PGMs contamination initially occurs in airborne particulate matter (PM), roadside dust, soil, sludge and water, etc., which finally results in uptake and bioaccumulation of these elements in the living organisms. The high demand for on-site monitoring of these metals in environmental matrices has attracted much attention from researchers since the concentration is still very low at present [16]. Moreover, in samples of complex composition, the determination of these elements is still a special challenge to trace element analysis [17–19].

There is a need for dynamic analytical techniques able to routinely measure parameters that are reliable predictors of bio-availability and bio-uptake and, thus of the potential ecotoxicological risk [20]. Ideally such measurements should be made *in situ* to avoid sampling and sample handling artefacts. This is a considerable challenge, techniques must be sufficiently robust to withstand field deployment, provide adequate sensitivity and furnish a signal that can be quantitatively interpreted in terms of chemical species present [20].

The measurements of total metal concentrations alone do not yield sufficient information on the ecotoxicological impact and fate of metallic elements. The measurement of specific species or groups of homologous species is therefore essential. The major existing techniques for trace-metals analyses are spectroscopic (in particular, graphite furnace atomic absorption spectroscopy (GF-AAS) and inductively coupled plasma mass spectroscopy (ICP-MS), neutron activation analysis (NAA) and voltammetric and chronopotentiometric analysis) [21].

Briefly, the advantage of NAA, GF-AAS and ICP-MS compared to voltammetric techniques is that they are applicable to a large number of elements. Their major drawbacks are their much higher cost and, above all, the fact that they allow measurements of total concentration only. Consequently, speciation measurements, using these detection techniques, are feasible only by coupling them with separation and extraction procedures. However, such steps significantly increase the risk of contaminations or chemical species modifications during sample storage or sample handling, and dramatically increase the cost of analyses. This is a major barrier to their application to routine speciation measurements on large sample sets, even though it would be the only means to interpret correctly the environmental impact of metals [21].

In contrast, the characteristics of voltammetric techniques make them particularly well suited for automatic (thus low cost) *in situ* speciation measurements, with no or minimum sample change, i.e., under conditions that dramatically minimize contaminations by reagents or losses by adsorption on containers. It has some intrinsic advantageous features like quick analysis speed, high precision and accuracy, relatively portable and inexpensive instrumentations and can be used 'on-site' measurements, biomedical, environmental and industrial applications [22–24].

The main objective of this work was aimed to apply ICP-MS analysis and to evaluate and compare the established simple method for analysis of road dust and roadside soil samples, with a voltammetric technique. Numerous studies of PGMs have been done in South Africa using ICP analysis and focusing on Pt analysis [25]. The only PGMs studies with stripping voltammetry data have been reported by van der Horst et al. [26]. This study supplied baseline information to enable future studies and detection of PGMs in soil and dust samples. Accordingly, since catalytic car converters may increasingly contribute as a main source of these environmental pollutants. Hence, there was a need for the present study so as to compare the concentrations of PGMs with studies in other parts of the world to ascertain the impact of catalytic car converters on environment quality. Furthermore, this work is also a contribution to the monitoring of PGMs in the South African environment since very little is known about the influence of these elements on organisms exposed to this type of contamination.

2. Experimental

2.1. Study area and site description

The study was conducted on the analysis of road dust and roadside soil samples, collected on busy roads of the Stellenbosch area in the Western Cape Province (South Africa). Soil and dust samples were collected on the Bottelary road that joins the R304 road into Stellenbosch (**Figure 1**). Sampling sites and corresponding samples collected on the Bottelary Road were labelled BOT1 to BOT4. A second set of samples were collected at sampling sites on the Old Paarl road that joins the R44 (or Adam Tas Road) that also runs into Stellenbosch. These sampling sites and corresponding samples were labelled OP1 to OP4 (**Figure 1**).

The overall sampling sites were categorized based on the level of traffic flow. **Figure 1** represents a schematic map of the study area and the sampling points.

2.2. Sample collection

Samples of soil and dust were collected from the eight sampling locations highlighted (**Figure 1**), by demarcating a square segment of about 10 cm^2 in the top soil. The grass on the top of some soil samples was removed, while the soil around the grassroots was shaken off to form part of the sample collected.

To avoid contamination during the collection and processing of sample material, nitrile gloves were worn at all times. It was hypothesised that soil and dust inputs from the surrounding area would be a primary source of particulates containing PGMs to settle on road surfaces. The OP sites are categorized as high-traffic zones. Site OP4 and BOT4 are the junctions that receive heavy traffic from R44 to R304, respectively. All sampling devices were cleaned by rinsing with pure water, 25% HNO_3 and 10% HCl, followed by drying for several days before sampling. Road dust and roadside soil were collected using hand auger and brush.

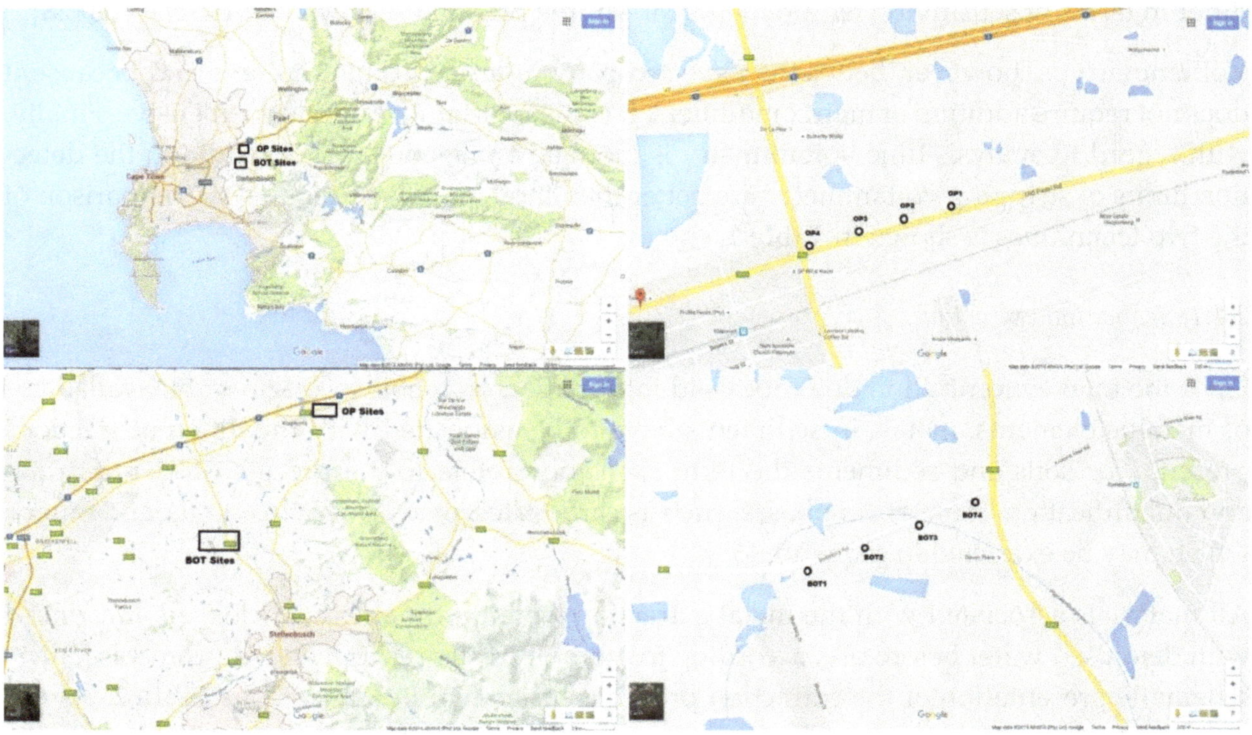

Figure 1. Schematic map of the study area near Stellenbosch (outside Cape Town), showing the sampling sites on Bottelary and Old Paarl Roads.

The samples were placed into double zip-locked plastic bags labelled BOT1–BOT4, OP1–OP4 to represent Bottelary and Old Paarl sites, respectively. The samples were transported to the laboratory on ice and transferred to clean storage containers, before drying was done. Soil samples were air-dried, grinded using mortar and pestle, and then sieved through a 1-mm mesh sieve to remove debris and small stones. The samples were then transferred into a zip-lock bag and stored in the fridge at 4°C until analysis. The working environment was also kept clean to avoid any cross-contamination [27].

2.3. Reagents

All chemical reagents used for calibration plots and design of electrodes were of analytical grades purchased from Aldrich (Germany). All other reagents used were provided by Merck (South Africa) and included sodium acetate, ammonia with a purity of ca. 25% ammonium chloride, hydrochloric acid (purity 32%) and nitric acid (purity 55%). Glacial acetic acid and ethanol (95%) were purchased from Kimix (South Africa). The electro-analytical measurements were performed in 0.2 M sodium acetate buffer (pH = 5.2) solution and a MilliQ deionized water (resistivity of 18.2 MΩ cm) was used for the preparation of the reagent solutions.

2.4. Instrumentation and analysis

In most inorganic analytical laboratories, PGMs analysis is usually performed using inductively coupled plasma-mass spectroscopy (ICP-MS). For the purpose of this study, both voltammetric and ICP-MS techniques were used. The two techniques were used to compare results, considering aspects of both analytical and technical details. These analytical techniques differ in terms of sensitivity, requirements for sample preparation and costs of analysis [28].

Voltammetry is, however, better than spectroscopy in terms of interferences and because it does not require addition of matrix modifiers. It is, also, certainly much less expensive. Finally, with regard to analysis time voltammetry is more advantageous [28]. In addition the detection limits of stripping voltammetry are better than those of ICP-MS analysis. Comparison of the two techniques is shown in **Table 1**.

2.4.1. Sequential extraction

Since the total concentration does not yield information on mobility, origin or bioavailability of metallic elements, metals in sediments have to be associated with the different fractions present. For soils and sediments the term speciation refers to the process of identification and quantification of metals in phases, such as carbonates, oxides, organic matter and others, which may be extracted in sequence [35].

All materials associated with the metal extraction were thoroughly acid-cleaned and rinsed with deionised water before use, according to internationally recommended protocols [36]. A schematic presentation of the extraction procedure is shown in **Figure 2**. In **Table 2**, a summary of the stepwise sequential extraction procedure with pH solution values is shown, which involved a carbonate-bound fraction, an iron manganese-bound fraction and an organic-bound fraction [36–39].

Parameters	ICP-MS	AdDPCSV	Ref.
Detection limit	ng L^{-1}	pg L^{-1}	[29–30]
Sample mineralization	Yes	Yes	[28]
Matrix modifier	Yes	No	[28]
Interferences	High	Very few	[31]
Sample volume required	Very small	Very small	[30]
Ease to use	Moderate easy	Very easy	[30]
Method development	Difficult	Very easy	[30, 32]
Capital costs	Very high	Very low	[21, 28, 30]
Sensitivity	Very high	Extremely high	[21, 33]
Sample preparation	Destructive	Non-destructive	[33, 34]

Table 1. Comparison of elemental techniques for the analysis of PGMs.

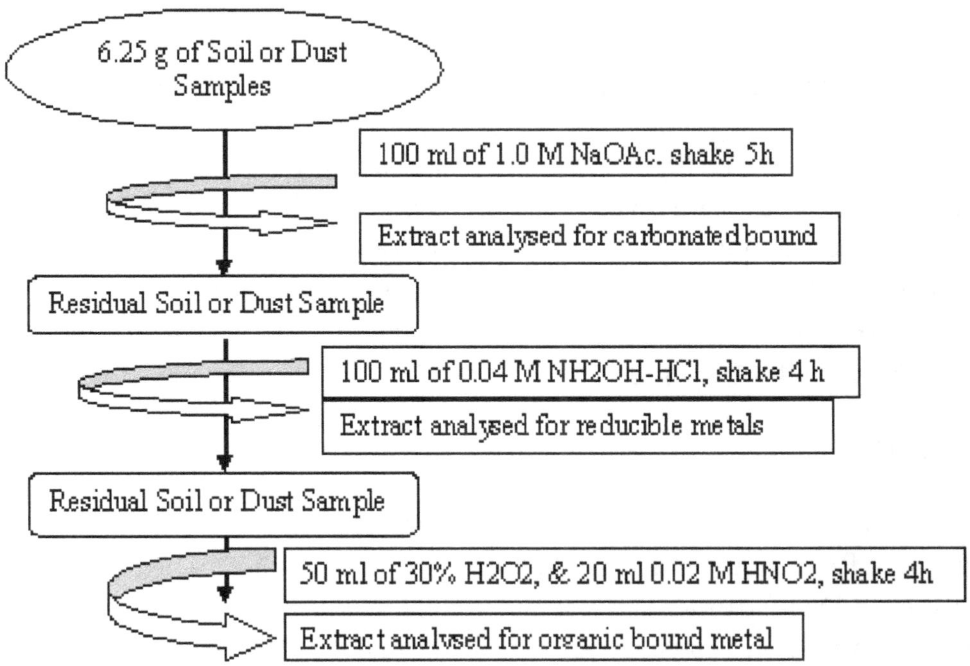

Figure 2. Sequential extraction procedure for trace metals in sediment, soil and dust samples.

Extraction fraction	Reactant concentration; pH
Carbonated bound	1.0 M NaOAc (sodium acetate); 2.0 pH = 4
Fe-Mn oxide bound	0.04 M NH_2OH–HCl (hydroxylamine hydrochloride), in 25% acetic acid; pH ≤ 2
Organic bound	A hydrogen peroxide (H_2O_2, 30%) solution including a 0.02-M HNO_3 solution; pH ≤ 2

Table 2. Summary and description of extracts used for each extraction step and the extraction phases of the sediments in the sequential extraction procedure [36].

Sample introduction			
Sample uptake rate		55 s	
Nebulizer type		Micromist	
Nebulizer gas		Argon, 0.9 L min^{-1}	
ICP-MS		**Data acquisition**	
RF power	1590 W	**Acquisition mode**	Peak hopping
Pulse stage voltage	15.70 V	**Stabilization time**	25 ms
Lens voltage	Omega Lens used	**Integration time**	0.5 s
Plasma gas	Argon, 0.98 L min^{-1}		
Auxiliary gas	Argon, 0.11 L min^{-1}		
Cones	Platinum		

Table 3. ICP-MS operating conditions for the determination of PGMs.

2.4.2. ICP-MS analysis

The analysis of the extracts produced by a sequential extraction procedure was achieved by using ICP-MS (Agilent Technologies, 7500 CX, Chemetrix, Midrand, RSA). Typical instrument operating conditions for the ICP-MS are listed in **Table 3**. The effects of plasma power, gas flow and sample depth were investigated in terms of PGM signal and interference formation. The standard solution of each of the metal ion was made in glass volumetric flask using double de-ionised water (18.2 mΩ cm) and the environmental samples were filtered through a 0.1 mm millipore filter. The concentration of the metals was then determined by means of extrapolation from a calibration plot.

2.4.3. Voltammetric analyses

Concentration of Pd, Pt and Rh in different speciation fractions were determined by adsorptive differential pulse cathodic stripping voltammetry (AdDPCSV). The instrument used was Epsilon analyser (BASi Instruments, West Lafayette, IN, USA) equipped with a three electrode system. The electrode set-up consisted of a conventional three electrode configuration, which comprised a glassy carbon (GC) working electrode, a Pt wire as counter electrode and silver/silver chloride (Ag/AgCl) as a reference electrode. Alumina micro polish and polishing pads (Buehler, IL, USA) were used for electrode polishing. A 0.2 M NaOAc buffer (pH = 5.2) solution was used as the supporting electrolyte. The volume of the voltammetric cell was 25 ml.

The sample containing each of the selected PGMs was conditioned by the addition of 0.2 M NaOAc buffer (pH = 5.2) solution and 1×10^{-5} M dimethylglyoxime (DMG) chelating reagent solution. Afterwards, 1 ml of sequentially extracted sample was placed into the voltammetric cell and the buffer was added. The solution was de-oxygenated for 150 s with high purity nitrogen and the adsorption study of PGMs was done using a pre-concentration potential of −1.2 V (vs. Ag/AgCl), under stirring conditions of 200 rpm for 120 s. The parameters measured in this technique are recorded in **Table 4**.

AdDPCSV	
Potential window	0.6–1.5 V
Electrodeposition potential	−1.2 V
Reduction time	120 s
Speed	200 rpm
Rest time	10 s
Time of deaeration with nitrogen	150 s

Table 4. Main parameters used in the adsorptive differential pulse cathodic stripping voltammetry.

Metal contents were determined by comparing voltammograms peak height from the sample with heights corresponding to standards on calibration graphs. The standard addition was used as a way of checking applicability of the calibration graph. All determinations were performed with five repetitions. From that series, basic statistical information was collected: mean values and standard deviations were calculated.

3. Results and discussion

3.1. Quality assurance

The analytical precision and quality assurance for the overall procedure was done by testing three sub-samples for each of the dust and soil samples. The AdDPCSV measurement showed good recoveries for all PGMs ions, with percentage recoveries between 89 and 112%. A blank sample consisting of an electrolyte (0.2 M sodium acetate buffer (pH = 5.2)) was used as a correction factor, the true concentration of the samples were taken as the difference between the measured concentrations of the samples to those of the blanks for each metal.

The procedural blanks were routinely analysed for every five samples. Since the certificate samples are not available, the analytical procedure was checked for accuracy by analysing enriched samples prepared by us. For metal contents lower than the detection limits a known quantity of Pd, Pt and Rh standards to samples of soil and dust was spiked [27]. The relative standard deviations on the metals measurements of recovery were found to be less than 15%.

3.2. Study of interferences

Precise and selective measurement of PGMs present in real sample matrices is a challenging task, as there are other commonly encountered cations and anions normally present in the real samples along with PGMs, posing a serious problem for electrochemical analysis. After obtaining the limits of detection, quantification and study of recovery, some possible interference in soil samples were investigated. Possible interfering ions of iron (Fe^{2+}), nickel (Ni^{2+}), cobalt (Co^{2+}), sodium (Na^{+}) and copper (Cu^{2+}) were tested using a 4:1 ratio of each metal ion possibly co-existing with selected PGMs. Interferences arising from oxygen containing inorganic ions of phosphate (PO_4^{3-}) and sulphate (SO_4^{2-}) that are expected to co-exist in PGMs were also evaluated [26, 40].

The results obtained have shown that although some of the cations appeared to have an interfering effect towards stripping analysis of PGMs complexes, the interfering was still at tolerable levels, as these ions only lead to approximately 5% decline in stripping peak current signals.

3.3. Analysis of soil samples by voltammetric and spectroscopic techniques

The concentration of the PGMs in soil samples was evaluated by AdDPCSV and ICP-MS analysis using the optimised parameters that are described in **Tables 1** and **2**, respectively.

As already mentioned in the introduction, concentration levels of PGM in environmental samples are very low, although a tendency to higher values is given due to anthropogenic impact. Analysis of the results in **Table 5** for sites OP1–OP4 and BOT1–BOT4 has been conducted to understand the prevalence of the PGMs at these sampling sites.

Table 5 shows mean concentrations and standard deviations for all the investigated sites. Analysis of the AdDPCSV and ICP-MS results that are shown in **Table 5** for Pd suggests that at OP sites, the major concentrations of Pd are associated with the Fe-Mn oxide-bound fraction, while at BOT sites, they are associated with the carbonated-bound fraction. The carbonated-bound fraction is considered to be introduced by land use activities (anthropogenic), which are considered to be weakly bound and, when in contact with the aqueous phase, will become more bio-available [41].

The Fe-Mn oxide-bound fraction has a scavenging effect and sometimes provides a sink of heavy metals, which can be unlocked under the correct redox potential and pH conditions [41]. The ICP-MS results obtained for Pt at the OP sites revealed that major concentrations are found in the Fe-Mn oxide-bound fraction. The BOT sites contained major concentrations of Pt in the carbonated-bound fraction. The results for the AdDPCSV analysis suggests that major concentration of Pt were found at the OP sites, while for the BOT sites the Pt concentrations were found in organic-bound fraction.

The ICP-MS results obtained for Rh at the OP sites revealed that the major concentrations were found in the Fe-Mn oxide-bound fraction, while for the BOT sites the Rh concentrations were obtained in carbonate-bound fraction.

Voltammetric results for Rh(III) analysis suggests that no major concentrations were found in extracted fractions of OP and BOT sampling sites. However, only for site BOT3 it was found that there were high concentrations of Rh in the organic-bound fraction. The results obtained for the samples collected at these sites detected by AdDPCSV analysis for Pd(II) ranged between 0.92 and 4.0 ng kg^{-1}. For Pt(II), the concentrations ranged between 0.84 and 0.99 ng kg^{-1}. For Rh(III), concentrations ranged between 0.42 and 1.21 ng kg^{-1}.

The results obtained for the samples collected at the OP and BOT sites, showed results for the ICP-MS analysis for Pd ranging between 0.01 and 0.34 μg kg^{-1}. For Pt the concentrations ranged between 0.004 and 0.07 μg kg^{-1}. For Rh, concentrations ranged between 0.002 and 0.26 μg kg^{-1}.

Soil samples							
Sites	Fraction	AdDPCSV(ng g^{-1})			ICP-MS (µg g^{-1})		
		Pd(II)	Pt(II)	Rh(III)	Pd	Pt	Rh
OP1	Carbonated	0.96 ± 0.01	0.93 ± 0.01	0.42 ± 0.01	0.02	0.01	0.005
	Fe-Mn	1.14 ± 0.05	0.94 ± 0.01	0.94 ± 0.01	0.198	0.01	0.01
	Organic	0.99 ± 0.01	0.92 ± 0.01	0.98 ± 0.01	0.01	0.004	0.002
OP2	Carbonated	0.98 ± 0.01	0.96 ± 0.04	0.92 ± 0.02	0.03	0.02	0.01
	Fe-Mn	1.17 ± 0.01	0.98 ± 0.01	0.94 ± 0.03	0.17	0.03	0.02
	Organic	0.97 ± 0.03	0.97 ± 0.02	0.96 ± 0.01	0.02	0.01	0.01
OP3	Carbonated	0.96 ± 0.02	0.96 ± 0.03	0.99 ± 0.01	0.04	0.01	0.02
	Fe-Mn	1.17 ± 0.01	0.84 ± 0.06	0.98 ± 0.01	0.15	0.04	0.03
	Organic	0.98 ± 0.01	0.97 ± 0.01	0.98 ± 0.04	0.03	0.03	0.02
OP4	Carbonated	0.98 ± 0.03	0.97 ± 0.01	0.98 ± 0.04	0.06	0.05	0.03
	Fe-Mn	1.17 ± 0.03	0.97 ± 0.01	0.97 ± 0.01	0.13	0.06	0.01
	Organic	0.99 ± 0.01	0.97 ± 0.01	0.97 ± 0.01	0.04	0.03	0.004
BOT1	Carbonated	1.04 ± 0.01	0.94 ± 0.01	0.86 ± 0.03	0.03	0.01	0.004
	Fe-Mn	0.98 ± 0.01	0.96 ± 0.01	0.96 ± 0.06	0.01	0.004	0.003
	Organic	0.94 ± 0.01	0.96 ± 0.01	0.76 ± 0.01	0.01	0.01	0.005
BOT2	Carbonated	1.31 ± 0.01	0.99 ± 0.04	0.99 ± 0.01	0.06	0.03	0.02
	Fe-Mn	0.98 ± 0.01	0.97 ± 0.01	0.98 ± 0.03	003	0.02	0.01
	Organic	0.94 ± 0.01	0.98 ± 0.01	0.75 ± 0.04	0.02	0.01	0.01
BOT3	Carbonated	4.00 ± 0.01	0.95 ± 0.01	0.95 ± 0.01	0.34	0.04	0.26
	Fe-Mn	0.98 ± 0.03	0.98 ± 0.01	0.97 ± 0.01	0.06	0.05	0.04
	Organic	0.97 ± 0.05	0.97 ± 0.03	1.21 ± 0.01	0.04	0.02	0.02
BOT4	Carbonated	0.92 ± 0.06	0.90 ± 0.01	0.86 ± 0.08	0.08	0.05	0.04
	Fe-Mn	0.98 ± 0.01	0.96 ± 0.02	0.86 ± 0.06	0.08	0.07	0.05
	Organic	0.98 ± 0.01	0.98 ± 0.01	0.75 ± 0.03	0.05	0.04	0.04

Table 5. Results obtained for the sequential extraction of PGM concentrations in soil samples obtained using AdDPCSV and ICP-MS analysis techniques.

Comparison of the above results has shown that better detection with AdDPCSV analysis was achieved, compared to the results obtained for ICP-MS analysis in **Table 5**.

Figure 3 displays the results obtained for the comparative analysis of PGMs by ICP-MS and AdDPCSV techniques in soil samples.

The results shown in **Figure 3** indicated that the results obtained for Pd seemed to be the highest at all of the investigated sampling sites, indicating that Pd(II) was the most mobile element in both spectroscopic and voltammetric results. The ICP-MS results showed the highest concentrations at site BOT3, followed by site OP1. AdDPCSV results showed the highest concentrations of Pd(II) at site BOT3, followed by site BOT2. Both the ICP-MS and AdDPCSV analysis were in agreement with the highest concentration of Pd found at site BOT3.

From the results shown in **Figure 3**, the order of abundance of PGMs in the soil samples was found to follow the order Pd > Pt > Rh. In this study, the concentration of Pd(II), Pt(II) and Rh(III) were successfully determined in the soil samples by AdDPCSV technique. The stripping peaks for Pd(II), Pt(II) and Rh(III) were obtained at average peak potentials of about

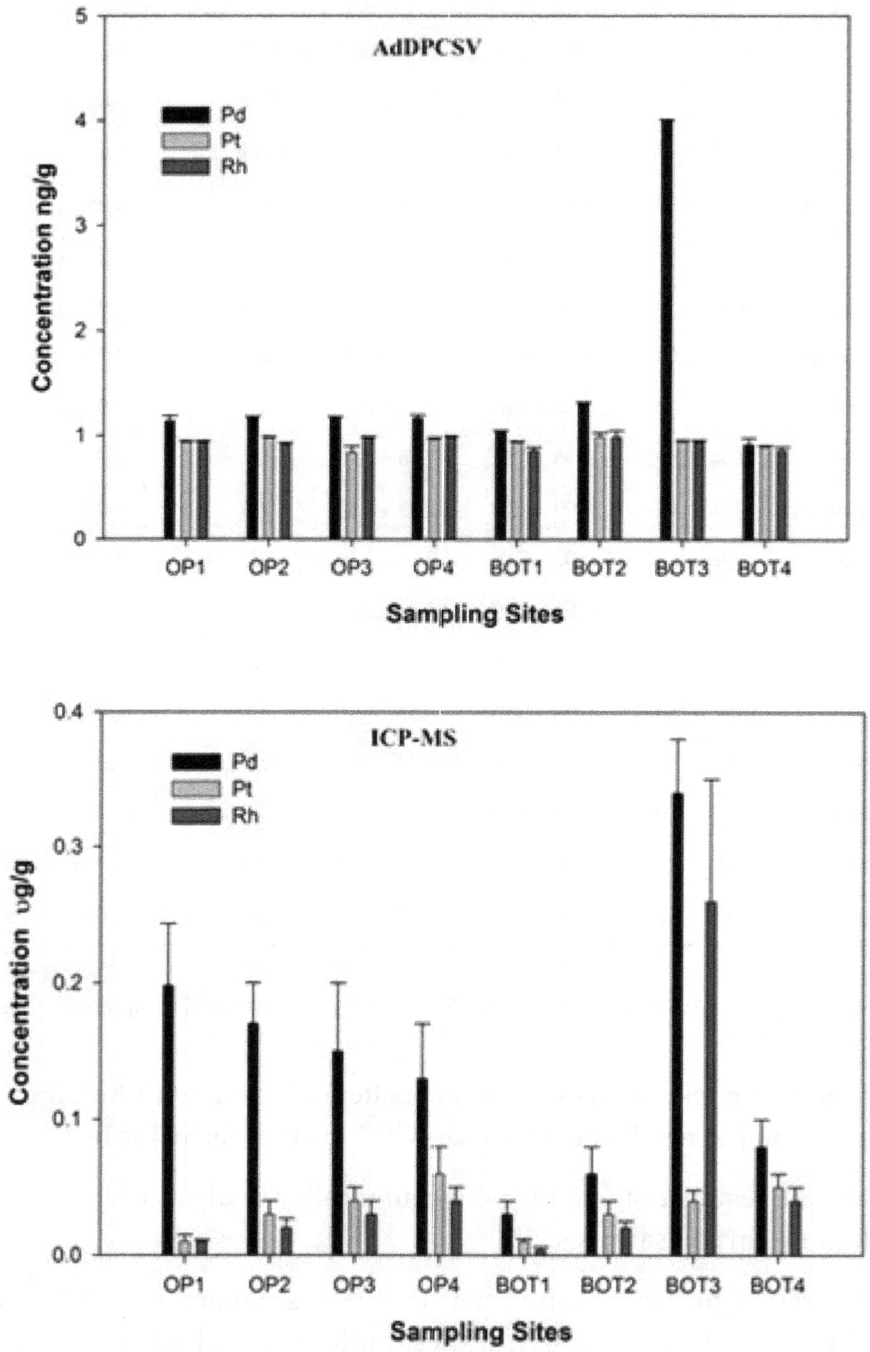

Figure 3. Comparative analysis of PGMs by ICP-MS and AdDPCSV techniques in soil samples.

−0.05, 0.04 and −0.25 V (vs. Ag/AgCl), respectively. These results showed a good comparison to the studies done by van der Horst et al. [26].

3.4. Analysis of dust samples by voltammetric and spectroscopic techniques

The PGMs levels in dust samples of the Bottelary (BOT) and Old Paarl (OP) Road samplings sites are outlined in **Table 6**. The experimental results for the AdDPCSV and ICP-MS analysis, displayed in **Table 4** revealed that for the OP sites, major concentrations of Pd was associated with the Fe-Mn oxide-bound fractions, while for the BOT sites it was associated with the carbonated-bound fractions.

The ICP-MS results obtained at sites OP4 and BOT4 revealed that major concentrations of Pt were found in the Fe-Mn oxide-bound fraction. The results obtained for the AdDPCSV analysis showed that major concentrations of Pt were found at sites OP2 and BOT2 in the carbonated-bound fraction.

The ICP-MS results obtained for Rh at the OP sites revealed no significant difference in concentrations obtained in the Fe-Mn oxide-bound and carbonated-bound fractions, while major concentrations at the BOT sites were obtained in the organic-bound fraction. The results for AdDPCSV analysis showed that no major concentration difference for the Rh concentrations were found in all of the fractions of all sites, except for site BOT3 that showed high Rh concentrations in the organic-bound fraction.

The results obtained for the samples collected at all sites detected by AdDPCSV analysis showed that the Pd(II) concentrations ranged between 0.96 and 1.35 ng kg^{-1}. The Pt(II) concentrations ranged between 0.87 and 0.99 ng kg^{-1}. The Rh(II) concentrations ranged between 0.10 and 0.95 ng kg^{-1}.

The results obtained for the samples collected at all sites detected by ICP-MS analysis showed that the Pd concentrations ranged between 0.02 and 0.15 $\mu g\ kg^{-1}$. The Pt concentrations ranged between 0.01 and 0.08 $\mu g\ kg^{-1}$. The Rh concentrations ranged between 0.002 and 0.26 $\mu g\ kg^{-1}$. These results obtained with AdDPCSV analysis showed better detection limits and were much lower values, compared to the results obtained for ICP-MS analysis (**Table 4**). It was therefore clear that the voltammetric technique was more sensitive than the spectroscopic technique investigated. Analysis of the results in **Figure 4** showed that the Pd concentrations were the highest at all of the investigated sites. It was again observed that Pd was the most mobile element in both spectroscopic and voltammetric results as observed in the results of **Figure 4**. The ICP-MS results showed the highest Pd concentrations at sites OP1 and OP3, followed by sites BOT3, OP4 and BOT2. Analysis of the AdDPCSV results showed the highest Pd concentrations at site BOT3, followed by site BOT4. The order of abundance of PGMs in the soil samples followed the order Pd > Pt > Rh.

3.5. Comparison between road soil and dust samples

Analysis of the PGMs concentrations was found to be generally a little bit higher in the soil samples than in the dust samples. A possible reason for this might be that the soil samples were taken from upper layer (0–2 cm), partly consisting of dust emitted from the roadway, containing particulate matter with PGMs.

Dust samples							
Sites	**Fraction**	**AdDPCSV(ng g^{-1})**			**ICP-MS (μg g^{-1})**		
		Pd	**Pt**	**Rh**	**Pd**	**Pt**	**Rh**
OP1	Carbonated	0.99 ± 0.01	0.93 ± 0.01	0.40 ± 0.01	0.04	0.02	0.01
	Fe-Mn	0.99 ± 0.01	0.98 ± 0.01	0.15 ± 0.01	0.15	0.02	0.01
	Organic	1.05 ± 0.01	0.99 ± 0.01	0.93 ± 0.02	0.02	0.01	0.004
OP2	Carbonated	1.35 ± 0.01	0.93 ± 0.02	0.70 ± 0.01	0.07	0.04	0.03
	Fe-Mn	0.99 ± 0.01	0.97 ± 0.05	0.42 ± 0.06	0.06	0.04	0.03
	Organic	1.05 ± 0.01	0.95 ± 0.01	0.89 ± 0.01	0.03	0.02	0.01
OP3	Carbonated	1.35 ± 0.01	0.92 ± 0.01	0.70 ± 0.01	0.09	0.05	0.04
	Fe-Mn	0.99 ± 0.03	0.98 ± 0.01	0.25 ± 0.01	0.15	0.06	0.05
	Organic	1.00 ± 0.02	0.98 ± 0.01	0.95 ± 0.01	0.05	0.05	0.03
OP4	Carbonated	1.30 ± 0.01	0.96 ± 0.01	0.72 ± 0.03	0.06	0.06	0.05
	Fe-Mn	0.97 ± 0.03	0.96 ± 0.02	0.34 ± 0.03	0.13	0.09	0.06
	Organic	1.00 ± 0.02	0.96± 0.02	0.95 ± 0.04	0.04	0.06	0.03
BOT1	Carbonated	0.98 ± 0.01	0.93 ± 0.01	0.80 ± 0.01	0.03	0.06	0.005
	Fe-Mn	0.98 ± 0.01	0.87 ± 0.04	0.24 ± 0.02	0.01	0.02	0.01
	Organic	1.01 ± 0.01	0.99 ± 0.02	0.92 ± 0.01	0.01	0.02	0.01
BOT2	Carbonated	0.99 ± 0.01	0.99 ± 0.01	0.80 ± 0.01	0.06	0.08	0.03
	Fe-Mn	0.98 ± 0.02	0.89 ± 0.03	0.13 ± 0.05	003	0.04	0.02
	Organic	1.12 ± 0.01	0.99 ± 0.01	0.30 ± 0.02	0.02	0.05	0.03
BOT3	Carbonated	0.97 ± 0.01	0.92 ± 0.01	0.17 ± 0.01	0.34	0.09	0.04
	Fe-Mn	0.97 ± 0.01	0.94 ± 0.01	0.40 ± 0.01	0.06	0.08	0.04
	Organic	1.20 ± 0.01	0.95 ± 0.01	0.60 ± 0.01	0.04	0.06	0.04
BOT4	Carbonated	0.96 ± 0.02	0.89 ± 0.02	0.10 ± 0.08	0.08	0.10	0.05
	Fe-Mn	0.97 ± 0.01	0.91 ± 0.02	0.88 ± 0.06	0.08	0.09	0.07
	Organic	1.18 ± 0.01	0.96 ± 0.01	0.47 ± 0.01	0.05	0.08	0.05

Table 6. Results obtained for the sequential extraction of PGM concentrations in dust samples obtained using AdDPCSV and ICP-MS analysis techniques.

Although the AdDPCSV and ICP-MS techniques have different levels of detection, it was found that both soil and dust samples from all sampling sites have relatively high concentrations of Pd. Therefore, the observed Pd levels in soil and dust may either reflect the earlier introduction of Pd base catalysts or be indicative of the more widespread shift away from Pt towards Pd as the main catalytic component [42, 43].

The Pt: Pd ratios in environmental samples and the results of solubility experiments have suggested that Pd is more soluble than Pt or Rh and thus more susceptible to aqueous transport, resulting in its possible higher levels in the environment [44].

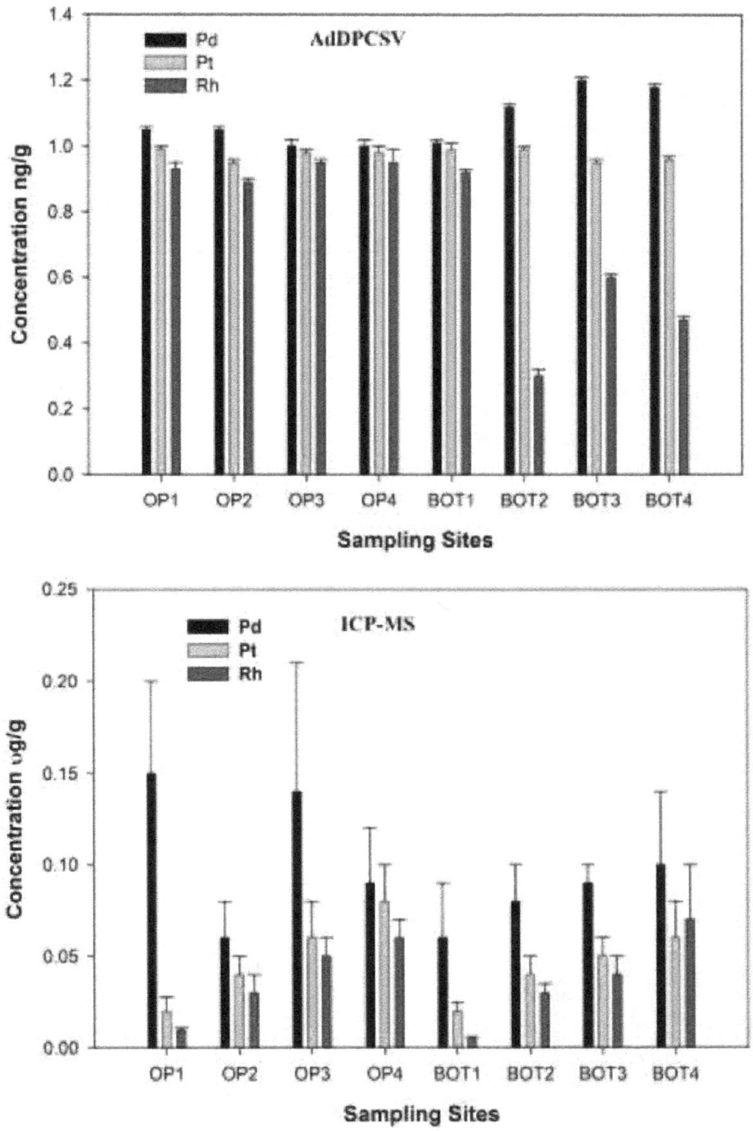

Figure 4. Comparative analysis of PGMs by ICP-MS and AdDPCSV techniques in dust samples.

The results obtained for the AdDPCSV and ICP-MS analysis in this study was also compared to the results obtained in other related studies. These results are displayed in **Tables 7–9**.

For comparison, the PGMs global concentration levels obtained using spectroscopic techniques are shown in **Table** 7. Analysis of the results in **Table** 7 has shown that the past studies in South Africa investigated only platinum using ICP analysis in roadside dust samples. This study shows the concentration of all the investigated PGMs detected up to ng g^{-1}. Although this is an interesting result, it has no confirmation in the literature about Pd and Rh in South Africa. Both the global concentration and the results from the present study reveal that the introduction of catalytic converters to automobiles in the process of reducing the harmful emissions from the exhaust emissions has resulted in the increase of PGMs in roadside dust. This study is therefore updating the results for South Africa and comparing the results available for the international scenario.

The results in **Table 8** shows the levels of PGMs obtained in global roadside soil samples. Analysis of the results in **Table 8** has shown that most of the global studies have been done on ICP analysis for the detection of platinum in roadside soils. Both the global concentration and the results from the present study reveal that the introduction of catalytic converter to automobiles in order to reduce the harmful emissions from car exhaust systems has resulted in the increase of PGMs in roadside dust matrices.

Roadside dust samples					
Country/city	**Pd**	**Pt**	**Rh**	**Unit**	**Reference**
South Africa/ Stellenbosch	20–150	10–80	4–74	ng g^{-1}	This study
South Africa/ Cape Town	–	4	–	ng g^{-1}	[25]
South Africa/Port Elizabeth	–	6	–	ng g^{-1}	[25]
South Africa/ Pretoria	–	23	–	ng g^{-1}	[25]
South Africa/ Rustenburg	–	223	–	ng g^{-1}	[25]
Ghana/Accra	–	39	–	ng g^{-1}	[45]
Germany/ Karlsruhe	–	112–169	–	ng g^{-1}	[44]
Germany/ Stuttgart	–	10	35	ng L^{-1}	[44]
Germany/ Frankfort	6	72	18	ng L^{-1}	[46]
Germany/ Frankfort	21.3	101.3	18.7	ng L^{-1}	[18]
Germany/ Unknown	1–146	5-8	–	ng g^{-1}	[47–49]
UK/Richmond	–	0.42–29.8	–	ng g^{-1}	[50]
UK/Nottingham	92.9	69.55	–	ng g^{-1}	[51]
Sweden/ Gothenburg	80	196	93	ng g^{-1}	[52, 53]
Sweden/ Unknown	213	56	74	ng g^{-1}	[47]
Poland/Bialystok	37.5	110.7	19.6	ng g^{-1}	[54]
Australia/Perth	37.5	110.7	19.6	ng g^{-1}	[55]
Austria/Styria	4	55	10.3	ng g^{-1}	[56]
Mexico/Mexico city	53.2–74	307.5–332.7	26–39.1	ng g^{-1}	[57]
Spain/Madrid	39–191	31–252	11–182.1	ng g^{-1}	[58]
Italy/Rome	102–504	14.4–62.2	1.9–11.1	ng g^{-1}	[59]
Greece/Ioannina	12.1–18.2	3.2–306	6.1–64.	ng g^{-1}	[60]

Table 7. Comparison of global concentration levels of PGMs in roadside dust samples by ICP spectroscopy analysis.

In **Table 9**, the results obtained for the voltammetric analysis of PGMs in roadside dust and soil samples are shown. Very few papers relevant to the voltammetric determination of PGMs in roadside soil and dust are reported in literature.

Roadside soil samples					
Country/City	**Pd**	**Pt**	**Rh**	**Units**	**Reference**
South Africa/ Stellenbosch	10–340	4–70	2–260	ng g^{-1}	This study
Ghana/Accra	–	15	–	ng g^{-1}	[45]
Germany/Hanau	–	23–112	–	ng g^{-1}	[18]
Germany/Mainz	7.2	87	–	ng g^{-1}	[61]
Germany/ Darmstadt	–	7.2–58.6	–	ng g^{-1}	[62]
Germany		15.6–31.7		ng g^{-1}	[63]
Germany		0.9–200		ng g^{-1}	[44]
Germany		27		ng g^{-1}	[64]
UK/Richmond		0.3–8		ng g^{-1}	[50]
UK/Nottingham		0.19–1.33		ng g^{-1}	[50]
UK/Birmingham		0.05–4.45		ng g^{-1}	[50]
Mexico city	307.5–332.7	53.2–74	26–39.1	ng g^{-1}	[58]

Table 8. Comparison of global concentration levels of PGMs in roadside soil samples by ICP spectroscopy analysis.

Country/city	**Pd**	**Pt**	**Rh**	**Units**	**Reference**
Roadside dust samples					
South Africa/ Stellenbosch	0.96–1.35	0.87–0.99	0.10–0.95	ng g^{-1}	This study
South Africa/ Stellenbosch	1.13–3.62	1.41–5.46	3.32–8.70	ng L^{-1}	[26]
Saudi Arabia/Jeddah city	192.3	–	–	μg L^{-1}	[65]
Roadside soil samples					
South Africa/ Stellenbosch	0.92–4.0	0.84–0.99	0.42–1.21	ng g^{-1}	This study
South Africa/ Stellenbosch	0.94–2.5	2.80–4.15	3.65–7.10	ng L^{-1}	[26]

Table 9. Comparison of global concentration levels of PGMs in roadside soil and road dust by voltammetric techniques.

Analysis of the results in **Table 9** has shown a lack of data in global concentrations of PGMs using voltammetric studies. One of the first investigations on roadside dust PGMs levels in Saudi Arabia was undertaken by Ba-Shami et al. [65]. This proves that voltammetric methods for detection of PGMs in road soil and dust samples are still in early stages globally. Most countries of the world have used spectroscopic studies to determine the concentration levels of PGMs in roadside dust and soil. The study by Ba-Shami et al. [65] has shown that voltammetric techniques can be applied to analyse roadside soil and dust samples for Pd(II) levels. In the current study, it was shown that voltammetric studies are sensitive enough to determine Pd(II), Pt(II) and Rh(II) in roadside soil and dust samples.

4. Conclusion

The present study showed the use of AdDPCSV analysis with a GCE/rGO-SbNPs sensor for the voltammetric determination of PGMs in roadside soil and dust samples. At all the sampling sites evidence of Pd, Pt and Rh were observed in different roadside media by AdDPCSV and ICP-MS analysis. The results showed that automobile catalysts are a minor source of PGMs into the South African environment with relatively low concentrations at high volume traffic sites near Stellenbosch, Western Cape. Useful analytical data on platinum group metal concentrations in soil and dust samples have been collected. Matrices effects in the samples were minimised by diluting the samples without the need of prior treatment for voltammetric analysis. The maximum concentrations of PGMs in road soil samples sampled near Stellenbosch using voltammetry were as follows: Pd(II) = 4 ng g^{-1}, Pt(II) = 0.99 ng g^{-1} and Rh(III) = 1.21 ng g^{-1}. The maximum concentration of PGMs in road soil samples near Stellenbosch using spectroscopy were as follows: Pd = 0.34 μg g^{-1}, Pt = 0.07 μg g^{-1} and Rh = 0.26 μg g^{-1}. The maximum concentrations of PGMs in roadside dust samples of Stellenbosch using voltammetry were as follows: Pd(II) = 1.35 ng g^{-1}, Pt(II) = 0.99 ng g^{-1} and Rh(III) = 0.95 ng g^{-1}. The maximum concentration of PGMs in roadside dust samples of Stellenbosch using spectroscopy were as follows: Pd = 0.34 μg g^{-1}, Pt = 0.10 μg g^{-1} and Rh = 0.06 μg g^{-1}. It was interesting to note that the data obtained in this study and compared to the globally available data revealed that Pt was the PGMs being investigated the most using spectroscopy as technique in the past. Unfortunately, not a lot of data are available concerning the voltammetric studies of PGMs in soil and dust samples. The lack of data in this field enables us to make only a few general comments. The improvement of LODs due to the development of novel analytical techniques has recently enabled detection of PGMs concentrations in the pg g^{-1} range. The results of our investigations showed voltammetry as more sensitive than spectroscopic methods. The research study confirmed the significant role of traffic-related activities as a source of particles of anthropogenic origin to urban road surfaces and surrounding environment. These particles contain heavy metals that may become available for transport with storm water runoff and eventually impact on receiving water quality and the aquatic ecosystem. The outcome of this study provides further evidence of the need for voltammetric methods for detection of PGMs pollution. Although environmental levels of PGMs are still relatively low, recent reports of a progressive increase in these concentrations makes further research essential.

Acknowledgements

The authors would like to thank the National Research Foundation (South Africa) and the CSIR (NRE, Stellenbosch) for supporting this research. We further acknowledge the support of the SensorLab, Chemistry Department, University of the Western Cape, Bellville, South Africa. A word of thanks also to the Chemistry Department, Faculty of Applied Sciences, Cape Peninsula University of Technology, Bellville, South Africa.

Author details

Bongiwe Silwana[1,3], Charlton van der Horst[1], Emmanuel Iwuoha[1] and Vernon Somerset[2*]

*Address all correspondence to: somersetv@cput.ac.za

1 SensorLab, Department of Chemistry, University of the Western Cape, Bellville, South Africa

2 Department of Chemistry, Faculty of Applied Sciences, Cape Peninsula University of Technology, Bellville, South Africa

3 Department of Chemistry, Durham University, Durham, United Kingdom

References

[1] Opasola, O.A., Sawyerr, H.O., Bolaji, A.S., Iromini, F.A., Akande, I.O., Femi-Adepoju, A.G., Adepoju, A. O., Fatukasi, B.A., Odeniyi, M.A., 2015. Studies on metal-resistant bacteria isolated from spent-engine oil contaminated soil, in Ogbomoso, Oyo State, Nigeria. Int. J. Environ. Sci. Technol. 5(2), 62–67.

[2] Alloway, B.J., 1995. The origin of heavy metals in soils. In B.J. Alloway (Ed.), Heavy metals in soils. London, UK: Blackie Academic and Professional, pp. 38–57.

[3] Schmid, M., Zimmerman, S., Krug, F., Sures, B., 2007. Influence of platinum, palladium and rhodium as compared with cadmium, nickel and chromium on cell viability and oxidative stress in human bronchial epithelial cell. Environ. Int. 33, 385–390.

[4] Okoro, K.I., Igene, J.O., Ebabhamiegbebho, P.A., Evivie, S.E., 2015. Lead (Pb) and Cadmium (Cd) levels in fresh and smoke-dried grasscutter (Thryonomys swinderianus Temminck) meat. Afr. J. Agric. Res. 10(32), 3116–2122.

[5] Van der Horst, C., Silwana, B., Iwuoha, E., Somerset, V., 2012. Stripping voltammetric determination of palladium, platinum and rhodium in freshwater and sediment samples from South African water resources. J. Environ. Sci. Health, A, 47(13), 2084–2093.

[6] Silwana, B., van der Horst, C., Iwuoha, E., Somerset, V. 2014. Screen-printed carbon electrodes modified with a bismuth film for stripping voltammetric analysis of platinum group metals in environmental samples. Electrochim. Acta. 128, 119–127.

[7] Zimmermann, S., Messerschmidt, J., Von, B.A., Taraschewski, H., Sures, B., 2002. Biological availability of traffic related platinum group elements (palladium, platinum and rhodium) to the zebra mussel in water containing road dust. Environ. Toxicol. Chem. 21, 2713–2718.

[8] Palacios, M. A., Gomez, M. M., Maldovan, M., Morrisson, G., Rauch, S. and McLead, C., 2000. Platinum group elements: quantification in collected exhaust fumes and studies of catalyst surfaces. Sci. Total Environ. 257, 1–15.

[9] Moldovan, M., Gomez, M.M., Palacios, M., 1999. Determination of platinum, rhodium and palladium in car exhaust fumes. J Anal. At. Spectrom. 14, 1163–1169.

[10] Ravindra, K., Bencs, L., Grieken, R. 2004. Platinum group elements in the environment and their health risk. Sci. Total Environ. 318, 1–43.

[11] Silvana A., Ramíreza, Gabriel, J., Gordillo., 2009. Adsorption and reduction of palladium–dimethylglyoxime complex. J. Electroanal. Chem. 629, 147–151.

[12] Cicchella, D., De Vivo, B., Lima, A., Albanese, S., Mc Gill, R.A.R., Parrish, R.R., 2008. Heavy metal pollution and Pb isotopes in urban soils of Napoli, Italy. Geochem. Explor. Environ. Anal. 8 (1), 103–112.

[13] Limbeck, A., Puls, C., Handler, M., 2007. Platinum and palladium emissions from on-road vehicles in the Kaisermühlen Tunnel (Vienna, Austria). Environ. Sci. Technol. 41 (14), 4938–4945.

[14] Kanicky, V., Otruba, V., Mermet, J.M., 1999. Comparison of some analytical performance characteristics in inductively coupled plasma spectrometry of platinum group metals and gold. Talanta. 48 (4–5), 859–866.

[15] Barbante, C., Veysseyre, A., Ferrari, C., Van de Velde, K., Morel, C., Capodaglio, G., 2001. Greenland snow evidence of large scale atmospheric contamination for platinum, palladium and rhodium. Environ. Sci. Technol. 35, 835–839.

[16] Bujdo, M., Hagarova, I., Matu, P., Anecka, L., Kubova, J., 2012. Optimization of determination of platinum group elements in airborne particulate matter by inductively coupled plasma mass spectrometry. Acta Chim. Slov. 59, 124–128.

[17] Jarvis, K.E, Parry, S.J., Piper, J.M., 2001. Temporal and spatial studies of autocatalyst-derived platinum, rhodium and palladium and selected vehicle-derived trace elements in the environment. Environ. Sci. Technol. 35, 1031–1036.

[18] Zereini, F., Dirksen, F., Skerstupp, B., Urban, H., 1998. Sources of anthropogenic platinum-group elements (PGE): automotive catalysts versus PGE-processing industries. Environ. Sci. Pollut. Res. 5, 223–230.

[19] Barefoot, R.R., Van Loon, J.C., 1999. Recent advances in the determination of the platinum group elements and gold. Talanta. 49, 1–14.

[20] Sigg, L., Black, F., Buffle, J., Cao., J., Cleven, R., Davison, W., Galceran, J., Gunkel, P., Kalis, E., Kistler, D., Martin, M., Noel, S., Nur, Y., Odzak, N., Puy, J., van Riemsdijk, W., Temminghoff, E., Tercier-Waeber, M-L., Toepperwien, S., Town, R.M., Unsworth, E., Warnken, K.W.,

Weng, L., Xue, H., Zhang, H., 2006. Comparison of analytical techniques for dynamic trace metal speciation in natural freshwaters. Int. J. Environ. Sci. Technol. 40(6), 1934–1941.

[21] Buffle, J., Tercier-Waeber, M.L., 2005. Voltammetric environmental trace metal analysis and speciation: from laboratory to in situ measurements. Trends Anal. Chem. 24 (3), 172–191.

[22] Wang, Z., Liu, G., Zhang, L., Wang, H., 2012. Bismuth modified hybrid binder carbon paste electrode for electrochemical stripping detection of trace heavy metals in soil. Int. J. Electrochem. Sci. 7, 12326–12339

[23] Economou, A., 2010. Recent developments in on-line electrochemical stripping analysis—An overview of the last 12 years. Anal. Chim. Acta. 683 (1), 38–51.

[24] Tesarova, E., Baldrianova, L., Hocevar, S.B., I. Svancara, I., Vytras, K., Ogorevc, B., 2009. Anodic stripping voltammetric measurement of trace heavy metals at antimony film carbon paste electrode. Electrochim. Acta. 54, 1506–1510.

[25] Rauch. S., Fatoki, O. S. 2010. Platinum and lead in South African road dust. In: Rauch et al. (eds)., Highway and Urban Environment. Springer, The Netherlands. ISBN 978-90-481-3043-6.

[26] Van der Horst, C., Silwana, B., Iwuoha, E., Somerset, V. 2015. Bismuth–silver bimetallic nanosensor application for the voltammetric analysis of dust and soil samples. J. Electroanal. Chem. 752, 1–11.

[27] Orecchio, S., Amorello, D. 2010. Platinum and rhodium associated with the leaves of *Nerium oleander* L.; analytical method using voltammetry; assessment of air quality in the Palermo (Italy) area. J. Hazard. Mater. 174, 720–727.

[28] Locatelli, C., 2006. Simultaneous square wave stripping voltammetric determination of platinum group metals (PGMs) and lead at trace and ultra-trace concentration level, application to surface water. Anal. Chim. Acta. 557, 70–77.

[29] Locatelli, C., Melucci, D., Torsi, G., 2005. Determination of platinum-group metals and lead in vegetable environmental bio-monitors by voltammetric and spectroscopic techniques: critical comparison. Anal. Bioanal. Chem. 382, 1567–1573.

[30] Silwana, B., van der Horst, C., Iwuoha, E., Somerset, V., 2015. Synthesis, characterization and electrochemical evaluation of reduced graphene oxide modified antimony nanoparticles. Thin Solid Films. 592, 124–134.

[31] Markert, B., 1995. Sample preparation for trace element analysis in plant matrices. Sci. Total Environ. 176, 45–61.

[32] Gupta, J.G., Bertrand, N.B. 1995. Direct ICP-MS of trace and ultratrace elements in geological materials after decomposition in a microwave oven. Talanta. 42, 1947–1957.

[33] Muynck, D., Vanhaecke, F., 2009. Development of method based on ICP-MS for the simultaneous determination of elements in bone and dental tissue. Spectrochim. Acta, B. 64, 408–415.

[34] Brown, R.J.C., Milton, M.J.T., 2005 Analytical techniques for trace element analysis: an overview. TrAC Trends Anal. Chem. 24 (3), 266–274.

[35] Jignesh, S., Vineeta, K., Abhay, S., Vilasrao, K., 2012. Analytical methods for estimation of metals. Int. J. Res. Pharm. Chem., 2 (1), 146–163.

[36] Somerset, V., Van der Horst, C., Silwana, B., Walters, C., Iwuoha, E., 2015. Biomonitoring and evaluation of metal concentrations in sediment and crab samples from the North-West Province of South Africa. Water, Air, & Soil Pollut. 226(3), 2329–2329.

[37] Olujimi, O.O., Oputu, O., Fatoki, O., Opatoyinbo, O.E., Aroyewun, O.A., Baruani, J., 2015. Heavy metals speciation and human health risk assessment at an illegal gold mining site in Igun, Osun State, Nigeria. J. Health Poll. 8, 19–32.

[38] Okoro, H.K., Fatoki, O.S., Adekola, F.A., Ximba, B.J., Snyman, R.G., 2012. A review of sequential extraction procedures for heavy metals speciation in soil and sediments. Open Access Scientific Reports. 1(3): 1–9.

[39] Li, L., Xu, Z., Wu, J., Tian, G., 2010. Bioaccumulation of heavy metals in the earthworm Eisenia fetida in relation to bioavailable metal concentrations in pig manure. Bioresource Technol. 101, 3430–3436.

[40] Silwana, B., van der Horst, C., Iwuoha, E., Somerset, V., 2016. Reduced graphene oxide impregnated antimony nanoparticle sensor for electroanalysis of platinum group metals. Electroanalysis, 28: 1597–1607.

[41] Pradhanang, S., 2014. Distribution and Fractionation of Heavy Metals in Sediments of Karra River, Hetauda, Nepal. J. Inst. Sci. Technol. 19 (2), 123–128.

[42] Morera, M.T., Echeverría, J., Garrido, J., 2001. Bioavailability of heavy metals in soils amended with sewage sludge. Can J. Soil Sci. 82, 433–438.

[43] Schuster, M., Schwarzer, M., Risse, G., 2000. Determination of palladium in environmental samples. In: Zereini F, Alt F, editors. Anthropogenic platinum group element emissions and their impact on man and environment. Berlin: Springer, pp. 173 –182.

[44] Schafer, J., Eckhardt, J.D., Berner, Z.A. Stüben, D., 1999. Time-dependent increase of traffic-emitted platinum group elements (PGE) in different environmental compartments. Environ. Sci. Technol. 33, 3166–3170.

[45] Kylander, M.E., Rauch, S., Morrison, G.M., Andam, K., 2003. Impact of automobile emissions on the levels of platinum and lead in Accra, Ghana. J. Environ. Monit. 5, 91–95.

[46] Schafer, J., Puchelt, H., 1998. Platinum group metals (PGM) emitted from automobile catalytic converters and their distribution in roadside soils. J Geochem. Explor. 64, 301–314.

[47] Lesniewska, B.A., Godlewska-Zylkiewicz, B., Bocca, B., Caimi, S., Hulanicki, A., 2004. Platinum, palladium and rhodium content in road dust, tunnel dust and common grass in Bialystok area (Poland): a pilot study. Sci. Total Environ. 321, 93–104.

[48] Sures, B., Zimmermann, S., Messerschmidt, J., von Bohlen, A., Alt, F., 2001. First report on the uptake of automobile catalyst emitted palladium by European eels (Anguilla anguilla) following experimental exposure to road dust. Environ. Pollut. 113, 341–345.

[49] Helmers, E., Mergel, N., Barchet R., 1994. Platinum in ash from sewage sludge incinerators and in grass, UWSF – Z Umweltchem Ökotox. 6, 130–134.

[50] Farago, M.E., Kavanagh, P., Blanks, R., Kelly, J., Kazantzis, G., Thornton, I., Simpson, P.R., Cook, J.M., Parry, S., Hall, G.M., 1996. Platinum metal concentrations in urban road dust and soil in the United Kingdom. Fresenius J. Anal. Chem. 354, 660–663.

[51] Hutchinson, E.J., Farago, M.E, Simpson, P.R., 2000. Changes in platinum concentrations in soils and dusts from UK cities. In: Zereini F, Alt F, editors. Anthropogenic Platinum-Group Element Emissions. Their Impact on Man and Environment. Berlin: Springer-Verlag, pp. 57–64.

[52] Motelica-Heino, M., Rauch, S., Morrison, G.M, Donard. O.F.X., 2001. Determination of palladium and rhodium concentrations in urban road sediments by laser ablation-ICP-MS. Anal. Chim. Acta. 436, 233–244.

[53] Pearson, D.G., Woodland, S.J., 2000. Solvent extraction/anion exchange separation and determination of PGEs (Os, Ir, Pt, Pd, Ru) and Re–Os isotopes in geological samples by isotope dilution ICP-MS. Chem. Geol. 165, 87–107.

[54] Rauch, S., Morrison, G.M., 1999. Platinum uptake by the freshwater isopod Asellus aquaticus in urban rivers. Sci. Total Environ. 235, 261–268.

[55] Whitely, J.D., Murray, F., 2003. Anthropogenic platinum group elements (Pt, Rh, Pd) concentrations in road dust and road side soils from Perth, Western Australia. Sci. Total Environ. 317, 121–135.

[56] Godlewska-Zylkiewicz, B., Zaleska, M., 2002. Preconcentration of palladium in a flow-through electrochemical cell for determination by graphite furnace atomic absorption spectrometry. Anal. Chim. Acta. 462, 305–312.

[57] Morton, O., Puchelt, H., Hernández, E., Lounejeva, E., 2001. Traffic-related platinum group elements (PGE) in soils from Mexico City. J. Geochem. Explor. 72, 223–227.

[58] Gomez, B., Gómez, M., Sanchez, J.L., Fernández, R., Palacios, M.A., 2001. Platinum and rhodium in airborne particulate matter and road dust. Sci. Total Environ. 269, 131–144.

[59] Petrucci, F., Bocca, B., Alimonti, A., Caroli, S., 2000. Determination of Pd, Pt and Rh in airborne particulate and road dust by high resolution ICP-MS: a preliminary investigation of the emission from automotive catalysts in the urban area of Rome. J. Anal. At. Spectros. 15, 525–528.

[60] Tsogas, G.Z., Giokas, D.L., Vlessidis, A.G., Aloupi, M., Angelidis, M.O., 2009. Survey of the distribution and time-dependent increase of platinum-group element accumulation along urban roads in Ioannina (NW Greece). Water, Air Soil Pollut. 201, 265–281.

[61] Müller, M., Heumann, K.G., 2000. Isotope dilution inductively coupled plasma quadrupole mass spectrometry in connection with a chromatographic separation for ultra-trace determinations of platinum group elements (Pt, Pd, Ru, Ir) in environmental samples. Fresenius J. Anal. Chem. 368, 109–115.

[62] Patel, K.S., Sharma, P.C., Hoffmann, P., 2000. Graphite furnace atomic absorption spectrometric determination of palladium in soil. Fresenius J. Anal. Chem. 367, 738–741.

[63] Alt, F., Eschnaur, H.R., Mergler, B., Messerschmidt, J., Tolg, G., 1997. A contribution to the ecology and oenology of platinum. Fresenius J Anal. Chem. 357, 1013–1019.

[64] Zereini, F., Zientek, C., Urban, H., 1993. Concentration and distribution of platinum group elements (PGE) in soil–platinum metal emission by abrasion of catalytic converter materials. UWSF – Z Umweltchem Ökotox. 5, 130–134.

[65] Ba-Shami, R.M., Gazzaz, H., Bashammakh, A.S., Al-Sibaai, A.A., El-Shahawi, M.S., 2014. "Redox behavior and adsorptive cathodic stripping voltammetric determination of nanomolar levels of palladium using a novel Schiff base reagent containing a squaric acid moiety," Anal. Methods. 6(17), 6997–7005.

5

Graphene against Other Two-Dimensional Materials: A Comparative Study on the Basis of Photonic Applications

Rafael Vargas-Bernal

Abstract

Two-dimensional materials represent the basis of technological development to produce applications with high added value for nanoelectronics, photonics, and optoelectronics. In first decades of this century, these materials are impelling this development through materials based on carbon, silicon, germanium, tin, phosphorus, arsenic, antimony, and boron. 2D materials for photonic applications used until now are graphene, silicene, germanene, stanene, phosphorene, arsenene, antimonene, and borophene. In this work, the main strategies to modify optical properties of 2D materials are studied for achieving photodetection, transportation, and emitting of light. Optical properties analyzed here are refractive index, extinction coefficient, relative permittivity, absorption coefficient, chromatic dispersion, group index, and transmittance. The transmittance spectra of various two-dimensional materials are presented here with the aim of classifying them from photonic point-of-view. A performance comparison between graphene and other two-dimensional materials is done to help the designer choose the best design material for photonic applications. In next three decades, a lot of scientific research will be realized to completely exploit the use of 2D materials either as single monolayers or as stacked multilayers in several fields of knowledge with a special emphasis in the optoelectronics and photonic industry in benefit of the industry and ultimately to our society.

Keywords: graphene, 2D materials, photonic devices, optical properties, nanophotonics, optoelectronics

1. Introduction

The urgent need for photonic materials with useful and fascinating properties is leading researchers around the world to exploit technologically Van der Waals materials.Two-dimensional (2D) materials due to their unique optical and electronic properties have an enormous potential in photonics and optoelectronics. These two-dimensional materials are stable, simple, and consisting of layers of few-atom-thick. Particularly, materials such as graphene and its derivatives, two-dimensional materials based on silicon, germanium, tin, phosphorus, arsenic, antimony, bismuth, and boron, as well as transition metal dichalcogenides are being synthesized for the development of innovative applications in the ultraviolet, visible, and infrared and terahertz frequency ranges.

Optoelectronic devices can be defined as electronic devices capable of generating, detecting, interacting with, or controlling light [1, 2]. Main applications of flexible electronics that can be impelled using two-dimensional materials are large arrays of solar cells, wearable electronics, and transparent displays. The presence of a direct band gap in two-dimensional monolayer materials makes them completely attractive for optoelectronic applications. Physical properties of two-dimensional materials that are attractive for the development of photonic applications are mechanical flexibility, electrical conductivity, and optical transparency. Solid-state properties such as morphology and crystallinity directly influence the optical properties of two-dimensional materials. 2D materials have the potential to impact technologies of modulation, transmission, and detection of photonic signals [3–5]. These materials are attractive for photonics and optoelectronic devices, such as displays, touch screens, photodetectors, light-emitting diodes (LEDs), and solar cells. Two-dimensional transition metal dichalcogenides (TMDCs) provide higher optical efficiency than their bulk counterparts due to that these nanomaterials present a direct band gap, which is not found in the volumetric versions [6]. Heterostructures based on TDMCs and other materials such as graphene, *h*-BN, and phosphorene present a band gap that can cover a wide spectral range from visible to mid-IR. Integrated photonic systems based on 2D materials can improve network performance while reducing costs.

Despite all the progress made by researchers so far in the development of two-dimensional materials, it has not been possible to understand all conditions that can be used to tune the optical properties, which influence the development of photonic devices and systems. This work has as aim studying the optical properties of two-dimensional materials for the development of applications in photonics and optoelectronics. The transmittance spectra of various two-dimensional materials are presented here with the aim of classifying these in the different types from photonic point-of-view. A performance comparison between graphene and other two-dimensional materials is done to help to the designer to choose the best design material for photonic applications. This document can be considered as a technical resource for academics, researchers, and engineers working in the photonics and optoelectronics and developing novel technologies based on emerging materials. Additionally, the document provides an overview of the state-of-the-art of the photonic applications based on graphene and other two-dimensional materials. A direct comparison of the performance and qualities that different types of two-dimensional nanomaterials have in the photonic applications is made with the aim of determining the advantages and disadvantages that the designer can exploit technologically in each of these materials. This chapter has been divided as follows: Section 2 introduces basic concepts about the optical properties that can be found in two-dimensional materials and a comparison of optical performance between graphene and other materials is performed. Main applications of two-dimensional materials in optoelectronics and photonics are described in Section 3. Finally, conclusions about of the work are given in Section 4.

2. Optical properties of two-dimensional materials

Light is an electromagnetic radiation within a certain portion or a set of wavelengths of the electromagnetic spectrum. All types of light are emitted, transported, and absorbed in tiny packets called photons and exhibit properties of both wave and particle. The study of light and the interaction of light and matter are known as optics, and it continues being an important research area in modern physics. With the introduction of two-dimensional materials and the control of their optical properties was possible to extend the development of photonic applications to new and novel technological applications of high added value [2].

Main optical properties that allow describing the performance of an engineering material are the following: (1) refractive index, (2) extinction coefficient, (3) relative permittivity, (4) absorption coefficient, (5) chromatic dispersion, and (6) group index. To understand what role each of these optical properties has in the performance of optical materials, the subsequent paragraphs will describe each of them.

Reflection is the change in direction of the beam of light at an interface (boundary between different substances) between two different media so that the wave returns into the medium from which it was generated. This effect can be either specular (mirror-like) or diffuse (retaining the energy, but losing the image) depending on the nature of the interface, as visualized in **Figure 1**. Incident light (illustrated in red) travels in the medium 1 and strikes against a surface, a fraction of the light is reflected in the same medium (shown in blue), and other

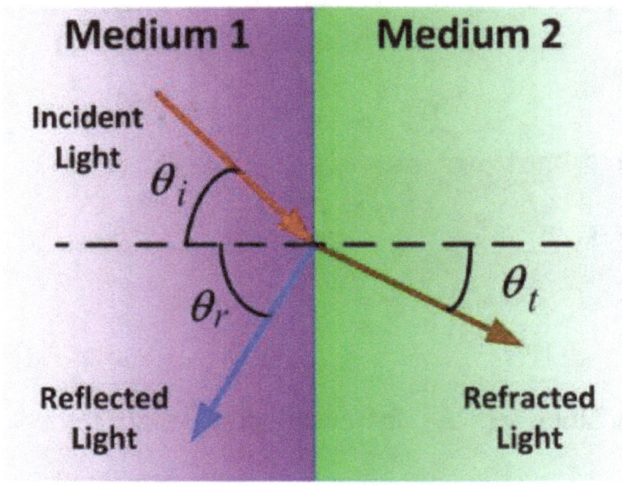

Figure 1. Basic optical phenomena between a pair of physical media.

fraction of the light is transmitted into or refracted into medium 2 (depicted in brown). When light reflects off a material denser than the external medium, it undergoes a polarity inversion (change of direction). In contrast, a less dense, lower refractive index material will refract or transmit the incident light that was applied to it.

Light propagation in absorbing materials is described using a complex-valued refractive index. The imaginary part is related with the attenuation, while the real part accounts for refraction. The refractive index or index of refraction n of a material can be defined as a dimensionless number that describes how light propagates through that medium. This property represents the rate to which the speed of light in vacuum travels through a specific medium. In other words, it determines how much light is refracted when it enters a material. The refraction is presented by the change in direction of propagation of an electromagnetic wave, oscillation of electrical and magnetic fields with varying or fixed wavelength, due to a change in its transmission medium. The wavelength (λ) is the spatial period of a wave, that is, the distance over which the wave's shape repeats, and thus the inverse of the spatial frequency. A transmission medium is a solid material that can propagate light either infrared, visible and/or ultraviolet. In addition, refractive index varies with respect to the wavelength and/or photon energy of the light applied to the material. Visible light is usually defined as that having wavelengths in the range of 400–700 nm (430–750 THz), and it is between the infrared (with longer wavelengths) and the ultraviolet (with shorter wavelengths). Infrared light is an invisible radiant energy with wavelengths between 700 nm (430 THz) and 1,000,000 nm (300 GHz). And, ultraviolet light is an electromagnetic radiation with a wavelength from 10 nm (30 PHz) to 400 nm (750 THz). The electromagnetic spectrum of light is shown in **Figure 2**.

Optical properties determine the response of materials to incident electromagnetic radiation. The optical properties of a material are influenced by the nature of its surface and its electronic structure. The most known properties are reflection, refraction, transmission, and absorption. For each material, the incident radiation is partially transmitted, partially reflected, and partially absorbed.

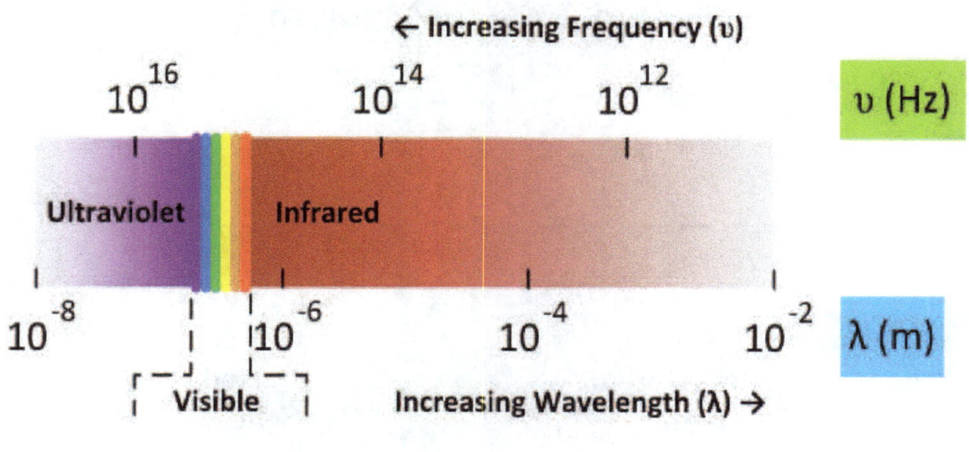

Figure 2. The electromagnetic spectrum of light.

When light passes through a medium, some part of it will always be attenuated. This can be conveniently considered by defining a complex refractive index $n_c = n + j\kappa$, where the real part (n) is the refractive index and indicates the phase velocity, whereas the imaginary part (κ) is called the extinction coefficient or mass attenuation coefficient and it indicates the amount of attenuation when the electromagnetic wave propagates through the material.

The electric permittivity ε and magnetic permeability μ of a medium determine the phase velocity $v = c/n$ of electromagnetic radiation through that medium. These are related by means of the relationship $\varepsilon\mu = 1/v^2$, where c is the speed of light in vacuum and n is the refractive index of the optical medium. The response of materials to alternating electric fields is characterized by a complex permittivity defined as $\varepsilon = \varepsilon' + j\varepsilon''$, where ε' is the real part of the permittivity, which is related to the stored energy within the medium; and ε'' is the imaginary part of the permittivity, which is related to the dissipation (or loss) of energy within the medium. In addition, complex permittivity ε must have poles only for positive imaginary parts, and therefore satisfies the Kramers-Kronig relations. The imaginary part of ε leads to absorption loss if it is positive and absorption gain if it is negative.

Absorption coefficient, attenuation coefficient, or narrow beam attenuation coefficient of a material characterizes how easily it can be penetrated by a beam of light. A large attenuation coefficient means that the beam is quickly "attenuated" (weakened) as it passes through the medium, and a small attenuation coefficient means that the medium is relatively transparent to the beam. The SI unit of attenuation coefficient is the reciprocal meter (m^{-1}). A small attenuation coefficient suggests that the material is relatively transparent, while a larger value suggests greater degrees of opacity. The attenuation coefficient depends upon the type of material and the energy of the radiation. In addition, for electromagnetic radiation, the higher the energy of the incident light and the lower dense the material to be characterized, the lower the corresponding attenuation coefficient will be.

The chromatic dispersion of an optical medium is the phenomenon where the phase velocity and group velocity of light propagating in a transparent medium depend on the optical frequency. It is defined as $dn/d\lambda$. A related quantitative measure is the group velocity dispersion. The group index or group velocity refractive index (n_g) is defined as $n_g = c/v_g$ where v_g is the group velocity. When the dispersion is small, the group velocity can be linked to the phase velocity by the relationship $v_g = v - \lambda(dv/d\lambda)$, where λ is the wavelength in the medium. In this case, the group index can be written in terms of the wavelength dependence of the refractive index as $n_g = n/(1 + (\lambda/n)(dn/d\lambda))$.

A summary of the optical properties of the two-dimensional materials is provided in **Table 1**. These properties cannot be used individually to evaluate the optical performance of a material. The relative permittivity of the material can predict, for example, whether the material can be transparent or not by the sign and value of the real and imaginary parts, but it is not the only argument to conclude about this quality of the material. Relative permittivity of different two-dimensional materials used in nanophotonics and optoelectronics are shown in **Figure 3**.

The refractive index of photonic materials such as two-dimensional materials varies by the photon energy and/or wavelength of the light applied. Using the refractiveindex.info database,

Material	Refractive index (n)	Extinction coefficient (κ)	Relative permittivity (ε)	Absorption coefficient (μ) (cm^{-1})	Chromatic dispersion ($dn/d\lambda$) (μm^{-1})	Group index (n_g)
Graphene	1.0575	2.2962	$\varepsilon' = -4.1541$ $\varepsilon'' = 4.8564$	1.1621E06	25.520	−5.2791
Borophene	2.3465	1.5154	$\varepsilon' = 3.2098$ $\varepsilon'' = 7.1119$	1.0521E06	4.1925	1.5877
Indiene	0.7000	4.7000	$\varepsilon' = -21.600$ $\varepsilon'' = 6.5800$	1.0739E06	1.9000	−0.3450
Stanene	2.1600	6.3500	$\varepsilon' = -35.657$ $\varepsilon'' = 27.432$	1.0931E06	3.1429	−0.13429
Bismuthene	4.0890	4.0890	$\varepsilon' = -2.8729$ $\varepsilon'' = 36.199$	0.0525E06	5.2612	−1.4879
InAs	3.7973	0.49254	$\varepsilon' = 14.177$ $\varepsilon'' = 3.7407$	0.0848E06	−0.94298	4.4857
MoO_3	2.0576	0.59358	$\varepsilon' = 3.8813$ $\varepsilon'' = 2.4427$	0.3004E06	−0.5988	2.2063
MoS_2	2.3156	2.6594	$\varepsilon' = -1.7101$ $\varepsilon'' = 12.316$	1.3459E06	18.621	−2.3079
$MoSe_2$	1.9136	2.3809	$\varepsilon' = -2.0069$ $\varepsilon'' = 9.1125$	1.2050E06	9.7345	−0.50345
$MoTe_2$	1.8903	2.3354	$\varepsilon' = -1.8809$ $\varepsilon'' = 8.8292$	1.1819E06	18.017	−2.5834

Table 1. Optical properties for a small set of two-dimensional materials.

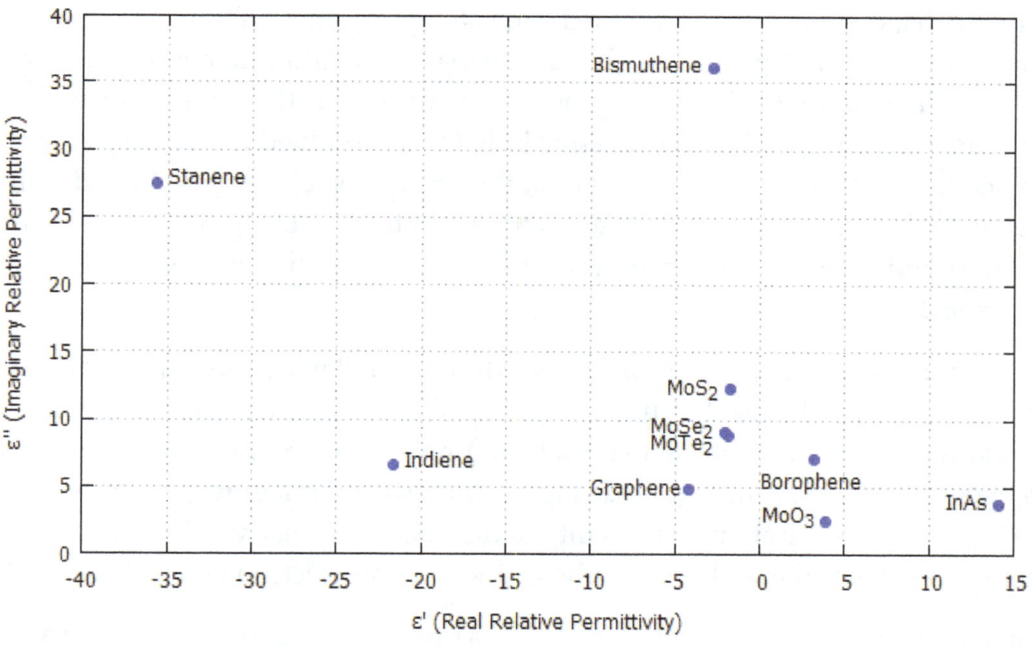

Figure 3. Relative permittivity of different two-dimensional materials.

a pair of graphs has been obtained to know the performance of two-dimensional materials (see **Figures 4** and **5**). Materials with very high or very low refractive index do not have the greatest transparency necessarily, as visualized through the set of graphs obtained. A wide range of optical properties can be obtained through two-dimensional materials.

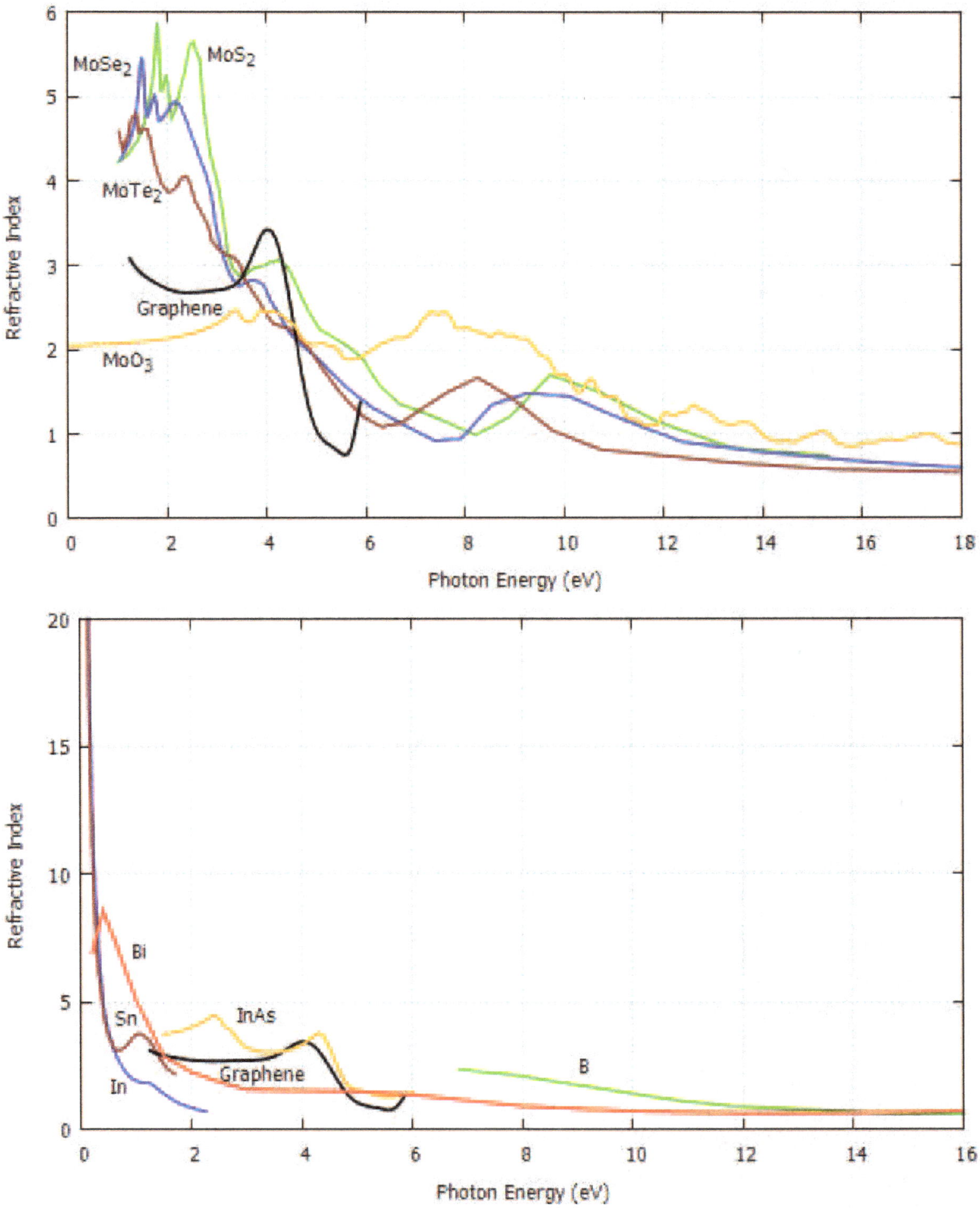

Figure 4. Refractive index versus photon energy for different two-dimensional materials.

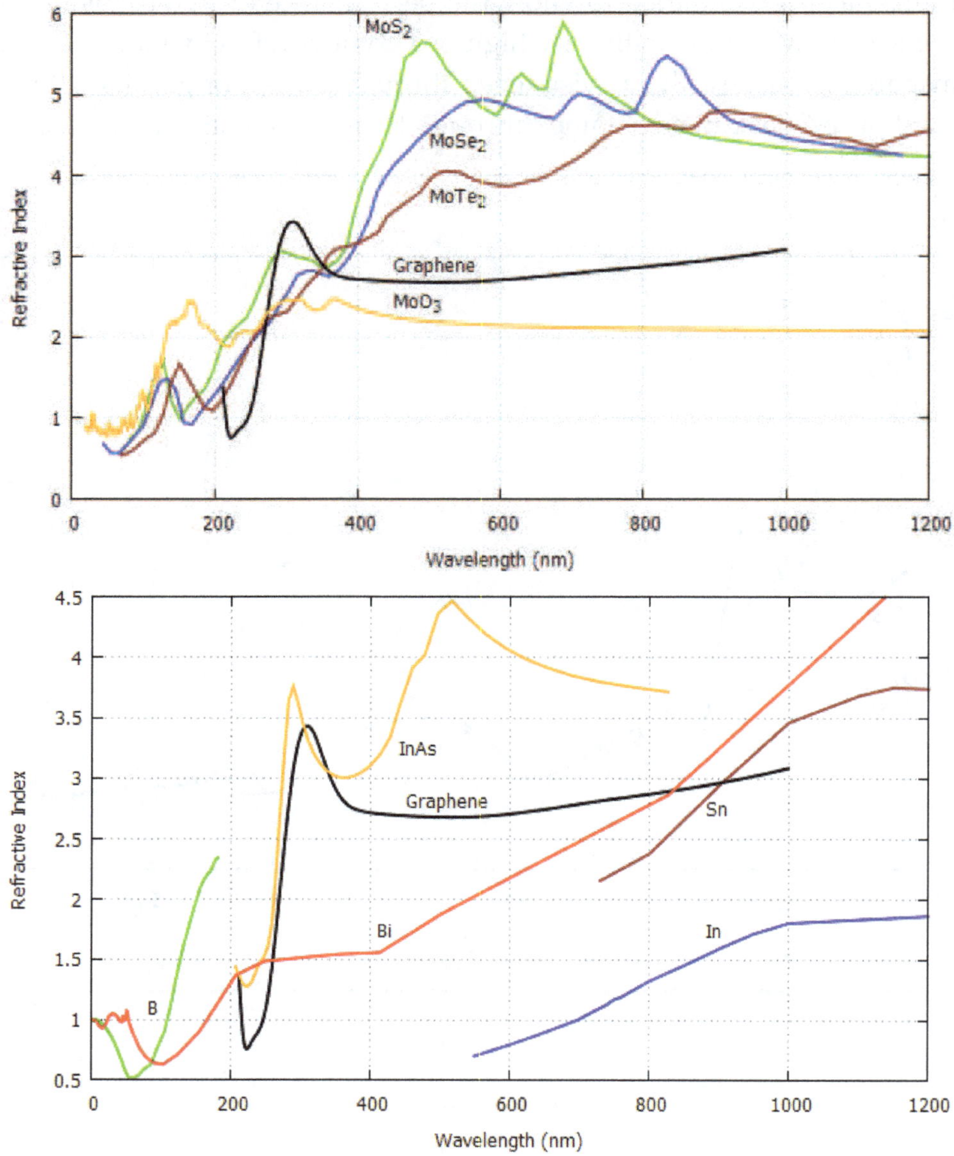

Figure 5. Refractive index versus wavelength for different two-dimensional materials.

Transmittance of the surface of a material is its effectiveness in transmitting radiant energy. It can be defined as the fraction of incident electromagnetic power that is transmitted through a material. Internal transmittance refers to energy loss by absorption, whereas total transmittance is due to absorption, scattering, reflection, etc. A set of transmittance graphs have been obtained through the refractiveindex.info database. These graphs are based on a spectral line or wavelength laser with value 0.2483 μm (KrF) for different two-dimensional materials, and the results are outlined in **Figures 6–15**. The graphs contain the transmittance behavior for different thicknesses of two-dimensional materials ranging from 0.1 to 10 nm. Indiene and stanene present low transmittance and other two-dimensional materials have high transmittance in all optical spectra, as shown in **Figures 6–15**. The value of the transmittance varies widely throughout the optical range, so the designer should check in which wavelength range his device or system must operate to properly choose the two-dimensional material will be used in its photonic or optoelectronic design.

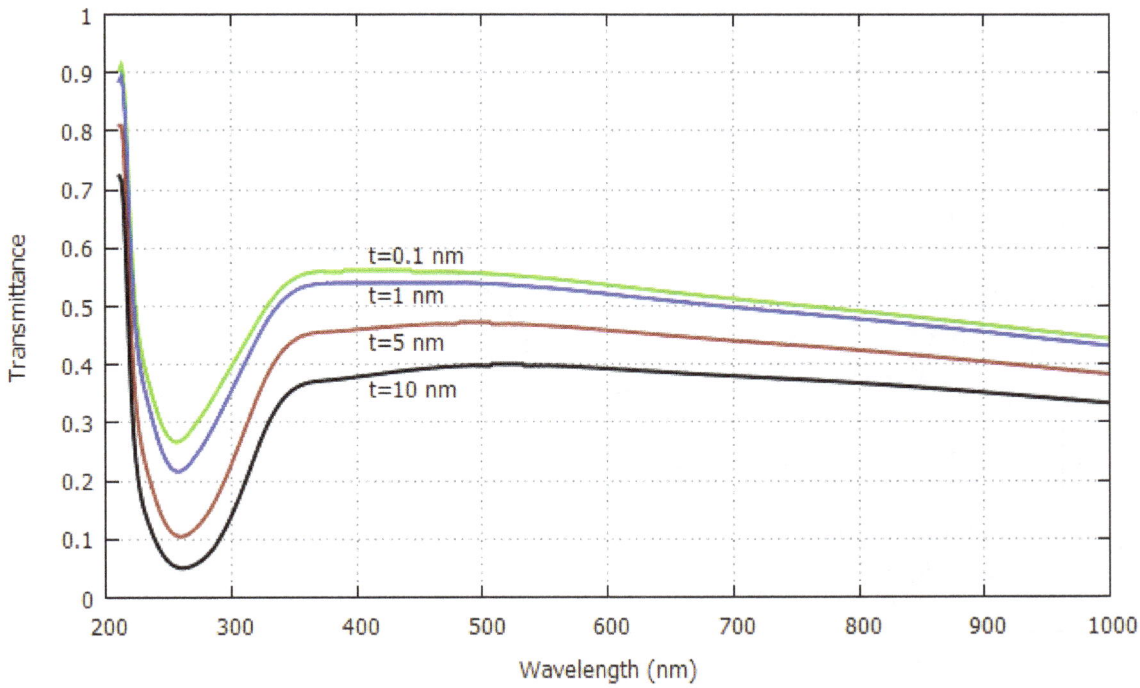

Figure 6. Total transmittance versus wavelength in graphene for different thicknesses.

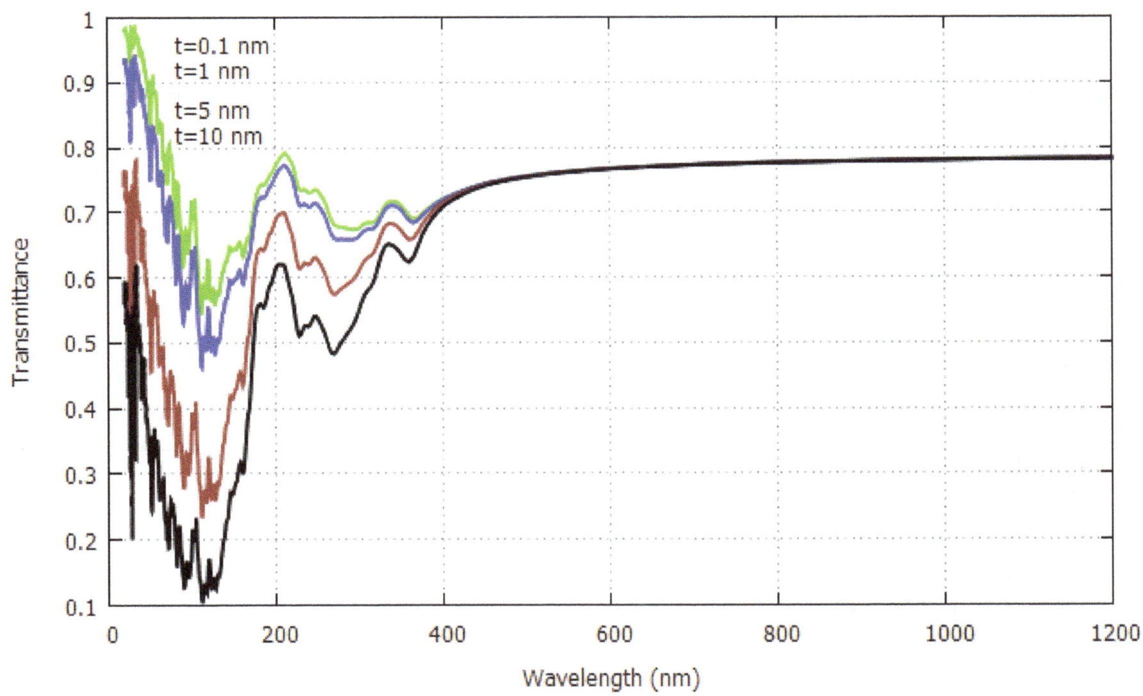

Figure 7. Total transmittance versus wavelength in tungsten(VI) oxide for different thicknesses.

After reviewing the different graphs obtained, it is possible to affirm that the two-dimensional materials can allow the development of emitting materials, waveguides, and light detectors throughout the optical spectrum.

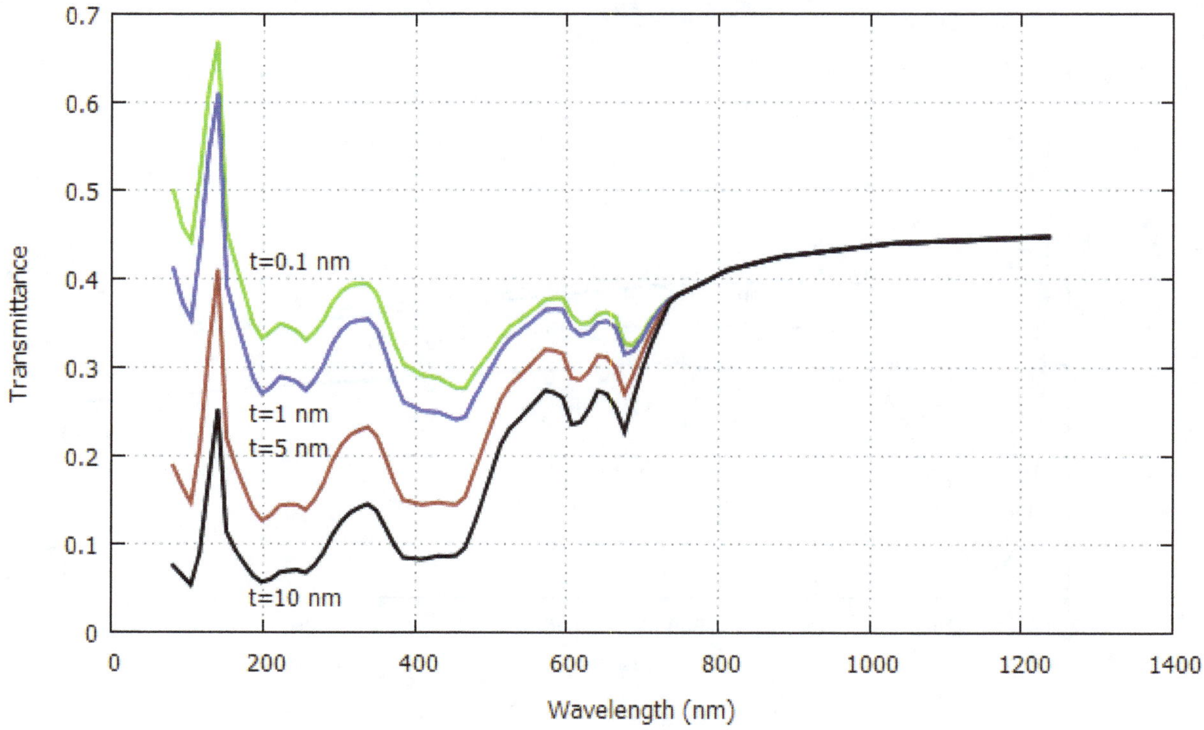

Figure 8. Total transmittance versus wavelength in molybdenum disulfide for different thicknesses.

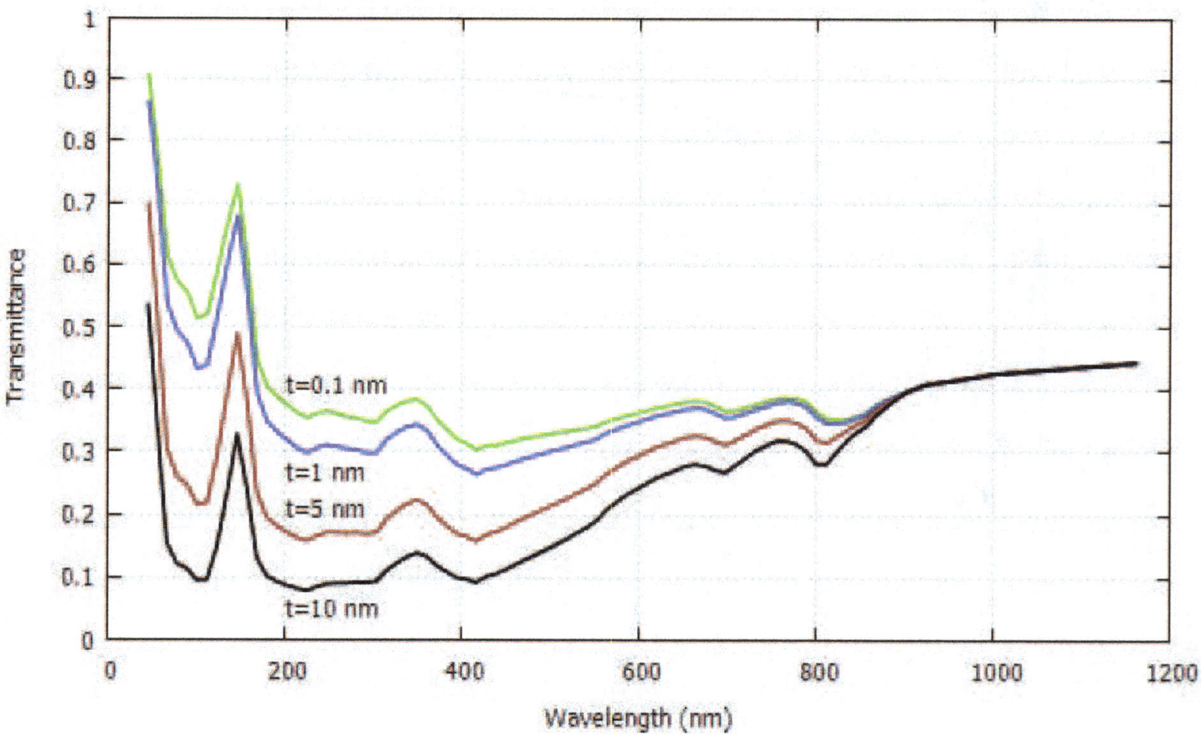

Figure 9. Total transmittance versus wavelength in molybdenum diselenide for different thicknesses.

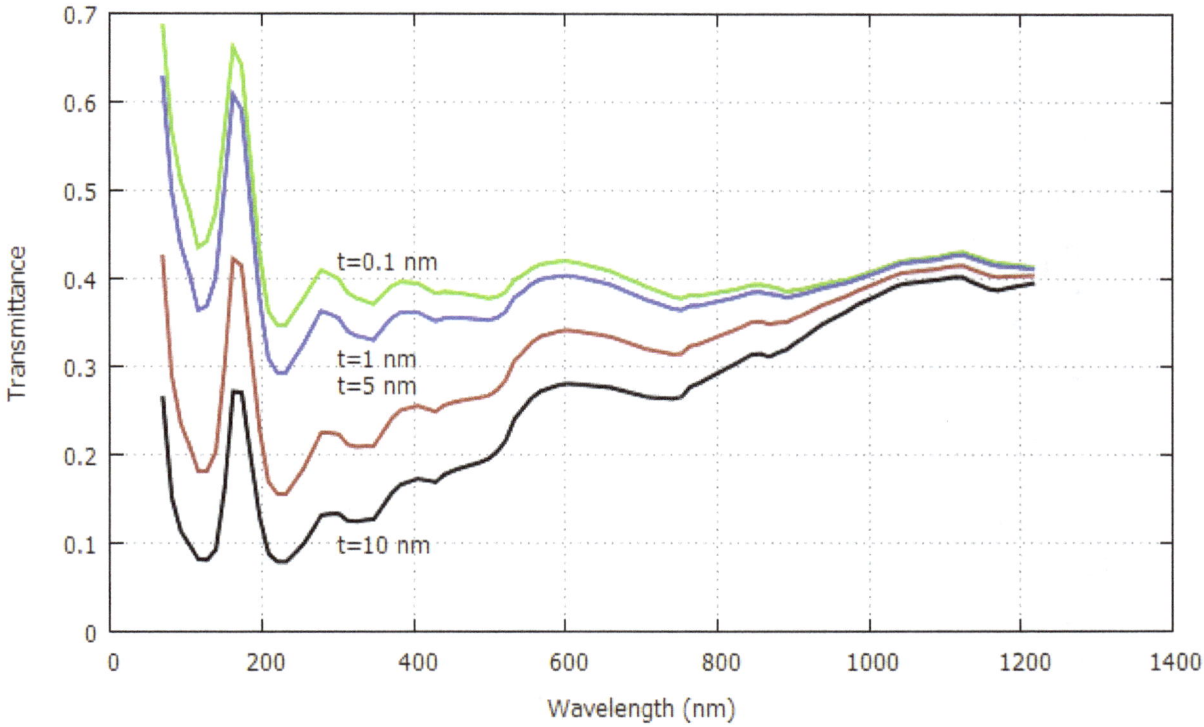

Figure 10. Total transmittance versus wavelength in molybdenum telluride for different thicknesses.

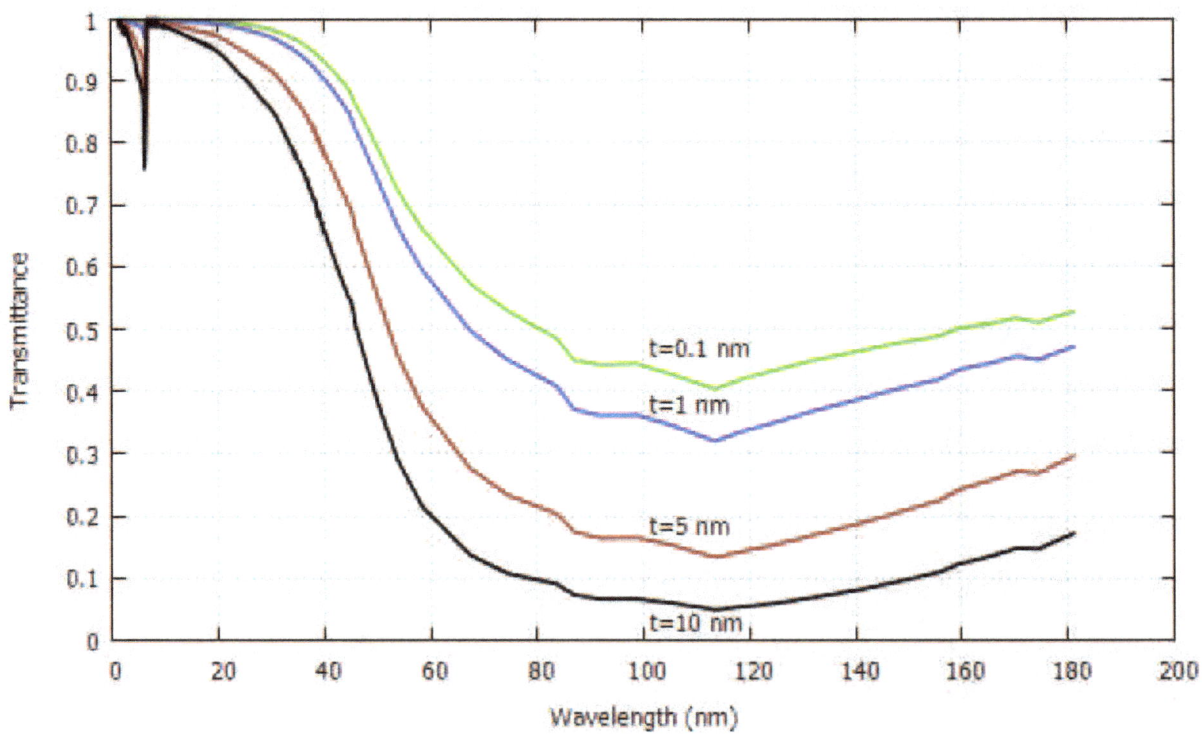

Figure 11. Total transmittance versus wavelength in borophene for different thicknesses.

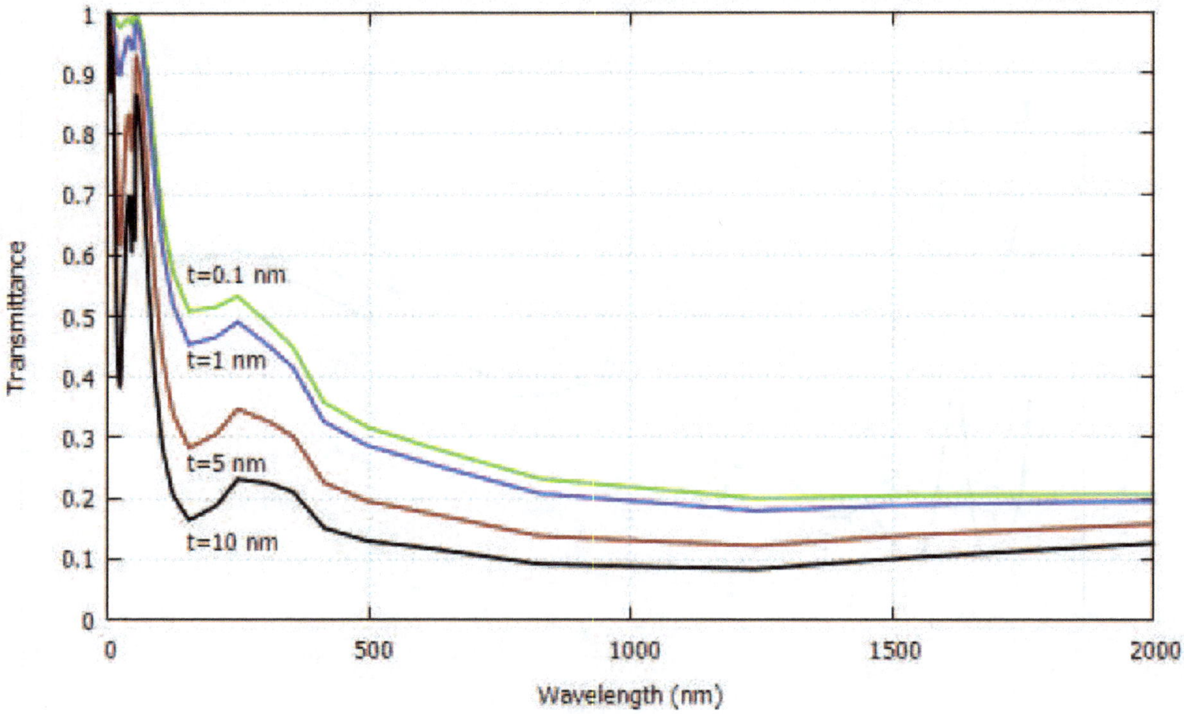

Figure 12. Total transmittance versus wavelength in bismuthene for different thicknesses.

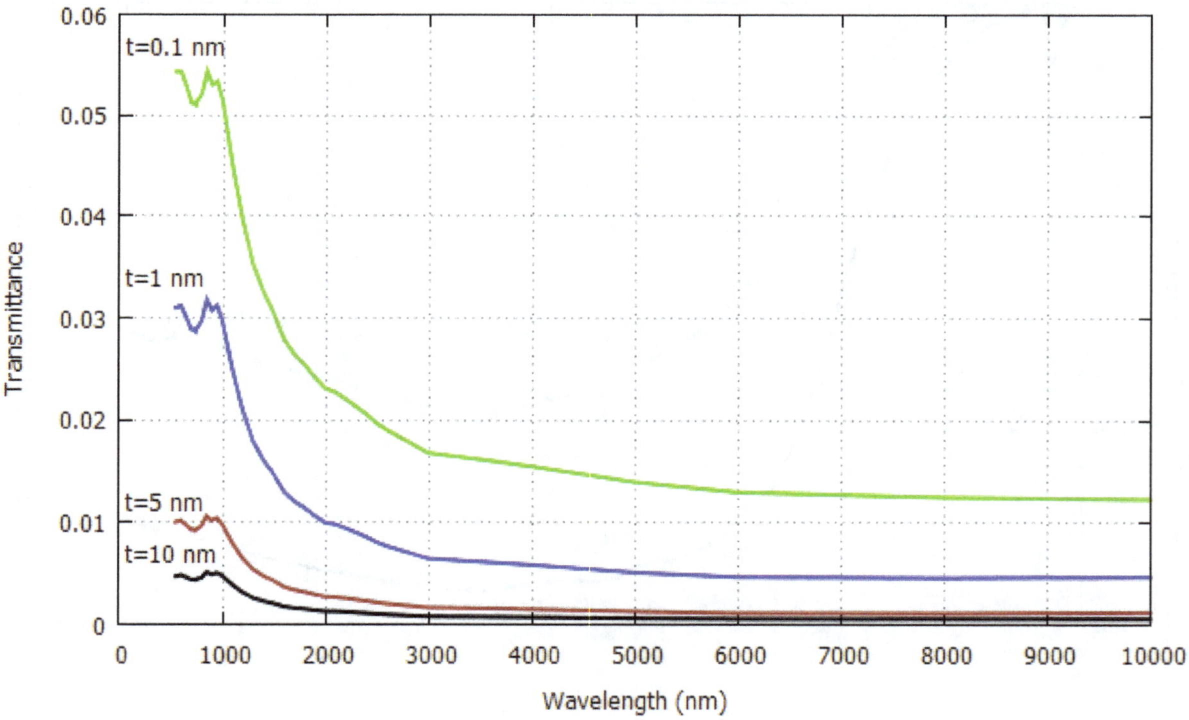

Figure 13. Total transmittance versus wavelength in indiene for different thicknesses.

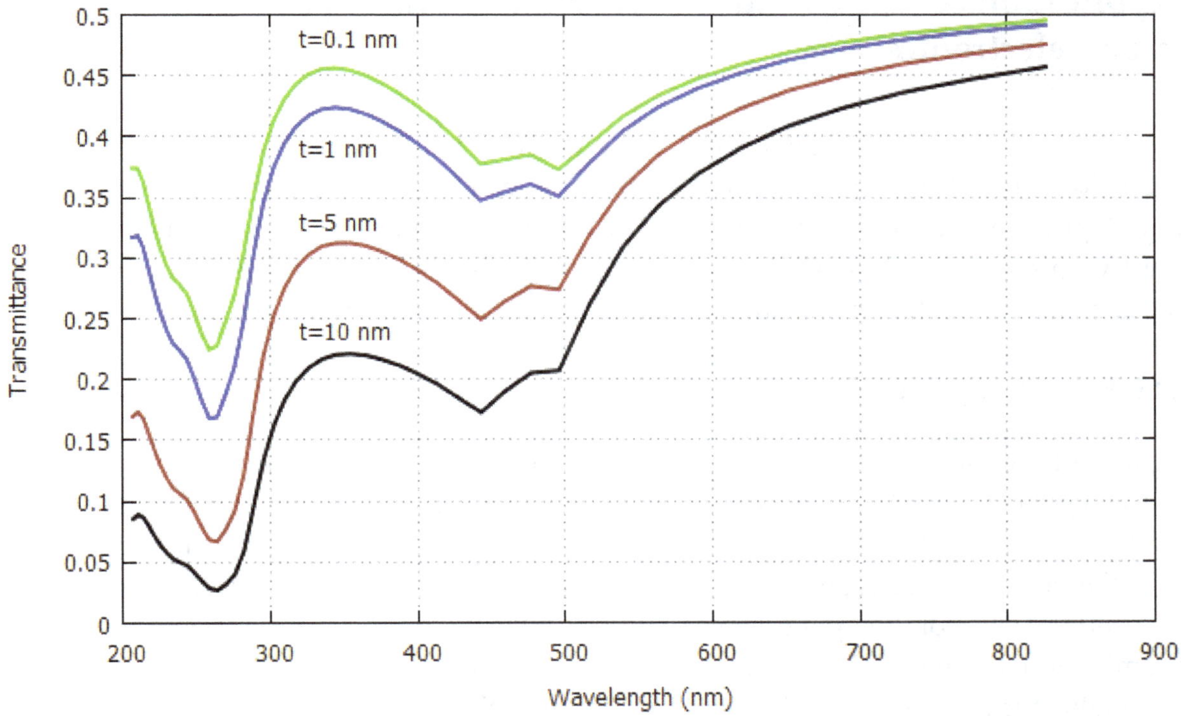

Figure 14. Total transmittance versus wavelength in arsenide for different thicknesses.

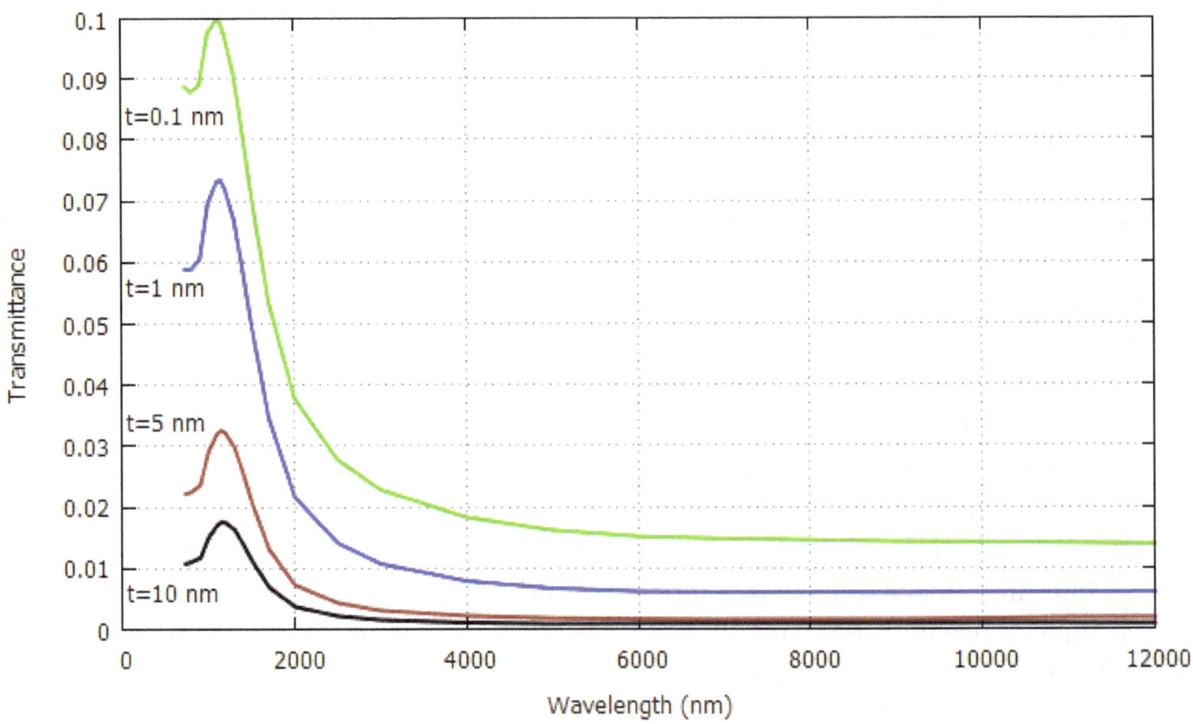

Figure 15. Total transmittance versus wavelength in stanene for different thicknesses.

3. Applications of the two-dimensional materials in photonics and optoelectronics

Two-dimensional materials due to their semiconductor behavior with direct band gap allow the development of numerous optoelectronic applications thanks to the access to spin and valley degrees of freedom. Among the main photonic applications that can be developed using two-dimensional materials are light-emitting diodes, laser diodes, photoluminescent materials, photodetectors, waveguides, optical modulators, and in valleytronics.

3.1. Light-emitting diodes

A light-emitting diode can be defined as a two-terminal semiconductor light source, which is based in a *p-n* junction diode that emits light when it is activated by means of an electric field [2]. When an electrical voltage is applied to the terminals, electrons can recombine with holes within the device, releasing light in the form of photons. This effect is called electroluminescence, and the color of the light (visible wavelength) is determined by the energy band gap of the semiconductor material used in the manufacture of the emitter. In addition, these emitters are available for ultraviolet, visible, and infrared wavelengths.

The typical two-dimensional semiconductors, such as MoS_2, $MoSe_2$, WS_2, WSe_2, and phosphorene, have band gaps that are smaller than 2.0 eV reducing their photoresponse in the blue and UV range [7, 8]. Arsenene and antimonene present wide band gaps and high stability since they have an indirect bandgap as two-dimensional monolayers, and these under small biaxial strain can transform their electronic behavior from indirect into direct bang gap semiconductors. The values of refractive index are 2.3 (α-Sb) and 1.5 (β-Sb) at the zero energy limit and scale up to 3.6 in the ultraviolet region [8].

The electroluminescence has been observed in two-dimensional monolayer field-effect transistors or *p-n* light-emitting diodes made of MoS_2 [9, 10], WSe_2 [11], and WS_2 [12]. Single layer molybdenum disulfide (MoS_2) field-effect transistors can emit electroluminescence (visible light) at 1.8 eV thanks to its direct band gap [9]. Electrostatic gating must be used to improve control and efficiency of light emission. A light-emitting diode based on a two-dimensional monolayer MoS_2 deposited on a heavily *p*-type doped silicon substrate [10]. This device emits a high energy exciton peak in 2.255 eV through of a direct-exciton transition at room temperature which opened the possibility of controlling valley and spin excitation. The optimization of lighting, displays, optical interconnects, and optical sensors depends on the development of light-emitting diodes with high efficiency, well-defined spectral properties, compactness, and integrability [11]. The use of a thin boron nitride (*h*-BN) support as a dielectric layer in a *p-n* light-emitting diode (LED) with a monolayer WSe_2-induced electrostatically yields bright electroluminescence with 1000 times smaller injection current and 10 times smaller linewidth than in MoS_2 monolayers. This quality can be obtained thanks to the control of the voltage applied to the LED. Ambipolar ionic liquid gated field-effect transistors based on WS_2 mono- and bilayers have electroluminescence at 2.14 and 1.82 eV, respectively [12]. It was observed that a much broader range of carrier density is achieved.

3.2. Laser diodes

A laser diode can be defined as an electrically pumped semiconductor laser where active laser medium is formed by a *p-n* junction of a semiconductor diode [2]. These devices regularly operate as directional lighting sources. These devices emit coherent light and they are supplied either by a forward electrical bias or an optical pumping. Optically pumped semiconductor lasers use a semiconductor material as the gain medium, or other laser (often another diode laser) as the pump source. This type allows a better wavelength selection and does not present interference from internal electrode structures.

A continuous-wave nanolaser can operate in the visible regime using a monolayer based on a two-dimensional material as a gain medium with the aim of confining the direct band excitons within one nanometer of the surface of a photonic crystal cavity (PCC) [13]. An optical pumping threshold, as low as 27 nW at 130 K, must be used to stimulate the spontaneous emission of photons. The use of surface-gain geometry allows an unprecedented accessibility, and thus the ability to tailor gain properties by means of external controls such as electrostatic gating and current injection, and therefore it can be operated using an electrical pumping. This topology is scalable and compatible with integrated photonics for on-chip communication technologies.

Advances in two-dimensional materials have allowed the implementation of a two-dimensional excitonic laser, by embedding a WS_2 monolayer in a microdisk resonator, with a high-quality factor and optical confinement presenting a bright light at visible wavelengths [14]. This laser device has high quantum yield, small footprint, and low power consumption. This device allows the development of two-dimensional on-chip optoelectronics for high-performance optical communication and computing applications.

3.3. Photoluminescence

Photoluminescence can be defined as light emission from any form of matter after the absorption of photons [2]. The quality of tuning the optical properties of a material is important for the development of good photonic devices [1]. Two-dimensional materials present photoluminescence, which is defined as the light emission from any form of matter after the absorption of photons. Two-dimensional materials, such as graphene [15, 16], MoS_2 [17–19], $MoTe_2$ [20, 21], WS_2 [22], MoS_2-WS_2 [23], $MoSe_2$ [24, 25], and phosphorene [26], have been researched as photoluminescent materials. Luminescence can be achieved thanks to the reduction of thickness of materials, since it produces a transition from indirect band gap to a direct band gap. The doping or defects as well as strain associated with the presence of the boundaries in two-dimensional materials lead to strong photoluminescence quenching or enhancement [1]. Therefore, photoluminescence properties of two-dimensional materials can be tuned through chemical doping when dopants cover its surface [27]. Pressure produces a shift in the band gap energy of the material and therefore widens its spectrum although it reduces the photoluminescence intensity [28]. Uniaxial tensile strain increases the photoluminescence peak intensity in a two-dimensional material, due to that it guarantees the direct band gap [29]. Electrical gating of two-dimensional materials increases photoluminescence intensity, whereas the photoluminescence wavelength remains nearly constant [30]. The photoluminescence of a two-dimensional material is strongly affected by the type of substrate where it is deposited [31]. Piezoelectric substrates such as $SrTiO_3$ have a higher emission intensity than dielectric substrates such as SiO_2 which reduces the emission intensity. In addition, the photoluminescence intensity from transition metal dichalcogenide alloys is drastically increased for monolayer samples composed of MoS_2-$MoSe_2$ and MoS_2-WS_2 with respect to multilayer samples [32].

Monolayer phosphorene operates as an exciton and trion emitter whose performance is comparable with other 2D transition metal dichalcogenide (TMD) semiconductors such as MoS_2, although semiconductors such as WSe_2 and $MoSe_2$ present higher carrier lifetimes as shown in **Figure 16** [33]. An exciton is the main mechanism for light emission in semiconductors at low temperature. The existence of exciton states may be inferred from the absorption of light associated with their excitation. Typically, excitons are observed just below the band gap. A trion is a localized exciton which consists of three charged quasiparticles. A negative trion consists of two electrons and one hole and a positive trion consists of two holes and one electron. Trion states were predicted theoretically and then observed experimentally in various optically excited semiconductors, especially in quantum dots and quantum well structures. The strong anisotropic atomic structure of single-layer black phosphorus makes that optical response be sensitive to the magnitude and the orientation of the applied strain to the material [34]. Biaxial strain can tune the optical band gap of phosphorene from 0.38 eV (at −8% strain) to 2.07 eV (at 5.5%).

3.4. Photodetectors

A photodetector or photosensor is an electronic device that converts the photons of light into electrical current [2]. The operating principle of a photodetector can be basically described as

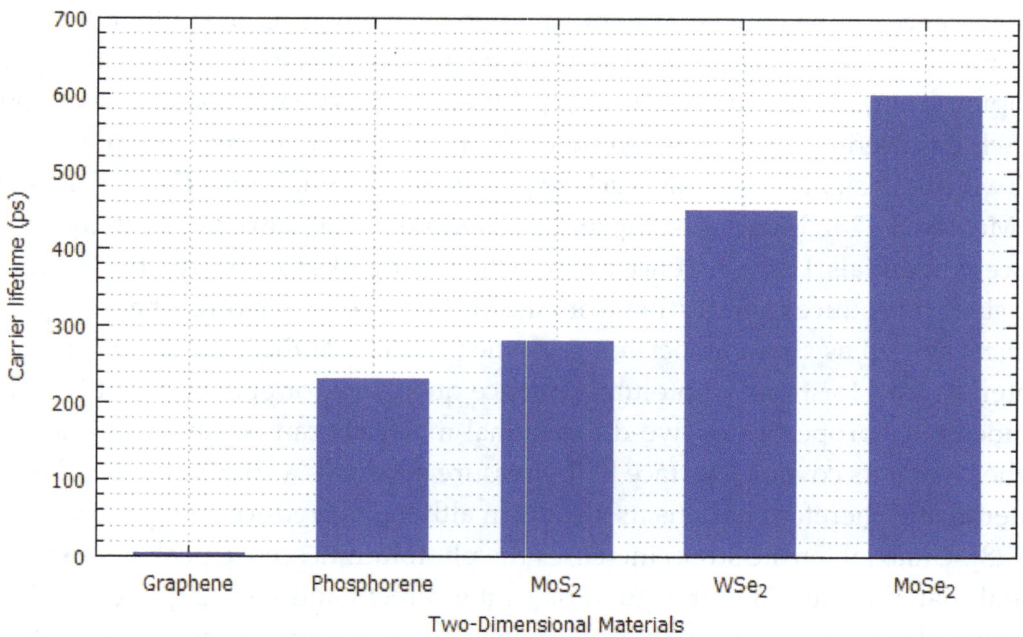

Figure 16. Carrier lifetimes in different two-dimensional materials.

follows. When the photons of light incident on a semiconductor material have energy greater than the forbidden band of this material, they create electron-hole pairs (excitons) or free carriers depend on the exciton binding energy in the material. Bound excitons separated by an applied or built-in electric field can generate a photocurrent. Photodetectors based on semiconductor materials can be classified into two large groups: photodiodes and phototransistors. The former has a much faster response and the latter have better sensitivities.

A graphene photodetector working in the photoconductor mode and based on photothermoelectric effect and bolometric effect (zero and nonzero biased operations, respectively) was integrated on silicon nitride (Si_3N_4) waveguide [35]. This combination of materials facilitates unprecedented nonlinear and optoelectronic applications. The graphene has a behavior as a highly *p*-doped semiconductor that produces a photocurrent generation.

The ultrashort lifetime of photocarriers, caused by the fast recombination of graphene, results in the weak response of light and limits its application in photodetection [36]. With the aim of improving the performance of the graphene in photodetection, it must be coupled with photonic structures or building vertical heterostructures by stacking graphene with other layered materials and increasing the doping level considering the band slop and the absorption at certain wavelengths.

Monolayer WSe_2 and MoS_2 heterostructures present strong photoluminescence intensity, which suggest that there is a strong interlayer coupling of the charge carriers [5].

3.5. Solar cells

A solar cell or photovoltaic cell can be defined as an electrical device that converts the energy of light directly into electricity by the photovoltaic effect, that is, a physical and chemical phenomenon where light is absorbed causing excitation of an electron or other charge carrier to a higher energy state [2]. Electrical properties, such as current, voltage, or resistance, vary when solar cells are exposed to light.

Thanks to their semiconducting character, graphene and monolayer transition metal dichalcogenides can be considered as potential sunlight absorbers in thicknesses less than 1 nm, which represents a value with 1 order of magnitude higher than for GaAs and Si [37]. Two stacked monolayers have been proposed: (1) a Schottky barrier solar cell between MoS_2 and graphene and (2) an excitonic solar cell based on a MoS_2/WS_2 bilayer. These structures can reach power conversion efficiencies of up to ~1%, corresponding to values of approximately 1 to 3 orders of magnitude higher than to power densities of the best existing ultrathin solar cells.

The ultrathin transition metal dichalcogenide films and their Van der Waals heterostructures have been used to develop photovoltaic devices with the aim of offering superior performance, extreme flexibility, and long lifetime due to suitable band gap and excellent light absorption capability [38]. A *p*-Si/MoS_2 heterojunction-based solar cell produces an enormous quantity of photogenerated carriers and increases the photocurrent generated by them. The semiconductor heterostructures and superlattices represent a solar cell design platform without the limitation of lattice matching present in other materials [5].

3.6. Waveguides

An optical waveguide can be defined as a physical structure that guides or transports electromagnetic waves in the optical spectrum [2]. The most common types of optical waveguides are optical fibers and rectangular waveguides. These are commonly used as components in integrated optical circuits or as a transmission medium in optical communication systems. In addition, waveguides can be classified by their geometry (planar, strips, fiber waveguides), mode structure (single-mode, multiple-mode), refractive index distribution (step index, gradient index), and material (glass, polymer, semiconductor material).

The ultrashort optical pulses can be propagated through graphene-based nanoribbon waveguides in telecommunications [39]. The pulses can experiment attenuation, high-order dispersive effects and nonlinear effects when they pass through graphene. Fortunately, these effects can be controlled with the input signal power and the chemical potential of the graphene. These waveguides can be applied to other nanophotonic devices such as filters, modulators, antennas, switches, and other devices.

3.7. Optical modulators

An optical modulator can be defined as a device used to modulate a beam of light, which is carried over free space or propagated through an optical waveguide [2]. These devices can be classified by the parameter of a light beam that is manipulated into amplitude modulators, phase modulators, polarization modulators, etc. The easiest way to perform the modulation of light intensity is to modulate the electric current that drives the light source, as in a laser diode. In addition, optical modulators can be classified by the properties of the materials used to modulate the light beam into two groups: absorptive modulators and refractive modulators. In absorptive modulators, the absorption coefficient of the material is changed; and in refractive modulators, the refractive index of the material is changed.

Two-dimensional materials can realize various functions at once due to their diverse physical properties as it is the case of the multifunctional optical modulators where optical modulators and photodetectors or optical modulators and plasmon waveguides can be implemented with the same semiconductor material [40]. Main advantages that these materials have on optical modulators are a broad operation bandwidth from the visible to terahertz range, small footprint, low cost, easy integration, and large flexibility. This quality facilitates the integration of diverse electronic and optical components with an "all-in-one" solution. However, before of reaching the commercial success of the two-dimensional materials, a trade-off between performance and cost must be considered.

3.8. Valleytronics

Particularly, transition metal dichalcogenide monolayers present a strong spin-orbit coupling, which leads to a spin-orbit splitting of hundreds meV in the valence band and a few meV in the conduction band [1, 41]. This quality allows the control of the electron spin by tuning the excitation laser photon energy that is applied to the material. This quality allows controlling the electrical charge or spin degree of free carriers, and therefore, novel devices can be proposed. Among these devices, a WSe_2-based electrically switchable chiral light-emitting transistor has been designed [42]. This device operates as a spin LED that shows circularly polarized electroluminescence, but depends on the application of an external magnetic field to operate. Circular polarized light can be used on three-dimensional displays or as spin sources for spintronics and as information carriers in quantum computation.

4. Conclusions

A lot of promising applications in nanophotonics will be developed in the coming decades. The main challenge for large-scale applications of two-dimensional materials is the ability to control the synthesis processes in order to optimize the physical and chemical properties and to extend its use in a wide variety of novel photonic applications. Most applications in nanophotonics of two-dimensional materials are still in the research phase or stage of technological development. However, the progress reported in this work indicates that the practical implementation of devices and systems will be based on simple and heterogeneous structures. The main applications are focused on the development of photodetectors based on the principle of photoconductivity, transparent electrodes, waveguides, and solar cells based on the photoelectric effect. The large variety of available two-dimensional materials, their heterostructures and hybrid systems that are being researched will continue introducing novel properties to outperform competing technologies applied in electronics and photonics. Designers can make use of an innumerable amount of two-dimensional materials based on the different chemical elements that have been synthesized or designed in computer. Fortunately, graphene and its derivatives are no longer alone and several material choices, such as MoS_2, $MoSe_2$, WS_2, WSe_2, phosphorene, borophene, arsenene, antimonene, bismuthene, indiene, and stanene, can be exploited through the optical range.

Acknowledgements

The author wants to thank financial support from CONACYT (contract no. 152524, basic science), Tecnológico Nacional de México (contract no. M00.1/1671/2015), and Instituto Tecnológico Superior de Irapuato (ITESI).

Author details

Rafael Vargas-Bernal

Address all correspondence to: ravargas@itesi.edu.mx

Department of Materials Engineering, Higher Technological Institute of Irapuato, Irapuato, Guanajuato, Mexico

References

[1] Kolobov, A.V., Tominaga, J. Two-Dimensional Transition-Metal Dichalcogenides. Switzerland: Springer; 2016. 545 p. DOI: 10.1007/978-3-319-31450-1

[2] Liu, J.-M. Principles of Photonics. Cambridge, United Kingdom: Cambridge University Press; 2016. 260 p. DOI: 10.1017/CBO9781316687109

[3] Ponraj, J.S., Xu, Z.-Q., Dhanabalan, S.C., Mu, H., Wang, Y., Yuan, J., Li, P., Thakur, S., Ashrafi, M., Mccoubrey, K. Photonics and Optoelectronics of Two-Dimensional Materials beyond Graphene. Nanotechnology. 2016; 27(46): 462001. DOI: 10.1088/0957-4484/27/46/462001

[4] Xia, F., Wang, H., Xiao, D., Dubey, M., Ramasubramaniam A. Two-Dimensional Material Nanophotonics. Nature Photonics. 2014; 8(12): 899–907. DOI: 10.1038/nphoton.2014.271

[5] Duan, X., Wang, C., Pan, A., Yu, R., Duan, X. Two-Dimensional Transition Metal Dichalcogenides as Atomically Thin Semiconductors: Opportunities and Challenges. Chemical Society Reviews. 2015; 44(24): 8859–8876. DOI: 10.1039/c5cs00507h

[6] Tian, H., Chin, M.L., Najmaei, S., Guo, Q., Xia, F., Wang, H., Dubey, M. Optoelectronic Devices based on Two-Dimensional Transition Metal Dichalcogenides. Nano Research. 2016; 9(6): 1543–1560. DOI: 10.1007/s12274-016-1034-9

[7] Zhang, S., Yan, Z., Li, Y., Chen, Z., Zeng, H. Atomically Thin Arsenene and Antimonene: Semimetal-Semiconductor and Indirect-Direct Band-Gap Transitions. Angewandte Chemie International Edition. 2015; 54(10): 3112–3115. DOI: 10.1002/anie.201411246

[8] Singh, D., Gupta, S.K., Sonvane, Y., Lukačević, I. Antimonene: A Monolayer Material for Ultraviolet Optical Nanodevices. Journal of Materials Chemistry C. 2016; 4(26): 6386–6390. DOI: 10.1039/c6tc01913g

[9] Krane, N., Lotze, C., Läger, J.M., Reecht, G., Franke, K.J. Electronic Structure and Luminescence of Quasi-Freestanding MoS_2 Nanopatches on Au(111). Nano Letters. 2016; 16(8): 5163–5168. DOI: 10.1021/acs.nanolett.6b02101

[10] Ye, Y., Ye, Z., Gharghi, M., Zhu, H., Zhao, M., Wang, Y., Yin, X., Zhang, X. Exciton-dominant Electroluminescence from a Diode of Monolayer MoS_2. Applied Physics Letters. 2014; 104(19): 193508. DOI: 10.1063/1.4875959

[11] Ross, J.S., Klement, P., Jones, A.M., Ghimire, N.J., Yan, J., Mandrus, D.G., Taniguchi, T., Watanabe, K., Kitamura, K., Yao, W., Cobden, D.H., Xu, X. Electrically Tunable Excitonic Light-Emitting Diodes based on Monolayer WSe_2 p-n Junctions. Nature Nanotechnology. 2014; 9(4): 268–272. DOI: 10.1038/nnano.2014.26

[12] Jo, S., Ubrig, N., Berger, H., Kuzmenko, A.B., Morpurgo, A.F. Mono- and Bilayer WS_2 Light-Emitting Transistors. Nano Letters. 2014; 14(4): 2019–2025. DOI: 10.1021/nl500171v

[13] Wu, S., Buckley, S., Schaibley, J.R., Feng, L., Yan, J., Mandrus, D.G., Hatami, F., Yao, W., Vučković, J., Majumdar, A., Xu, X. Monolayer Semiconductor Nanocavity Lasers with Ultralow Thresholds. Nature. 2015; 520(7545): 69–72. DOI: 10.1038/nature14290

[14] Ye, Y., Wong, Z.J., Lu, X., Ni, X., Zhu, H., Chen, X., Wang, Y., Zhang, X. Monolayer Excitonic Laser. Nature Photonics. 2015; 9(11): 733–737. DOI: 10.1038/nphoton.2015.197

[15] Pal, S.K. Versatile Photoluminescence from Graphene and Its Derivatives. Carbon. 2015; 88: 86–112. DOI: 10.1016/j.carbon.2015.02.035

[16] Du, D., Song, H., Nie, Y., Sun, X., Chen, L., Ouyang, J. Photoluminescence of Graphene Oxide in Visible Range Arising from Excimer Formation. Journal of Physical Chemistry C. 2015; 119(34): 20085–20090. DOI: 10.1021/acs.jpcc.5b04529

[17] Gao, W., Lee, Y.H., Jiang, R., Wang, J., Liu, T., Ling, X.Y. Localized and Continuous Tuning of Monolayer MoS_2 Photoluminescence Using a Single Shape-Controlled Ag Nanoantenna. Advanced Materials. 2016; 28(4): 701–706. DOI: 10.1002/adma.201503905

[18] Fan, X., Xu, P., Li, Y.C., Zhou, D., Sun, Y., Nguyen, M.A.T., Terrones, M., Mallouk, T.E. Controlled Exfoliation of MoS_2 Crystals into Trilayer Nanosheets. Journal of the American Chemical Society. 2016; 138(15): 5143–5149. DOI: 10.1021/jacs.6b01502

[19] Andleeb, S., Singh, A.K., Eom, J. Chemical Doping of MoS_2 Multilayer by p-Toluene Sulfonic Acid. Science and Technology of Advanced Materials. 2015; 16(3): 035009. DOI: 10.1088/1468-6996/16/3/035009

[20] Lezama, I.G., Arora, A., Ubaldini, A., Barreteau, C., Giannini, E., Potemski, M., Morpurgo, A.F. Indirect-to-Direct Band Gap Crossover in Few-Layer $MoTe_2$. Nano Letters. 2015; 15(4): 2336–2342. DOI: 10.1021/nl5045007

[21] Ruppert, C., Aslan, O.B., Heinz, T.F. Optical Properties and Band Gap of Single- and Few-layer $MoTe_2$ Crystals. Nano Letters. 2014; 14(11): 6231–6236. DOI: 10.1021/nl502557g

[22] McCreary, K.M., Hanbicki, A.T., Jernigan, G.G., Culbertson, J.C., Jonker, B.T. Synthesis of Large-Area WS_2 Monolayers with Exceptional Photoluminescence. Scientific Reports. 2016; 6: 19159. DOI: 10.1038/srep19159

[23] Chen, Y., Xi, J., Dumcenco, D.O., Liu, Z., Suenaga, K., Wang, D., Shuai, Z., Huang, Y-S., Xie, L. Tunable Band Gap Photoluminescence from Atomically Thin Transition-Metal Dichalcogenide Alloys. ACS Nano. 2013; 7(5): 4610–4616. DOI: 10.1021/nn401420h

[24] Kang, J., Zhang, L., Wei, S.-H. A Unified Understanding of the Thickness-dependent Bandgap Transition in Hexagonal Two-Dimensional Semiconductors. Journal of Physical Chemistry Letters. 2016; 7(4): 597–602. DOI: 10.1021/acs.jpclett.5b02687

[25] Wang, G., Palleau, E., Amand, T., Tongay, S., Marie, X., Urbaszek, B. Polarization and Time-resolved Photoluminescence Spectroscopy of Excitons in $MoSe_2$ Monolayers. Applied Physics Letters. 2015; 106(11): 112101. DOI: 10.1063/1.4916089

[26] Zhang, S., Yang, J., Xu, R., Wang, F., Li, W., Ghufran, M., Zhang, Y.-W., Yu, Z., Zhang, G., Qin, Q., Lu, Y. Extraordinary Photoluminescence and Strong Temperature/Angle-dependent Raman Response in Few-Layer Phosphorene. ACS Nano. 2014; 8(9): 9590–9596. DOI: 10.1021/nn503893j

[27] Liu, X., Qu, D., Ryu, J., Ahmed, F., Yang, Z., Lee, D., Yoo, W.J. P-type Polar Transition of Chemically doped Multilayer MoS_2 Transistor. Advanced Materials. 2016; 28(12): 2345–2351. DOI: 10.1002/adma.201505154

[28] Nayak, A.P., Pandey, T., Voiry, D., Liu, J., Moran, S.T., Sharma, A., Tan, C., Chen, C.-H., Li, L.-J., Chhowalla, M., Lin, J.-F., Singh, A.K., Akinwande, D. Pressure-Dependent Optical and Vibrational Properties of Monolayer Molybdenum Disulfide. Nano Letters. 2015; 15(1): 346–353. DOI: 10.1021/nl5036397

[29] Wang, Y., Cong, C., Yang, W., Shang, J., Peimyoo, N., Chen, Y., Kang, J., Wang, J., Huang, W., Yu, T. Strain-induced Direct-Indirect Bandgap Transition and Phonon Modulation in Monolayer WSe_2. Nano Research. 2015; 8(8): 2562–2572. DOI: 10.1007/s12274-015-0762-6

[30] Oh, H.M., Jeong, H., Han, G.H., Kim, H., Kim, J.H., Lee, S.Y., Jeong, S.Y., Jeong, S., Park, D.J., Kim, K.K., Lee, Y.K., Jeong, M.S. Modulating Electronic Properties of Monolayer MoS_2 via Electron-withdrawing Functional Groups of Graphene Oxide. ACS Nano. 2016, 10(11): 10446–10453. DOI: 10.1021/acsnano.6b06319

[31] Li, Y., Qi, Z., Liu, M., Wang, Y., Cheng, X., Zhang, G., Sheng, L. Photoluminescence of Monolayer MoS_2 on $LaAlO_3$ and $SrTiO_3$ Substrates. Nanoscale. 2014; 6(24): 15248–15254. DOI: 10.1039/c4nr04602a

[32] Choudhary, N., Park, J., Hwang, J.Y., Chung, H.-S., Dumas, K.H., Khondaker, S.I., Choi, W., Jung, Y. Centimeter Scale Patterned Growth of Vertically Stacked Few Layer only 2D MoS_2/WS_2 van der Waals Heterostructure. Scientific Reports. 2016; 6: 25456. DOI: 10.1038/srep25456

[33] Yang, J., Xu, R., Pei, J., Myint, Y.W., Wang, F., Wang, Z., Zhang, S., Yu, Z., Lu, Y. Optical Tuning of Exciton and Trion Emissions in Monolayer Phosphorene. Light: Science & Applications. 2015; 4(7): e312. DOI: 10.1038/lsa.2015.85

[34] Zhang, W., Ye, C., Hong, L., Yang, Z., Zhou, R. Molecular Structure and Dynamics of Water on Pristine and Strained Phosphorene: Wetting and Diffusion at Nanoscale. Scientific Reports. 2016; 6: 38327. DOI: 10.1038/srep38327

[35] Wang, J., Cheng, Z., Chen, Z., Xu, J.-B., Tsang, H.K., Shu, C. Graphene Photodetector Integrated on Silicon Nitride Waveguide. Journal of Applied Physics. 2015; 117(14): 144504. DOI: 10.1063/1.4917378

[36] Lin, L., Liao, L., Yin, J., Peng, H., Liu, Z. Building Graphene p-n Junctions for Next-Generation Photodetection. Nano Today. 2015; 10: 701–716. DOI: 10.1016/j.nantod.2015.11.006

[37] Furchi, M.M., Zechmeister, A.A., Hoeller, F., Wachter, S., Pospischil, A., Mueller, T. Photovoltaics in Van der Waals Heterostructures. IEEE Journal of Selected Topics in Quantum Electronics. 2017; 23(1): 4100111. DOI: 10.1109/JSTQE.2016.2582318

[38] Pradhan, S.K., Xiao, B., Pradhan, A.K. Enhanced Photo-Response in p-Si/MoS_2 Heterojunction-based Solar Cells. Solar Energy Materials & Solar Cells. 2016; 144: 117–127. DOI: 10.1016/j.solmat.2015.08.021

[39] Lima, Jr. A.W., Mota, J.C.M., Sombra, A.S.B. Attenuation, Dispersion and Nonlinearity Effects in Graphene-based Waveguides. Beilstein Journal of Nanotechnology. 2015; 6: 1221–1228. DOI: 10.3762/bjnano.6.125

[40] Sun, Z., Martinez, A., Wang, F. Optical Modulators with 2D Layered Materials. Nature Photonics. 2016; 10(4): 227–238. DOI: 10.1038/nphoton.2016.15

[41] Xie, L., Cui, X. Manipulating Spin-Polarized Photocurrents in 2D Transition Metal Dichalcogenides. PNAS. 2016; 113(14): 3746–3750. DOI: 10.1073/pnas.1523012113

[42] Zhang, Y.J., Oka, T., Ye, J.T., Iwasa, Y. Electrically Switchable Chiral Light-Emitting Transistor. Science. 2014; 344(6185): 725–728. DOI: 10.1126/science.1251329

6

Fundamentals of Chemical Vapor Deposited Graphene and Emerging Applications

Golap Kalita and Masaki Tanemura

Abstract

Graphene, the atomically thin sheet of sp^2 hybridized carbon atoms arranged in honeycomb structure, is becoming the forefront of material research. The chemical vapor deposition (CVD) process has been explored significantly to synthesis large size single crystals and uniform films of monolayer and bilayer graphene. In this prospect, the nucleation and growth mechanism of graphene on a catalytic substrate play the fundamental role on the control growth of layers and large domain. The transition metals and their alloys have been recognized as the active catalyst for growth of monolayer and bilayer graphene, where the surface composition of such catalysts also plays critical role on graphene growth. CVD-synthesized graphene has been integrated with bulk semiconductors such as Si and GaN for the fabrication of solar cells, photodetectors, and light-emitting diodes. Furthermore, CVD graphene has been integrated with hexagonal boron nitride (hBN) and transition metal dichalcogenides (TMDCs) for the fabrication of van der Waals heterostructure for nanoelectronic, optoelectronic, energy devices, and other emerging technologies. The fundamental of the graphene growth process by a CVD technique and various emerging applications in heterostructure devices is discussed in detail.

Keywords: graphene, CVD synthesis, atmospheric and low pressure, catalyst, graphene heterostructure

1. Introduction

Graphene was first introduced in 2004 by Prof. Andre Geim and Prof. Konstantin Novoselov by exfoliation of graphite using scotch-tape [1]. Prior to this ground breaking work, Peierls and Landau have predicted that the two-dimensional (2D) layered materials are thermodynamically unstable and difficult to find in nature [2–4]. The advent of atomically thin graphene revolutionized the 2D materials in physics, chemistry, and engineering fields, opening enormous potential applications [5–10]. Considering the pioneering work on graphene, the 2010 Nobel Prize in Physics was awarded to Prof. Geim and Prof. Novoselov. Graphene can be considered as the main building block of various other forms of sp^2 hybridized carbon atoms (**Figure 1**). One of the most well-known allotrope is buckminsterfullerene also called as buckyball or fullerene (C_{60}), which was discovered in 1986 by R. F. Curl, H. W. Kroto and R. E. Smalley [11]. The soccer ball look-alike buckyball with a diameter only of 7.1 Å and consisting of 12 pentagons and 20 hexagons carbon rings. Its existence had been predicted earlier in 1970, by the Japanese theoretician Eiji Ozawa et al. [12]. Similarly, carbon nanotubes (CNTs), another form of carbon, have generated significant interest in the research fraternity due to their fascinating properties such as a one-dimensional (1D) structure [13]. In contrast to a buckyball, the CNTs can be considered as a rolled sheet of graphene forming 1D tubular structure. A rolled single sheet graphene will form single walled CNTs, while multiple rolled graphene sheet will create multi-walled CNTs. Graphene quantum dots (GQDs) and nanoribbons are also explored considering their unique characteristics [14, 15]. The discussed various forms of crystalline carbon in different dimensions (i.e., zero, one, two, and three dimensions) exhibit contrasting physical and chemical properties and are prominent materials for wide

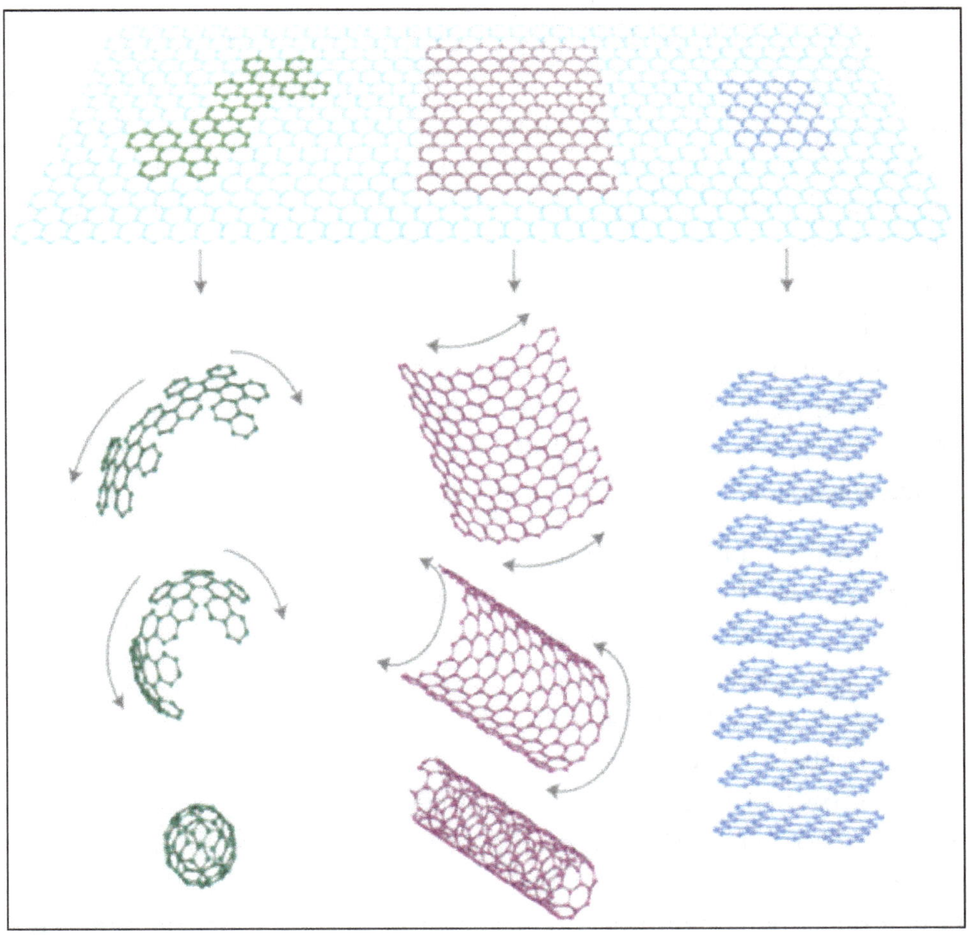

Figure 1. 2D graphene sheet is the building block of other forms of sp^2 hybridized carbon in all other dimensions. The graphene sheet can be wrapped to form fullerene (zero dimension), rolled into nanotubes (one dimension), or stacked into graphite (three dimension). [Reprinted with permission from Ref. [9]].

range of applications. Various physical phenomena and structure-related properties have been explored or discovered, while graphene can be ideal form of the materials to understand the various other forms. In the chapter, we focused on graphene growth process on catalyst substrates by the low and atmospheric pressure CVD process. Subsequently, we emphasize on different possible catalysts for graphene growth and their surface structures for selective growth of monolayer, bilayer, and large single crystal domains. Recent development and emerging applications of CVD graphene for the fabrication of heterostructures with bulk semiconductor and van der Waals heterostructures with other two-dimensional (2D) materials were introduced and explained.

2. Crystal structure and basic properties of graphene

The unit cell of graphene contains two carbon atoms and the graphene lattice can be considered to be made of by two sublattices, A and B (**Figure 2(a)**). The A and B sublattices are triangular Bravais lattices; therefore, the graphene honeycomb lattice can be viewed as a triangular Bravais lattice also called as hexagonal lattice with a two-atom basis (A and B). **Figure 2(a)** shows the honeycomb lattice of graphene. The vectors δ_1, δ_2, and δ_3 connect nearest neighboring carbon atoms, separated by a distance of a = 0.142 nm. The vectors a_1 and a_2 are basis vectors of the triangular Bravais lattice. The reciprocal lattice of the triangular lattice with primitive lattice vectors a^*_1 and a^*_2 is presented in **Figure 2(b)**. In the center, the shaded region represents the first Brillouin zone, with its center Γ, and the two inequivalent corners K (black squares) and K′ (white squares). The π-bands electronic dispersion relations for the 2D honeycomb crystal lattice of graphene under the tight binding (TB) approximation is shown in **Figure 2(c)**. Plots are shown for the electron energy dispersion for π and π^*-bands in the first and extended Brillouin zones as contour plots at equidistant energies and as pseudo-3D representations for the 2D structures. Electronic band structure can be simply presented by a Dirac cone, where the valance and conduction band touch at Dirac point [16]. The cone-shape energy band structures present linear electronic dispersion and density of states (DOS) of graphene, which significantly differ from conventional metals and semiconductors.

The quantum confinement of the electrons in absence of a third dimension provides graphene various exciting properties. Electrons in graphene behave as massless relativistic fermions at low temperatures, which is an unusual behavior for condensed-matter system [16]. Graphene shows an unusual (relativistic) quantum Hall effect with an applied perpendicular magnetic field at a temperature as high as room temperature [6]. The massless Dirac fermion (i.e., charge carrier) of graphene moves at ballistic speed in submicron length, close to relativistic speeds. It has been estimated that the intrinsic carrier mobility of graphene is as high as 200,000 $cm^2 V^{-1}s^{-1}$ [18]. The 2D graphene sheet is also known to be an excellent current conductor; the sustainability of current density is six orders higher than of normal Cu [9]. Graphene also shows exceptional mechanical strength, with breaking strength of ~40 N/m and the Young's modulus of ~1.0 TPa [8]. The thermal conductivity of a suspended graphene sheet at room temperature has been measured as in the range of 4.84×10^3–5.3×10^3 W/mK [19]. The absorption coefficient of graphene has been determined from the fine structure constant ($\alpha = e^2/\hbar c$), which signifies that the absorbance and

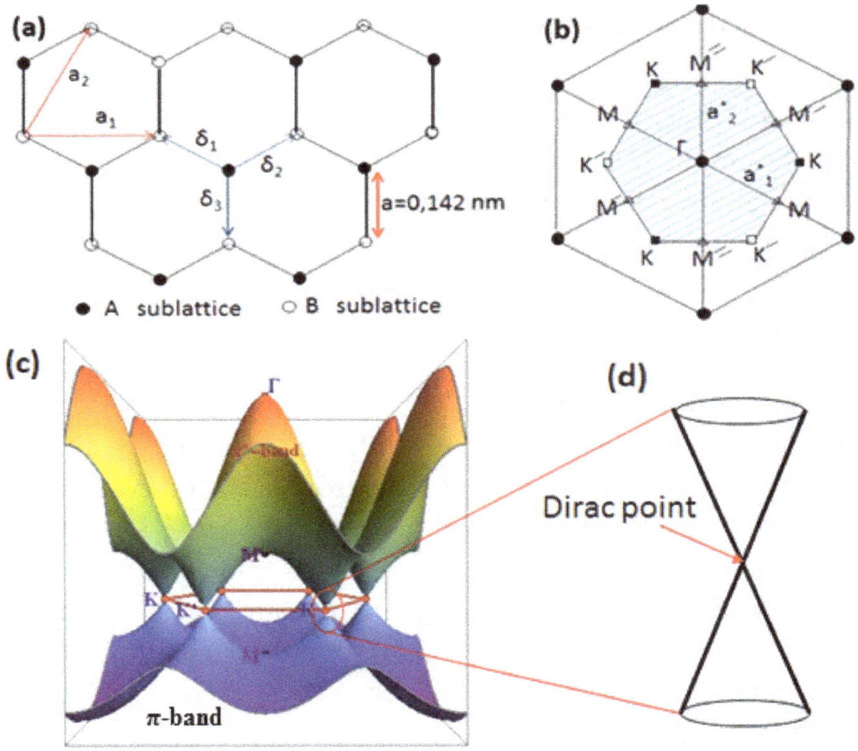

Figure 2. (a) Two graphene sub-lattices (A and B) and unit cell. (b) Reciprocal lattice of the graphene lattice, presenting the Brillouin zone, and (c) π-bands electronic dispersion relations for a 2D honeycomb crystal lattice of graphene under the tight binding (TB) approximation (http://demonstrations.wolfram.com/GrapheneBrillouinZoneAndElectronicEnergyDispersion). (d) Electronic band structure and presentation of Dirac point of 2D graphene crystal.

transparency of graphene is independent of wavelength. A monolayer graphene absorb 2.3 ± 0.1%; of the incident light with a negligible reflectance of <0.1% [7]. Graphene, with only one atomic layer, has a high surface area-to-volume ratio without affecting much of the mechanical properties. In the chemistry point of view, graphene sheets can be functionalized with other elements to achieve heterogeneous chemical and electronic structures. The basic properties of graphene-based material are summarized in **Table 1**.

Properties	Graphene	References
Electrical	• Band structure for Graphite • Carrier mobility~200,000 $cm^2 V^{-1}s^{-1}$ • Unusual (relativistic) quantum Hall effect • Excellent conductor (Can sustain 6 orders higher current than Cu)	Wallace [17] Morozov et al. [18] Zhang et al. [6] Geim and Novoselov [9]
Optical	• Graphene absorb 2.3 ± 0.1% of light (From fine structure constant), independent of wavelength	Nair et al. [7]
Thermal	Thermal conductivity~4.84 × 10^3 to 5.3 × 10^3 W/mK	Balandin et al. [19]
Mechanical	Breaking strength of ~40 N/m and Young's modulus of ~1.0 TPa	Lee et al. [8]
Chemical	Functionalization with various functional groups	Loh et al. [20]

Table 1. Properties of graphene 2D crystals.

3. Synthesis of graphene-based materials

The first demonstrated process to derive graphene was exfoliation of graphite by using a simple scotch tape technique [21]. The repeated exfoliation on a substrate surface enabled to derive a very thin sheet consisting of few-layers graphene. The unique physical properties of the exfoliated 2D graphene sheet led to several outstanding innovations in graphene deriving and synthesis process. Mechanical exfoliated graphene has been widely studied to investigate physical properties of graphene and electronic device fabrication [8–12, 21]. The scotch tape technique produces only a few micron sheet, which has only limited application prospective. Again, the technique is time consuming and basically depends on trial and error approach with difficulties to locate with repeatability. The other well-known approach explored in the very beginning of the graphene research is epitaxial growth on silicon carbide (SiC). Graphene on SiC has been synthesized by high temperature annealing process (>1300°C) in high vacuum chamber [22, 23]. In the annealing process, top layer of SiC undergoes thermal decomposition, where Si atoms desorb and carbon atoms remain on the surface rearranging and bonding to form sp^2 hybridized graphene structure. The main drawback of graphene growth on SiC is small domain structures with the presence of steps and terrace edges. Again, the high cost of SiC substrate and high processing temperature in vacuum chamber make less industrial attractive. The other alternative approach is reducing of chemically exfoliated graphene oxide (GO) to derive large quantity of graphene flakes [20, 24]. GO has been synthesized at around room temperature by a modified Hummer method that involves rigorous oxidation of pure graphitic materials [25, 26]. The oxygen-related functional groups enhance the interlayer distance of graphene sheets of graphite, which make it possible to exfoliate readily with reduce van der Waals interaction between the layers. GO is an insulating graphene sheet containing epoxy and hydroxyl oxygen functional groups on the basal plane and at the edges [27–29]. GO-based materials are electronically hybrid material that features both π state of sp^2 carbon and the σ state of sp^3 bonded carbon with oxygen. Controlling the sp^2 and sp^3 hybridized carbon atoms ratio with oxidation and reduction process the desirable optical and electronic properties can be achieved [20]. However, significant structural defects are induced in the graphene lattice with reduction of GO, which considerably effect the mechanical and electrical properties. Although the chemical exfoliation and reduction process can be carried out at room temperature to synthesize a large quantity of graphene, it does not satisfy the norms for electronic grade materials. Nevertheless, there are various possibilities for composite and electrode materials for energy conversion and storage devices.

In the above-discussed point of views, the CVD process is one of the most prominent approaches for the synthesis of high quality graphene with a controlled number of layers in large area. In the CVD approach, hydrocarbons are decomposed and catalyzed to form sp^2 hybridized carbon for lateral growth on a substrate. Graphene can be synthesized with desired structures depending on the CVD process, catalyst substrates, nature of precursors, base pressure, and gas composition. Recent significant development of high-quality and single crystal graphene synthesis by the CVD technique on metal substrates opens up new possibilities for applications [32–37]. In the following, we discuss in detail about the CVD synthesis process and growth mechanism of graphene on various catalyst substrates.

4. Development of chemical vapor deposition process

Growth of the graphitic layer on a transition metal substrate owing to atomic carbon segregation has been well known for many years [30]. Synthesis of planar nanographene (PFLGs) was demonstrated by Somani et al. by a simple CVD technique [31]. The graphitic structure included about ~35 layers of graphene, due to high segregation of carbon atoms on the Ni foil. The process was further developed to achieve control growth of monolayer, bilayer, and few layer graphene on the Ni-deposited SiO_2/Si and Ni foil [32, 33]. The main challenge of CVD graphene structure is to remove and transfer on an insulating substrate for physical properties characterization. In the later stage, transfer process of CVD graphene onto an arbitrary substrate without disturbing the intrinsic properties has been developed [34]. Transferring of a CVD graphene can be achieved by wet or dry etching of the catalytic layer and placing on a desirable substrate surface. The CVD process has been significantly explored to grow large size single crystals and high quality wafer-scale monolayer and bilayer graphene. In this prospect, the nucleation and growth mechanism of graphene on a catalytic substrate play the fundamental role in the formation of large domains and number of layers. The transition metals, alloys, and liquid metals have been recognized as the catalyst for growth of monolayer and bilayer graphene, where the surface composition of solid catalyst substrates also plays critical role on graphene growth. The thermodynamic and kinetic parameters are also crucial in graphene growth process, where pressure and temperature are both key factors [32–39]. The graphene growth mechanisms are discussed for the atmospheric and low pressure CVD technique in the following section.

5. Atmospheric and low pressure CVD technique

The atmospheric pressure chemical vapor deposition (APCVD) system consists of horizontal quartz tube connected to gas inlet and outlet [40]. **Figure 3(a)** and **(b)** presents schematic of two different possible APCVD and low pressure CVD (LPCVD) system depending on the precursor materials. Gaseous precursors can be supplied from external sources, while solid and liquid precursors can be directly inserted in the CVD chamber for graphene growth. The growth chamber pressure in graphene growth has been significantly explored to control the nucleation density and number of layers [41–45]. It has been observed that the activation energy of nucleation dramatically affects by the pressure i.e., different for the APCVD and LPCVD techniques. The effect of chamber pressure on graphene nucleation has been explored on Cu catalyst substrate, considering the self-limiting behavior for monolayer graphene growth on Cu [43–45]. It has been observed that the graphene nucleation density affected by several fundamental processes occurred on the Cu surface. These are: (i) precursor adsorption, (ii) formation of active carbon species (dehydrogenation), (iii) diffusion of active carbon on the surface, and (iv) critical size nuclei formation that competes with (v) desorption. Now, most of these processes are affected by the background pressure of CVD and responsible for the difference between the low and atmospheric pressure growth of graphene [45]. The overall pressure significantly affects desorption of various species from the catalyst surface. It has been explained that due to the collisions with the buffer gas in the diffusion layer close to the surface, the desorbed species have a higher returning rate to the surface at higher pressures. The effect is inversely proportional to the background pressure, making the evaporation rates at atmospheric pressure more than 3 orders of the magnitude lower than at reduced pressures. In LPCVD process, diffusion of the gas decreases proportionally to the reciprocal of the pressure. The pressure for LPCVD system is typically ~10–1000 Pa, while the standard atmospheric pressure is 101,325 Pa. If the pressure in the CVD process is reduced from atmospheric pressure to about 100 Pa the diffusion will decrease by almost 1000 times [46]. Hence, the velocity of mass transport will reduce to the substrate surface inside growth chamber, allowing uniform and homogenous growth of graphene. Thus, graphene growth is less dependent on mass transport velocity in the LPCVD process [45, 47].

It has been reported that the carbon atoms adsorption energies on a Cu surface vary within 4.1–7.5 eV range, where graphene nucleation on the catalyst substrate significantly depends on desorption/etching of small active carbon species. It has been also reported that the Cu evaporation rate in vacuum is significantly high, which is around 4 μm/hour at 1000°C [48]. The high evaporation rate of Cu in a LPCVD process promotes desorption of carbon species adsorbed on top of the catalyst, leading to a lower nucleation density. The reported activation energy of nucleation, E_{nuc} ~ 3–4 eV at low pressures and temperature >950°C [45, 47]. On the other hand, sublimation of Cu in the APCVD process is significantly less, which lead increased activation energy for Cu atoms desorption as well as desorption energy of carbon atoms from the catalyst surface (E_{nuc} ~ 9 eV) [45]. Thus, graphene nucleation and growth significantly differ depending on the pressure of the CVD chamber affecting absorption and desorption of carbon atoms (**Figure 4**). Along with Cu, various other catalyst substrates have been explored for uniform monolayer or bilayer graphene growth in the CVD process as discussed below.

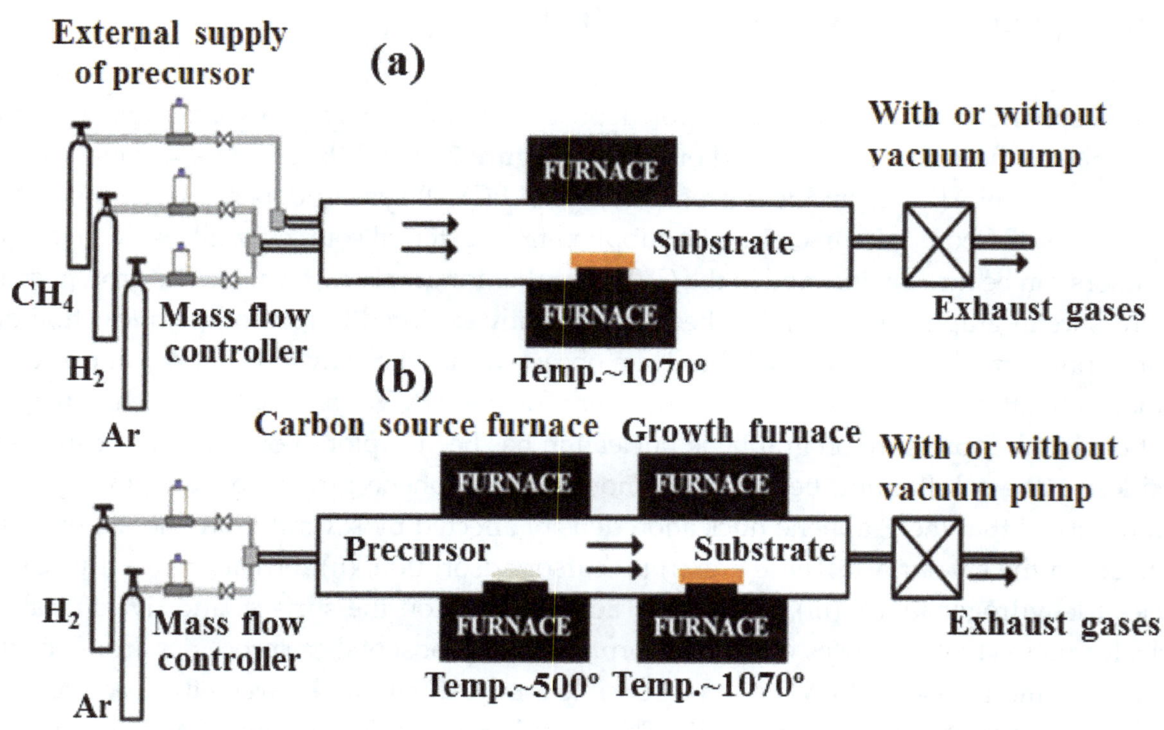

Figure 3. Two different possible CVD systems depending on the precursor materials. (a) External supply of gasses precursor and (b) Directly inserted solid and liquid precursors in APCVD and LPCVD modes.

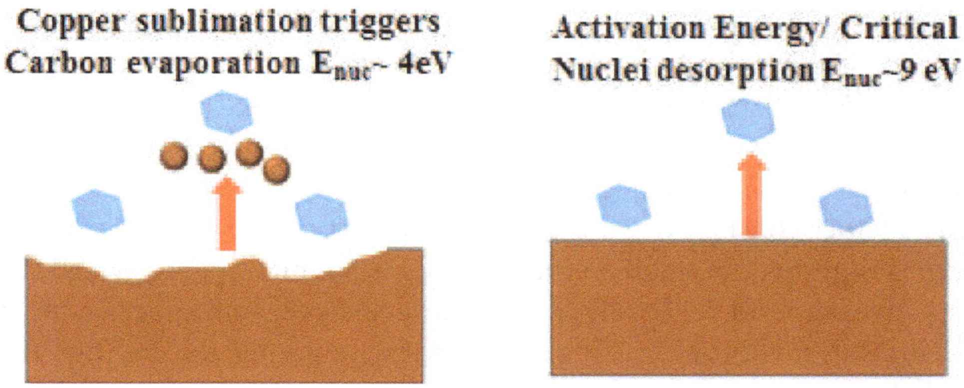

Figure 4. Activation energy differ for low and atmospheric pressure CVD. At low pressure, evaporation of Cu is significant which limits nucleation of graphene (E_{nuc} ~ 4 eV). Cu evaporation is less at atmospheric pressure CVD, where nucleation activation energy is higher (E_{nuc} ~ 9 eV). [Reprinted with permission from Ref. [45].

6. Catalysts for CVD process

6.1. Single transition metal catalyst

The CVD growth of graphene has been mostly investigated on transition metal catalysts (Ni, Cu, Co, Pt, Ag, Ru, Ir, Pd, etc.) [31–39, 49–53]. Growth of pyrolytic carbon films or layers of graphite have been first observed on Ni catalyst with introduction of hydrocarbon or evaporated carbon materials [30, 54]. Almost at the same time, the formation of graphitic layers on single crystal Pt and Ru was observed in catalysis experiments. The formation of graphitic layers can be explained with diffusion and segregation of carbon impurities on the metal surface during the annealing and cooling process. At an elevated temperature, hydrocarbons were thermally decomposed and surface absorption/desorption or segregation of carbon occur depending on carbon-metal solubility for graphene growth. The above-discussed (Section 5) absorption/desorption of carbon atoms for particular catalyst substrate can significantly differ, depending on carbon solubility on metal surface. Again, the metal catalyst can be polycrystalline or single crystalline in nature. The grain boundaries, crystallographic orientation of grain can significantly influence the absorption/desorption or segregation of carbon atoms, which leads diverse morphologies of graphene on different metal catalysts. In the following, we discussed two simple model of graphene growth process on highly carbon soluble Ni and low carbon soluble Cu catalyst. The stability and the reactivity of Ni at high temperatures, as well as its high carbon solubility (0.6 weight % at 1326°C), have been explored for graphene growth in the initial stage of CVD research [55]. The lattice constant of Ni is 3.52 Å and its first-neighbor distance in the bulk is 2.49 Å, which is almost identical to the lattice constant of graphene 2.46 Å [55, 56]. **Figure 5(a)** shows the growth model of graphene on the Ni catalyst. Bulk diffusion of carbon atoms can be dominant depending on the CVD growth conditions for high carbon soluble transition metal catalysts (**Figure 5(a)**). The nonequilibrium process leads to the carbon precipitation on the surface and the formation of graphene during the cooling down process [56–59]. It is difficult to obtain uniform graphene films with minimal microstructural defects on polycrystalline Ni, owing to multiple nucleation and unpredictable quantity of segregated carbon. Contrary to the Ni catalyst substrate, Cu has a filled 3D shell that results in a low carbon solubility (0.008 weight % at 1084°C) and reduces the tendency for adsorbing hydrocarbons onto the Cu surface [60–62]. This favors an extensive surface migration of carbon adatoms on the Cu surface and a minimum diffusion into the bulk of Cu. The low affinity of carbon for Cu is also shown in the absence of carbide formation. The lattice constant of Cu is 3.61 Å and its first-neighbor distance in the bulk is 2.56 Å, which is slightly different than the lattice constant of graphene which is 2.46 Å. The lattice mismatch is around 3.7%, larger than of between Ni and graphene (1.2%), which indicates favorable growth on the Cu surfaces but with easier transfer due to the weaker interactions between graphene and Cu substrate. The carbon source precursor molecules are decomposed before being adsorbed by the Cu catalyst where the dehydrogenation of the molecules takes place, followed by the surface migration and the growth (**Figure 5(b)**) [62–64]. Growth conditions and substrate grain orientations of Cu also influence the growth of the graphene crystals. Considering the effect of catalyst substrate polycrystalline nature, epitaxial growth of graphene on single crystal Ni (111) and Cu (111) has been also explored [65–67]. Graphene can be grown with a preferred orientation on Cu(111), and even seamless sticking of graphene domains on Cu(111) substrates has been obtained. Thus, unidirectional alignment of nucleated graphene crystals on a single-crystal metal

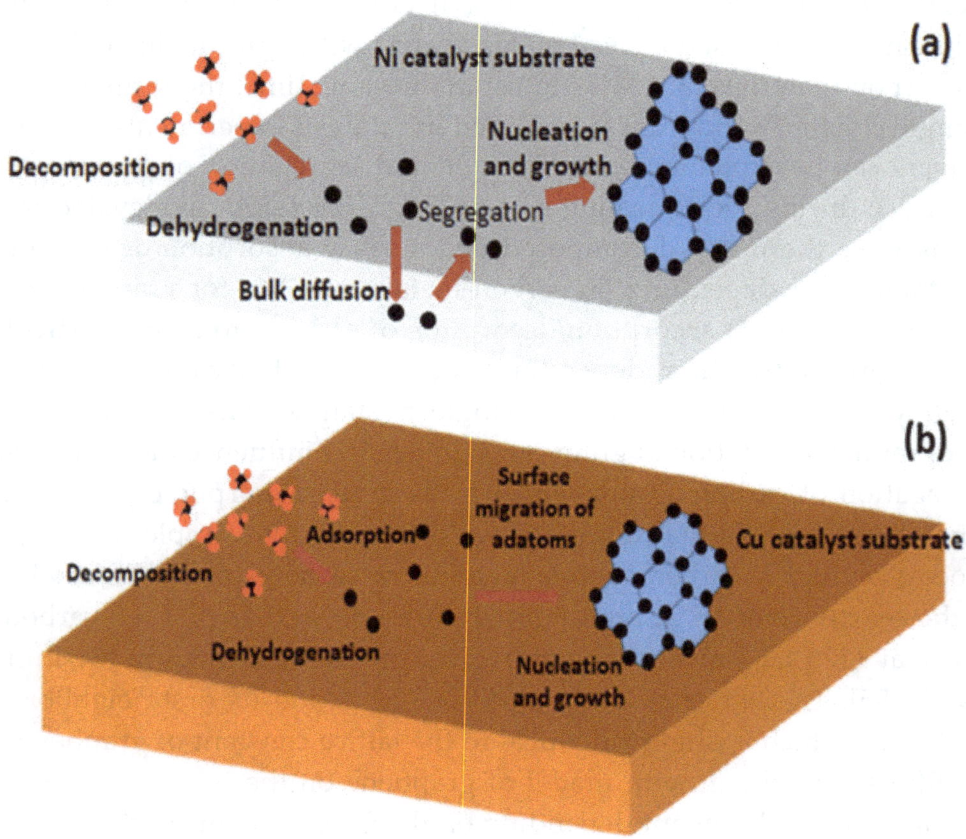

Figure 5. (a) Graphene growth process on highly carbon soluble metal substrate (e.g., Ni). Decomposition, dehydrogenation, bulk diffusion, and segregation process determined the graphene nucleation and growth. (b) Graphene growth process on a low carbon soluble metal substrate (e.g., Cu). Decomposition, dehydrogenation, and surface adsorption/desorption process determined the graphene nucleation and growth.

substrate can be the perfect solution to grow uniform graphene films. However, Synthesis of graphene on a single-crystal metal substrate is not suitable for large-scale production owing to the high cost and the difficulty of preparing single-crystal metal. Significant effort has been also made to synthesis large single crystal graphene on the metal catalyst to overcome the limitation of smaller graphene domain and polycrystalline nature [68–74]. In this prospect, modification of the Cu substrate surface by oxidation has enabled to reduce nucleation density, which allows large-area lateral growth of the same crystal (**Figure 6**) [72]. The processing of the Cu substrate surface is critical as well as the growth of kinetic parameters determine the size and shape of the single crystals. The main challenge toward this research prospect is to produce continuous large-area film with the same large size single domains in a less growth duration.

6.2. Binary transition metal catalyst

One of the important aspects, we learn from the above discussion, is that the traditional polycrystalline or single-crystalline metal catalysts have enabled the growth of uniform graphene film by complex pretreatments or precise parameter controlling. Rational designing of a binary catalyst can be significant to control nucleation density, number of layers, solubility of carbon and dopant, which will provide significant control in the CVD growth

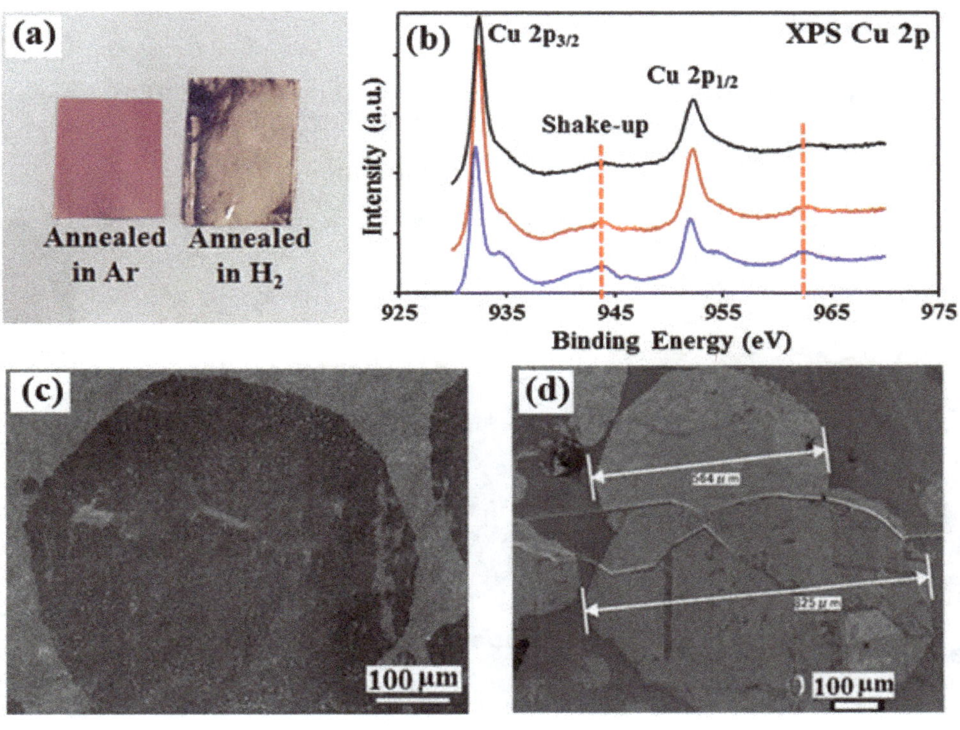

Figure 6. Oxidizing treatment of Cu for large domain growth. (a) Photographs of Cu foils after the annealing treatment in Ar and H_2 atmospheres. (b) XPS spectra of the Cu surface at different stages of the annealing. (c) SEM images of a large graphene domain (~510 µm). (d) Formation of millimeter scale structures with merging of two graphene domains (560 and 825 µm). [Reprinted with permission from Ref. [72].

process. Bilayer graphene with AA, AB stacking and rotation stacking fault is interesting prospect to observe novel physical properties (**Figure 7(a)**) [75–77]. It has been demonstrated that applying a vertical electric field in bilayer graphene-based field effect transistor (FET), a bandgap of the order of 0.2–0.3 eV can be observed [78]. Significant research effort has been given for control growth of bilayer graphene using a binary alloy catalyst to adjust the solubility of the carbon atoms. The solubility of carbon species at high temperature in Cu and Ni is significantly different, which leads to distinct growth behavior as discussed in Section 6.1. Growth of bilayer graphene can be achieved on Cu-Ni alloys with more than 90% coverage adjusting the Ni concentration in the alloy [79]. Synthesis of a highly uniform bilayer graphene film by the CVD process has been also achieved on an epitaxial Cu-Ni (111) binary alloy catalysts [80]. In the case of the epitaxial Cu-Ni alloy also the relative concentration of Ni and Cu as well as the other thermodynamic and kinetic parameters such as temperature, cooling profile, and gas composition strongly influence the uniformity of bilayer graphene growth. The other metal alloy such as Ni-Mo, Au-Ni, Co-Cu, etc. also can be significant to achieve control growth of graphene in a CVD process [81–84]. The present challenges toward this direction are selection of proper metal for alloy, composition, and alloy preparation techniques for further development. Significant effort has been also made to design alloy-based catalyst to achieve faster growth of larger graphene domains [85]. The binary alloy catalyst also facilitates to control the solubility of carbon and dopant atoms to achieve substitutional doping. Remi et al. have demonstrated that

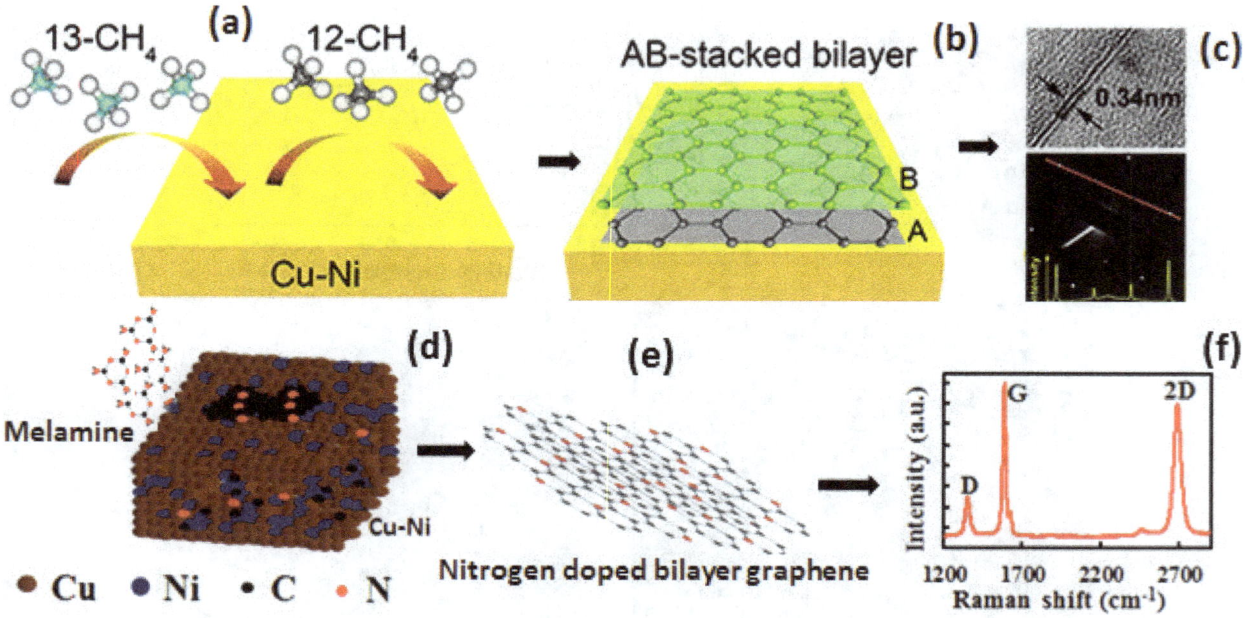

Figure 7. Graphene growth on the Cu-Ni binary alloy catalyst substrate by a CVD technique. (a) and (b) Schematics of the CVD growth process of AB stacked bilayer graphene on the Cu-Ni alloy. (c) Transmission electron microscopy (TEM) and diffraction analysis of the graphene. [Reprinted with permission from Ref. [76]]. (d) and (e) Schematic of nitrogen-doped bilayer graphene growth on the Cu-Ni binary alloy using a melamine solid precursor and (f) Raman spectra of the bilayer domain. [Reprinted with permission from Ref. [86]].

large nitrogen-doped bilayer graphene domains can be synthesized on the Cu-Ni binary alloy as shown in **Figure 7(b)** [86].

6.3. Liquid metal catalyst

Unidirectional nucleation and growth can be achieved on a liquid metal in the absence of a crystalline lattice and amorphous atomic structure, which is truly homogeneous [87, 88]. The quasi-atomically smooth surface of the liquid to avoid defects or grain boundaries such as those found in solid metals can be perfect choice for single layer graphene growth in the CVD process. The perfectly smooth liquid surface supports uniform nucleation and growth of graphene. The growth of graphene governed by a self-limited surface catalytic process and is robust to variations in CVD growth conditions [89, 90]. Homogenous monolayer graphene growth on liquid metals has been achieved using the CVD process. The p-block element, such as liquid Ga, has been also demonstrated as an attractive option for large high-quality graphene growth and expanding the catalyst family for graphene growth beyond d-block transition metals. Growth of monolayer graphene on liquid indium (In) and gallium (Ga) has been demonstrated, without the formation of sublayers as shown in **Figure 8** [90]. Thus, the development of various catalysts for the CVD process and achieving growth of graphene with desired numbers of layer, crystalline structure, and morphology with the highest quality will enable commercial applications. The comparison of various catalyst types for the graphene growth process is summarized in **Table 2**.

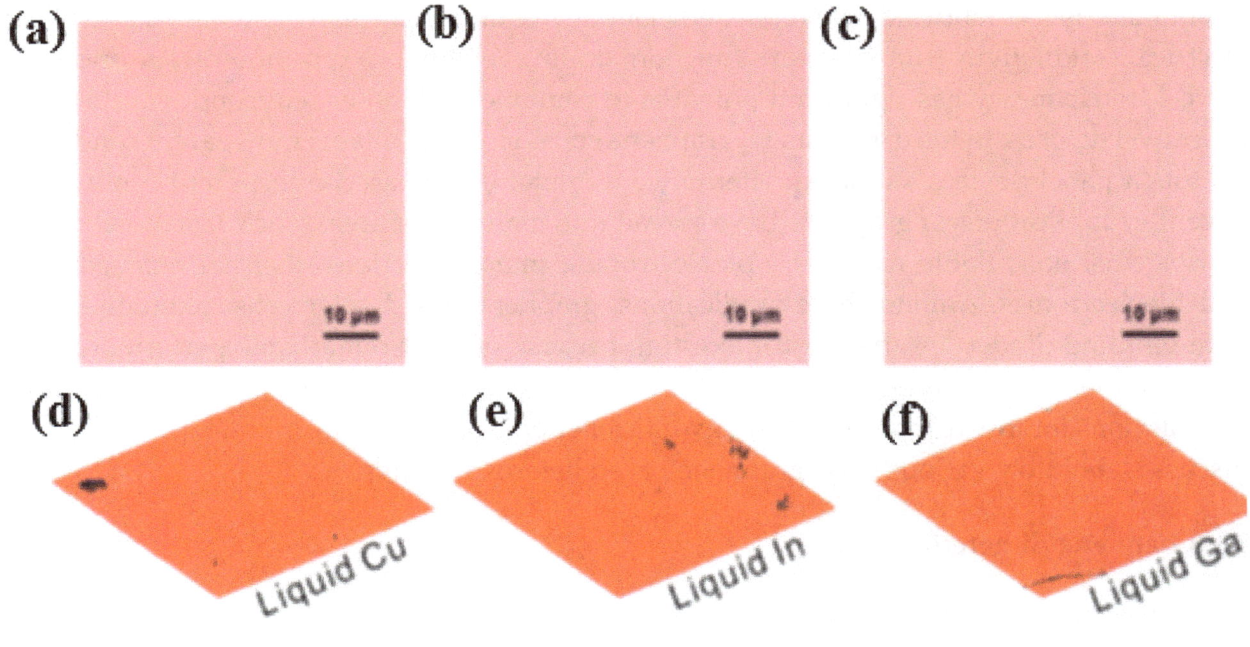

Figure 8. Typical growth results on liquid metal substrates. (a–c) Optical microscope images of graphene grown on liquid Cu, In and Ga respectively, which demonstrates the excellent uniformity of the single-layer graphene. (d-f) Layer distribution determined by RGB color analysis of the corresponding optical microscope images [Reprinted with permission from Chem. Mater. 2014; 26(12): 3637-3643].

Type of catalysts	Merits	Demerits
Polycrystalline metal	1. Low cost 2. Easy processability 3. Simple CVD approach	1. Nonuniform growth 2. Multilayer growth 3. Defect in grain boundaries
Single crystal metal	1. Unidirectional nucleation 2. Uniform monolayer	1. Difficult to produce large area graphene 2. High processing cost for single crystal metal preparation
Binary alloy	1. Control layer growth 2. Controllable doping of graphene	3. Alloy preparation technique 4. High processing cost
Liquid metal	1. Uniform single layer graphene	1. Handling and processability 2. Transfer process (for example, transferring from liquid Ga surface)

Table 2. A comparison of catalyst types for various graphene growth process.

7. Heterostructure of CVD graphene with other semiconductors

The heterostructure of CVD graphene with other bulk and two-dimensional (2D) semiconductors has been developing extremely fast. Novel heterostructure devices are in main focus for solar cell, light-emitting diodes, photodetectors, gas sensors, tunneling transistors, resonant tunneling diodes, etc. [91]. This has enabled to utilize and integrate the properties of different materials to fabricate new highly efficient device architecture. Graphene with excellent electrical conductivity, tunable work function, doping to obtain p-type and n-type and high optical transparency has significant potential to combine with other semiconductor to fabricate novel devices. In the earlier stage, graphene considered to suitable transparent conductor for touch panel devices and as an alternative to metal oxide-based transparent conductors. Recently, applications of graphene have spread into many other areas, such as nanomechanical systems, nonvolatile memory, optoelectronics, interconnections, thermal management, and bioelectronics. Among these applications, graphene-based solar cells, photodetectors, light-emitting diodes are the most interesting because of their remarkable performances as transparent electrodes and active layers for photoelectric devices, which make them promising solutions for fast-response and energy-efficient applications. In the following sections, we discuss elaborately various CVD graphene-based heterostructures.

7.1. Graphene-Si heterostructure

Heterojunction of CVD graphene and Si has been extensively studied for solar cells and photodetector applications [92]. The transparency of CVD graphene in a wide wavelength range makes it possible to fabricate a broad wavelength range photodetector and utilize much higher solar light to achieve high conversion efficiency. Again, the density of state (DOS) for graphene significantly differs than that of conventional metals. It can be observed that conventional metal has a finite DOS near Fermi energy, whereas graphene has a zero DOS at the Fermi energy (as discussed in Section 2). The unusual optoelectronic properties of graphene open new opportunities in device engineering than that of conventional metal-semiconductor-based devices. The CVD graphene on n-Si can create a Schottky junction with a large build-in potential due to its suitable work function as well as prospect to tune the barrier height. CVD synthesized a large-area monolayer or few-layer graphene has been transferred on the n-Si or p-Si substrate to create a heterojunction [94]. The CVD graphene film coated on the patterned Si substrate with an insulating and metallic layer attached perfectly with excellent contact (**Figure 9**). Excellent rectification diode characteristic can be observed for the graphene/n-Si device. The low leakage current can be attributed to uniform contact of the graphene film on Si surface. Photovoltaic action has been observed with illumination of white light. The exciton dissociation can be achieved to obtain a photovoltaic action by creating a suitable barrier height in the graphene/Si heterojunction [92–94]. Recent studies show more than 10% conversion efficiency of CVD graphene/n-Si Schottky junction solar cells [95]. The interesting fact is that by chemical or substitutional doping of graphene allows to tune the electronic state to make p-type or n-type as well as changing the work function. There are enormous potential in this device technology for various optoelectronics application. Such a simple device structure with easy fabrication process can be ideal to integrate for wafer scale production within the existing Si solar cell manufacturing process.

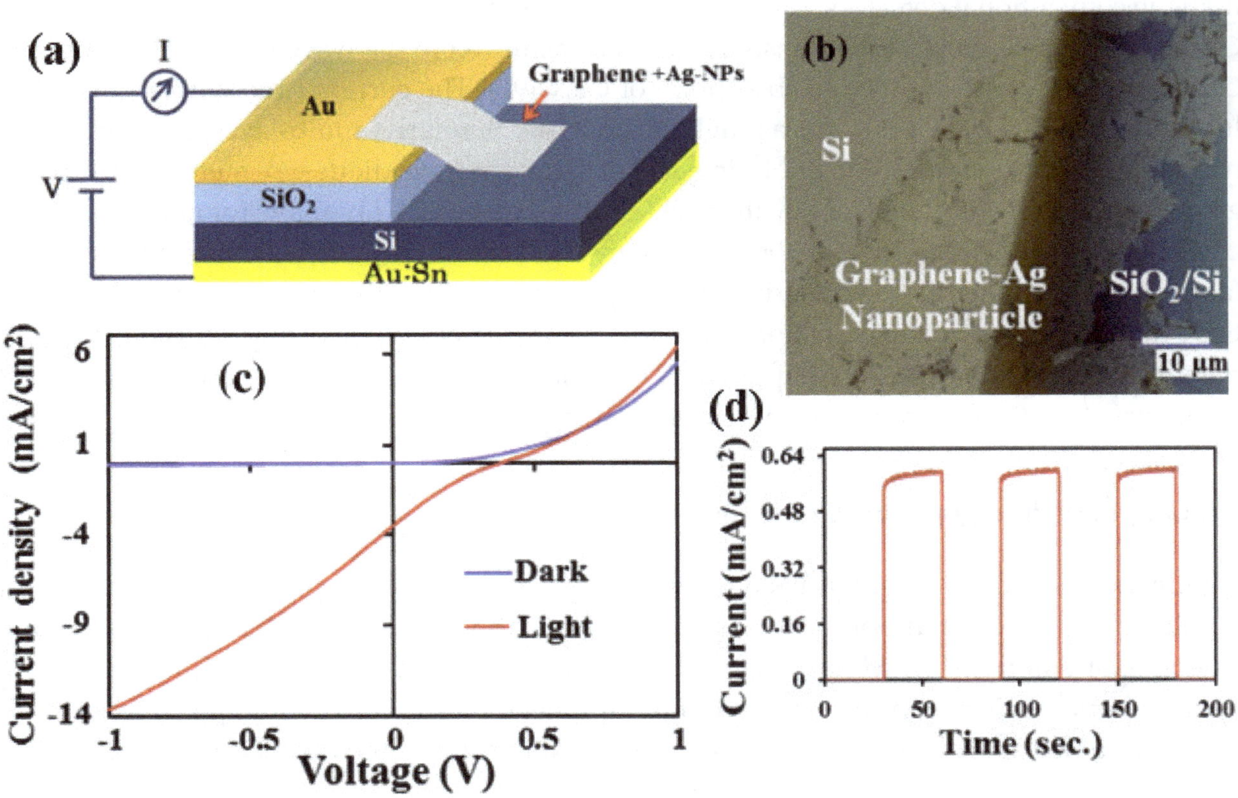

Figure 9. (a) Schematic of fabricated graphene-Si heterojunction. (b) Optical microscope image of silver nanoparticles (AgNPs) coated graphene/Si heterostructure. (c) Current-voltage curves for AgNPs-graphene/n-Si solar cell. (d) Photoresponsivity at 550 nm wavelength. [Reprinted with permission from Ref. [94]].

7.2. Graphene-GaN heterostructure

Integration of CVD graphene with wide bandgap semiconductors has also attracted significant attention for optoelectronic device applications. GaN is one of the most promising wide bandgap semiconductors for applications in optoelectronic and other electronic devices such as light-emitting diodes, laser diodes, solar cells, and high-electron-mobility transistors [95–99]. The transparency of metal oxide conductors is poor in the UV and near UV region, which affects the photodetectors efficiency and brightness of LEDs. In this prospect, a high transparent conductor in the UV region can be the only solution and CVD graphene is the ideal candidate. Another important aspect is low thermal conductivity of metal oxide transparent conductors, reduces the diffusion efficiency of the heat generated during the operation of LEDs. Graphene/GaN heterojunction can be most chemically robust and nondegradable under harsh conditions. The conventional metals like Ni, Pt, and Pd have been used for the fabrication of Schottky junction with GaN. However, diffusion of the metal into semiconductor at elevated temperatures enhances the tunneling of the carriers across the barrier, causing a decrease in thermionic emission current, i.e., lowering of the barrier and hence restricts the high temperature operation of the device. Graphene has been combined with n and p type GaN in lateral and vertical heterostructure to fabricate light emitting diode and photodetectors (**Figure 10**) [98]. The fabrication of light emitting diode has been demonstrated using a tunnel junction configuration such as graphene/n-$In_xGa_{1-x}N$/p-GaN. Similarly, a graphene/GaN-based Schottky UV photodetector has been investigated for higher and faster response than the conventional detectors. UV photodetector device has been fabricated with monolayer CVD graphene. High photoresponsivity has been achieved in the wavelength below 360 nm, corresponding to the band edge absorption of GaN [99]. The current ratio with and without luminescence has been achieved as high as 1.6×10^4. Similarly, multi-layer CVD graphene has been also investigated for photodetector or light-emitting diodes. A method of directly growing graphene films on GaN by the CVD process can be also favorable for fabrication process compatibility. Nevertheless, there are many challenges in graphene-GaN heterojunction fabrications and interface engineering. The important aspect is to understand the graphene/GaN interface physics and transport mechanism for carriers transporting across it. There is plenty of scope to develop 2D (graphene)-3D (Si, GaN, GaAs and other bulk semiconductors) heterostructure for fabrication of energy-efficient devices.

7.3. Graphene-hexagonal boron nitride heterostructure

Significant research interest has been given to develop atomically thin 2D heterostructure of graphene and hexagonal boron nitride (hBN) [100]. hBN can be considered as a structural analogue of graphene, which is composed of alternating boron and nitrogen atoms in a honeycomb lattice rather than the carbon atoms. Graphene with a zero bandgap is a semimetallic as discussed in Section 1, while hBN with a bandgap of ~6 eV is insulating [100–107]. Significant carrier mobility of graphene has been achieved on atomically flat hBN as a dielectric layer for the FET device. hBN with no dangling bond and charge traps make an ideal dielectric material for graphene electronics [104]. It has been also observed that a measurable bandgap can be induced in graphene using hBN as a substrate, opening a new prospect for the FET fabrication. Recent studies have also revealed that an obvious bandgap can be introduced by in-plane hBN and graphene heterostructures fabrication with a relatively high carrier mobility [106]. In these prospects of vertical heterostructures, the hBN and graphene layer can be transferred individually to fabricate a stack of layers, while a growth process is unavoidable for in-plane heterostructure fabrication. Various CVD approaches have been developed for the fabrication in-plane hBNC hybrid or hBN-graphene heterostructures on the catalyst substrate (e.g., Cu) [106, 107]. These approaches are one-batch growth and two-step patching growth models. Ajayan et al. have demonstrated the synthesis of hBNC film and observed significant influence on electronic property, such as the electrical conductivity and opening a bandgap attributing to the quantum confinement effect or spin polarization at specific C-BN boundaries (**Figure 11**) [106]. Subsequently, hydrogen-induced etching process of graphene or hBN layer has been developed for lateral epitaxial growth of h-BN/graphene structure with sharp interfaces [107]. Temperature-triggered chemical switching growth of in-plane and vertically stacked heterostructures has been also achieved in the same growth process [108]. There are significant potential to control the etched structure and domain formation in an hBN-graphene heterostructure to realize a considerable bandgap. The control in size, shape, interface, and domain structure will be critical for further development of both in-plane and vertical hBN/graphene heterostructure.

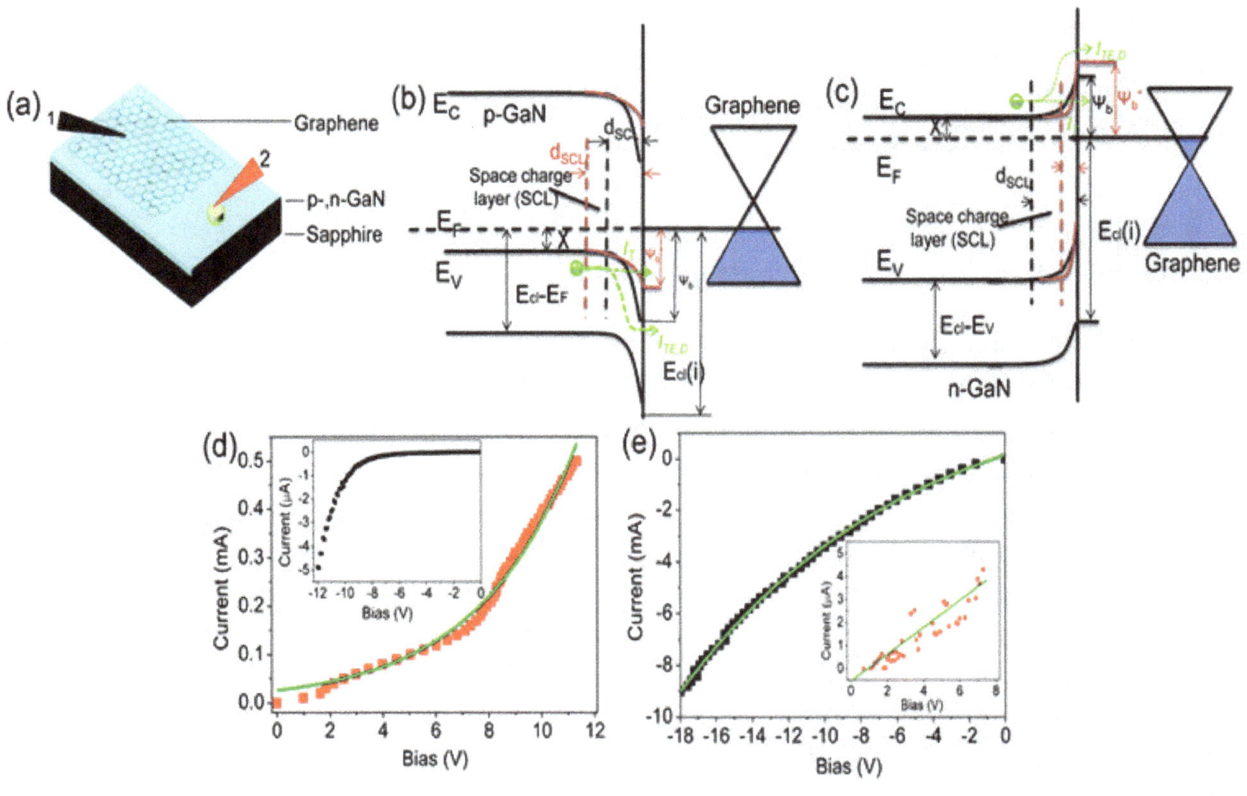

Figure 10. (a) Schematic diagram of graphene-GaN device. (b) and (c): Energy band diagram of graphene/p-GaN and graphene/n-GaN junctions. (d) and (e): Forward current-voltage (*I-V*) characteristics for p-, n-GaN/graphene junctions, respectively; insets of (d) and (e) present the corresponding reverse *I-V* curves. [Reprinted with permission from Ref. [98]].

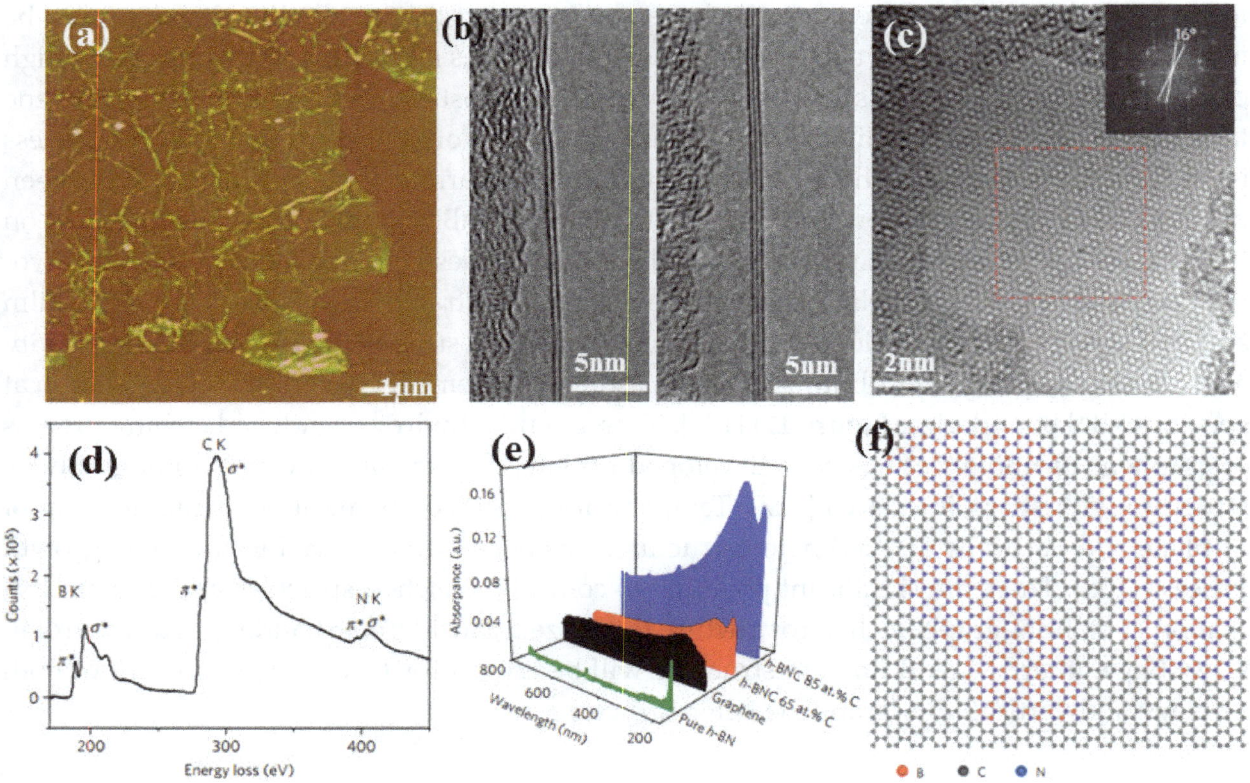

Figure 11. Synthesis of hBNC film by a CVD approach to introduce a bandgap in the hybrid structure. (a) AFM image of the uniform hBNC transferred onto SiO_2/Si substrates. (b) High resolution TEM images presenting the section view of the hBNC film. (c) Moiré patterns of a two layer packing in the hBNC film. (d) EELS spectrum of the hBNC films showing three K-shell excitation edges of B, C and N respectively. (e) UV-Vis absorption spectra of pure hBN, graphene, and hBNC hybrid. (f) Atomic models for the hBNC film. [Reprinted with permission from Ref. [106]].

7.4. Graphene-transition metal dichalcogenides heterostructure

The monolayer transition metal dichalcogenides (TMDC) materials have received significant attention due to the presence of a direct bandgap. TMDCs, such as molybdenum disulfide (MoS_2), tungsten disulfide (WS_2), rhenium disulfide (ReS_2), tungsten diselenide (WSe_2), etc., are the family of compound materials with the generalized formula MX_2, where M is transition metal and X is a chalcogen such as S, Se, or Te. The individual layer of TMDCs consists of three atomic layers in which the transition metal is sandwiched between two chalcogens. The electronic and optoelectronic components fabricated using TMDC-layered materials, such as FET, sensors, and photodetectors, have the potential to substitute conventional Si-based electronics and organic semiconductors [109–111]. TMDC-layered materials have been derived by mechanical exfoliation, liquid exfoliation, solvothermal process, and sulfurization of transition metal-based precursors [112–115]. Among various approaches, the CVD technique is one of the most suitable approaches to obtain wafer-scale uniformity for device fabrication [113, 115]. TMDC materials synthesis by the CVD process is independent of catalyst in contrast to the graphene and hBN CVD synthesis. TMDC layers have been combined with graphene and hBN for the fabrication of transistors, photodetectors, solar cell, sensors, etc. by exploiting their discrete physical properties [116, 117, 119]. Graphene and TMDCs, such as MoS_2, WS_2, etc., have been combined to fabricate a heterojunction photodetectors. Various other novel devices have been proposed or developed based on heterostructures formed between MoS_2 and graphene. Electronic logic and memory devices have already been constructed from graphene-MoS_2 hybrids [114]. The graphene-MoS_2 heterostructures have also been adopted to demonstrate an extremely high photo-gain and the ultrasensitive sensors fabrication [118, 119]. Considering the significant potential, efforts have been made to grow graphene-TMDCs hybrid structures by a CVD technique (**Figure 12**). Direct synthesis of TMDC layer on graphene has been developed; such large-area van der Waals heterostructures show a significant improvement in photoresponse. In-plane or vertically oriented growth of TMDCs on graphene also has great potential in the hydrogen evolution reaction and sensing device applications.

The beauty of these material systems is that we can design and develop accordingly to achieve something unexpected. Various other layered materials and their heterostructures are emerging with intriguing properties for possible practical nanoelectronics, optoelectronics and energy conversion and storage devices. We expect that the further development of the CVD method will be the forefront of research area to develop graphene and other 2D materials.

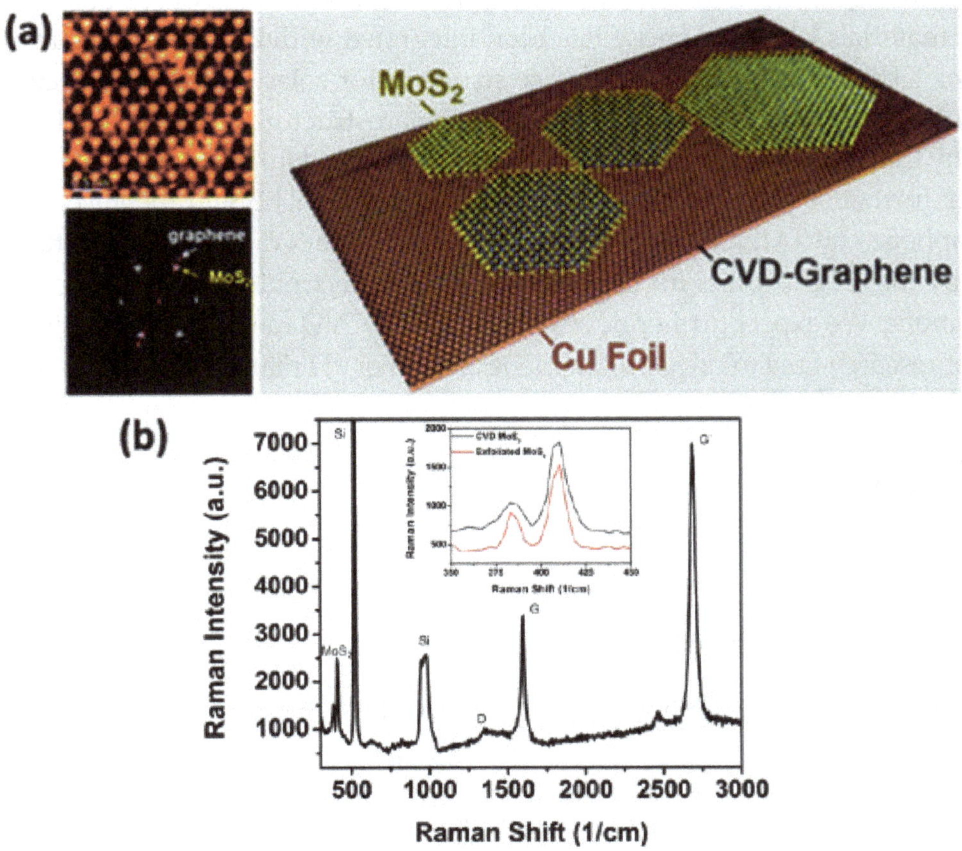

Figure 12. CVD synthesis of monolayer MoS_2 on graphene. (a) TEM images and electron diffraction pattern suggests the Van-der-Waals epitaxy of MoS_2/graphene layers. (b) Raman spectra of the MoS_2/graphene heterostructure. [Reprinted with permission from Ref. [119]].

8. Summary

In the very outset of the chapter, we have discussed the background of graphene research development leading to a new area of research in physics, chemistry, and engineering of 2D crystals. The crystal structure and basic optical, mechanical, electrical, and thermal properties of graphene were presented. Deriving or synthesis of graphene by various approaches was introduced. Among the synthesis processes, a CVD approach has been extensively explored for controlled growth of high quality and large-area graphene for electronic device applications. The thermodynamic and kinetic parameters have a critical role in graphene nucleation and growth process in the CVD method. The pressure and temperature are both key factors for graphene growth, where the kinetic factors, such as gas flow, precursor flow rate, and gas composition pressure also play important roles on graphene quality and structural morphology. We introduced the atmospheric and low pressure CVD technique and growth dynamics of graphene continues film and large single crystals. Catalyst also plays a significant role on the quality of graphene synthesized by the CVD process. We have discussed about the polycrystalline and single crystalline transition metal catalysts and their effect on graphene growth. The surface modification, such as oxidation, plays an important role to reduce nucleation density and thereby achieving centimeter scale single crystal graphene growth in a CVD approach. We also introduced the monolayer graphene growth behavior on a liquid metal catalyst. Next, we discussed about the emerging application of the CVD graphene and fabrication of various heterostructure and hybrid materials. CVD graphene has been integrated with bulk semiconductor such as Si and GaN for fabrication of advanced device structure for solar cell, photodetector, and light-emitting diode applications. In-plane and vertical heterostructures of hBN-graphene have been developed by the CVD process. The control in size, shape, and interface of domain structure will be critical for further development of both in-plane and vertical hBN/graphene heterostructures. Further, graphene and TMDCs (MoS_2, WS_2 etc.) based van der Waals heterostructures have been developed by the CVD process for nanoelectronic, optoelectronic, energy storage, and conversion applications. We expect further development of the CVD method, which will be one of the forefronts of research area to develop graphene and other 2D materials in the coming years.

Author details

Golap Kalita* and Masaki Tanemura

*Address all correspondence to: kalita.golap@nitech.ac.jp

Department of Applied Physics and Engineering, Nagoya Institute of Technology, Japan

References

[1] Novoselov KS, Geim AK, Morozov SV et al., Electric field effect in atomically thin carbon films. Science 2004; **306**: 666–669. DOI: 10.1126/science.1102896.

[2] Peierls RE, Quelques propriétés typiques des corpses solides. Ann. Inst. H. Poincaré 1935; **5**: 177–222.

[3] Landau LD, Lifshitz EM, Statistical Physics, Part I. Pergamon Press: Oxford; 1980. Sections 137 and 138.

[4] Mermin ND, Crystalline order in two dimensions, Phys. Rev. 1968; **176**: 250–254.

[5] Novoselov KS, Geim AK, Morozov SV et al., Two-dimensional gas of massless Dirac fermions in graphene, Nature 2005; **438**: 197–200. DOI:10.1038/nature04233.

[6] Zhang Y, Tan YW, Stormer HL et al., Experimental observation of the quantum Hall effect and Berry's phase in graphene. Nature 2005; **438**: 201–204. DOI:10.1038/nature04235.

[7] Nair RR, Blake P, Grigorenko AN et al., Fine structure constant defines visual transparency of graphene. Science 2008; **320**: 1308. DOI: 10.1126/science.1156965.

[8] Lee C, Wei X, Kysar J W et al. Measurement of the elastic properties and intrinsic strength. Science 2008; **321**: 385–388. DOI: 10.1126/science.1157996.

[9] Geim AK, Novoselov KS, The rise of graphene. Nature Mater. 2007; **6**: 183–191. DOI:10.1038/nmat1849.

[10] Berger C, Song Z, Li X, et al., Electronic confinement and coherence in patterned epitaxial graphene. Science 2006; **312**: 1191–1196. DOI: 10.1126/science.1125925.

[11] Kroto HW, Heath JR, Obrien SC et al. C_{60}: Buckminsterfullerene, Nature 1985; **318**: 162–163.

[12] Original paper in Japanese, translated in: Ozawa E, Kroto HW, Fowler PW, Wassermann E, Phil. Trans. R. Soc. (London) A. 1993; **1**: 343.

[13] Iijima S, Helical microtubules of graphitic carbon. Nature 1991; **354**: 56–58.

[14] Kosynkin DV, Higginbotham AL, Sinitskii A et al., Longitudinal unzipping of carbon nanotubes to form graphene nanoribbons. Nature 2009; **458**: 872–876.

[15] Xu X, Ray R, Gu Y. et al., Electrophoretic analysis and purification of fluorescent single-walled carbon nanotube fragments. J. Am. Chem. Soc. 2004; **126**(40): 12736–12737.

[16] Castro Neto AH, Guinea F, Peres NMR, Novoselov KS, Geim AK, The electronic properties of graphene. Rev. Mod. Phys 2009; **81**: 109–162. DOI: 10.1103/RevModPhys.81.109.

[17] Wallace PR, The band theory of graphite. Phys. Rev. 1947; **71**: 622.

[18] Morozov SV, Novoselov KS, Katsnelson MI et al. Giant intrinsic carrier mobilities in graphene and its bilayer. Phys. Rev. Lett. 2008; **100**. p. 016602.

[19] Balandin AA, Ghosh S, Bao W et al., Superior thermal conductivity of single-layer graphene. Nano Lett. 2008; **8**: 902–907.

[20] Loh KP, Bao Q, Eda G. et al. Graphene oxide as a chemically tunable platform for optical applications. Nature Chem. 2010; **2**: 1015–1024. DOI: 10.1103/PhysRevLett.100.016602.

[21] Novoselov KS, Jiang D, Schedin F et al., Two-dimensional atomic crystals. Proc. Natl. Acad. Sci. 2005; **102**: 10451–10453.

[22] Berger C, Song Z, Li T et al., Ultrathin epitaxial graphite: 2D electron gas properties and a route toward graphene-based nanoelectronics. J. Phys. Chem. B. 2004; **108**: 19912–19916.

[23] Lin YM, Dimitrakopoulos C, Jenkins KA, Farmer DB, Chiu HY, Grill A, Avouris P, 100-GHz transistors from wafer-scale epitaxial graphene. Science. 2010; **327**: 662. DOI: 10.1126/science.1184289.

[24] Stankovich S, Dikin DA, Dommett GHB et al., Graphene-based composite materials. Nature. 2006; **442**: 282–286.

[25] Hummers WS, Offeman RE, Preparation of graphitic oxide. J. Am. Chem. Soc. 1958; **80**: 1339–1339.

[26] Eda G, Fanchini G, Chhowalla M, Large-area ultrathin films of reduced graphene oxide as a transparent and flexible electronic material. Nature Nanotechnol. 2008; **3**: 270–274.

[27] Kalita G, Sharma S, Wakita K et al., A photoinduced charge transfer composite of graphene oxide and ferrocene. Phys. Chem. Chem. Phys. 2013; **15**: 1271–1274.

[28] Kalita G, Wakita K, Umeno M et al., Fabrication and characteristics of solution-processed graphene oxide–silicon heterojunction. Phys. Stat. Sol. RRL 2013; **7**: 340–343.

[29] Avouris P, Dimitrakopoulos C, Graphene: synthesis and application. Mater. Today 2012; **15**: 86–97.

[30] Eizenberg M, Blakely JM, Carbon monolayer phase condensation on Ni(111). Surf. Sci. 1979; **82**: 228–236.

[31] Somani P R, Somani S P, Umeno M, Planer nano-graphenes from camphor by CVD. Chem. Phys. Lett. 2006; **430**: 56–59.

[32] Yu Q, Lian J, Siriponglert S et al., Graphene segregated on Ni surfaces and transferred to insulators. Appl. Phys. Lett. 2008; **93**: 113103.

[33] Kalita G, Masahiro M, Uchida H et al., Few layers of graphene as transparent electrode from botanical derivative camphor. Mater. Lett. 2010; **64**: 2180–2183.

[34] Reina A, Jia X, John H et al., Large area, few-layer graphene films on arbitrary substrates by chemical vapor deposition. Nano Lett. 2009; **9**: 30–35.

[35] Kim K S, Zhao Y, Jang H et al., Large-scale pattern growth of graphene films for stretchable transparent electrodes. Nature 2009; **457**: 706–710.

[36] Li X S, Cai W W, An J H et al., Large-area synthesis of high-quality and uniform graphene films on copper foils. Science 2009; **324**: 1312–1314.

[37] McCarty K F, Feibelman P J, Loginov E, Bartelt N C, Kinetics and thermodynamics of carbon segregation and graphene growth on Ru(001). Carbon 2009; **47**(7): 1806–1813.

[38] Zhang W, Wu P, Li Z, Yang J, First-principles thermodynamics of graphene growth on Cu surfaces. J. Phys. Chem. C 2011; **115**: 17782–17787.

[39] Qi M, Ren Z, Jiao Y, Zhou Y, Xu X, Li W, Li J, Zheng X, Bai J, Hydrogen kinetics on scalable graphene growth by atmospheric pressure chemical vapor deposition with acetylene. J. Phys. Chem. C 2013; **117**(27): 14348–14353.

[40] Kalita G, Matsushima M, Uchida H et al., Graphene constructed carbon thin films as transparent electrodes for solar cell applications. J. Mater. Chem. 2010; **20**: 9713–9717.

[41] Vlassiouk I, Fulvio P, Meyer H et al., Large scale atmospheric pressure chemical vapor deposition of graphene. Carbon 2013; **54**: 58–67.

[42] Lin HC, Chen YZ, Wang YC, Chueh YL, The essential role of Cu vapor for the self-limit graphene via the Cu catalytic CVD method. J. Phys. Chem. C 2015; **119**(12): 6835–6842.

[43] Rosmi MS, Shinde SM, Rahman NDA et al., Synthesis of uniform monolayer graphene on re-solidified copper from waste chicken fat by low pressure chemical vapor deposition. Materials Research Bulletin 2016; **83**: 573–580.

[44] Robert MJ, Michael SA, Graphene growth dynamics on epitaxial copper thin films. Chem. Mater. 2013; **25**: 871–877.

[45] Ivan V, Sergei S, Murari R et al., Graphene nucleation density on copper: Fundamental role of background pressure. J. Phys. Chem. C 2013; **117**: 18919–18926.

[46] Stoffel A, Kovács A, Kronast W, Müller B, LPCVD against PECVD for micromechanical applications. J. Micromech. Microeng. 1996; **6**(1): 20–33.

[47] Kim H, Mattevi C, Calvo MR et al., Activation energy paths for graphene nucleation and growth on Cu. ACS Nano 2012; **6**: 3614–3623.

[48] Hersh H N, The vapor pressure of copper. J. Am. Chem. Soc. 1953, **75**: 1529–1531.

[49] Ayhan ME, Kalita G, Sharma S, Tanemura M, Chemical vapor deposition of graphene on silver foil as a tarnish-resistant coating. Phys. Status Solidi RRL 2013; **7**(12): 1076–1079.

[50] Sun J, Nam Y, Lindvall N et al., Growth mechanism of graphene on platinum: surface catalysis and carbon segregation. Appl. Phys. Lett. 2014; **104**: 4871978.

[51] Jin L, Fu Q, Zhang H, Mu R, Zhang Y, Tan D, Bao X, Tailoring the growth of graphene on Ru(0001) via engineering of the substrate surface. J. Phys. Chem. C 2012; **116**(4): 2988–2993.

[52] Hamilton J C, J. M. Blakely, Carbon segregation to single crystal surfaces of Pt, Pd and Co. Surf. Sci. 1980; **91**: 199–217.

[53] López G, Mittemeijer E, The solubility of C in solid Cu. Scripta Materialia 2004; **51**: 1–5.

[54] Acheson EG, United States Patent; 1896. No. 568323.

[55] Lahiri J, Miller T S, Ross A J, Adamska L, Oleynik I I, Batzill M, Graphene growth and stability at nickel surfaces. New J. Phys. 2011; **13**: 025001.

[56] Sharma S, Kalita G, Hirano R et al., Influence of gas composition on the formation of graphene domain synthesized from camphor. Materials Letters 2013; **93**: 258–262.

[57] Kalita G, Wakita K, Umeno M, Structural analysis and direct imaging of rotational stacking faults in few-layer graphene synthesized from solid botanical precursor. Jap. Jour. Appl. Phys. 2011; **50**: 070106.

[58] Kalita G, Wakita K, Takahashi M et al., Iodine doping in solid precursor-based CVD growth graphene film. J. Mater. Chem. 2011; **21**: 15209–15213.

[59] Zhang Y, Gomez L, Ishikawa F N et al., Comparison of graphene growth on single-crystalline and polycrystalline Ni by chemical vapor deposition. J. Phys. Chem. Lett. 2010; **1**: 3101–3107.

[60] Nie S, Wofford JM, Bartelt NC, Dubon OD, McCarty KF, Origin of the mosaicity in graphene grown on Cu(111). Physical Review B; 2011; **84**: 155425.

[61] Bhaviripudi S, Jia X, Dresselhaus MS, Kong J, Role of kinetic factors in chemical vapor deposition synthesis of uniform large area graphene using copper catalyst. Nano Letters 2010; **10**: 4128–4133.

[62] Hwang C, Yoo K, Kim SJ et al., Initial stage of graphene growth on a Cu substrate. J. Phys. Chem. C 2011; **115**: 22369–22374.

[63] Wood JD, Schmucker SW, Lyons AS et al., Effects of polycrystalline cu substrate on graphene growth by chemical vapor deposition. Nano Lett. 2011; **11**: 4547–4554.

[64] Bae S, Kim HK, Lee, Roll-to-roll production of 30-inch graphene films for transparent electrodes. Nat. Nanotechnol 2010; **5**: 574–578.

[65] Sutter PW, Flege JI, Sutter EA, Epitaxial graphene on ruthenium. Nat. Mater. 2008; **7**: 406–411.

[66] Addou R, Dahal A, Sutter P, Batzill M, Monolayer graphene growth on Ni(111) by low temperature chemical vapor deposition. Applied Physics Letters 2012; **100**: 3675481.

[67] Ago H, Ito Y, Mizuta N et al., Epitaxial chemical vapor deposition growth of single-layer graphene over cobalt film crystallized on sapphire. ACS Nano 2010; **4**: 7407–7414.

[68] Wu T, Ding G, Shen H et al., Triggering the continuous growth of graphene toward millimeter-sized grains. Adv. Funct. Mater. 2013; **23**: 198–203.

[69] Yan Z, Lin J, Peng Z et al., Toward the synthesis of wafer-scale single-crystal graphene on copper foils, ACS Nano 2012; **6**: 9110–9117.

[70] Sharma S, Kalita G, Ayhan ME et al., Synthesis of hexagonal graphene on polycrystalline Cu foil from solid camphor by atmospheric pressure chemical vapor deposition. J. Mater. Sci. 2013; **48**: 7036–7041.

[71] Sharma S, Kalita G, Papon R et al., Synthesis of graphene crystals from solid waste plastic by chemical vapor deposition, Carbon 2014; **72**: 66–73.

[72] Sharma KP, Shinde SM, Rosmi MS et al., Effect of copper foil annealing process on large graphene domain growth by solid source-based chemical vapor deposition. J. Mater. Sci. 2016; **51**(15): 7220–7228.

[73] Papon R, Pierlot C, Sharma S et al. Optimization of CVD parameters for graphene synthesis through design of experiments, Phys. Status Solidi B 2016; 1–7 DOI 10.1002/pssb.201600629

[74] Li J, Wang XY, Liu XR et al., Facile growth of centimeter-sized single-crystal graphene on copper foil at atmospheric pressure. J. Mater. Chem.2015; **C 3**: 3530.

[75] Warner JH, Rümmeli MH, Gemming T, Büchner B, Briggs G, Direct imaging of rotational stacking faults in few layer graphene. Nano Lett. 2009; **9**(1): 102–106.

[76] Wu Y, Chou H, Ji H et al., Growth mechanism and controlled synthesis of AB-stacked bilayer graphene on Cu–Ni alloy foils. ACS Nano 2012; **6**(9): 7731–7738.

[77] Tan L, Zeng M, Zhang T, Fu L, Design of catalytic substrates for uniform graphene films: from solid-metal to liquid-metal. Nanoscale. 2015; **7**: 9105.

[78] Oostinga JB, Heersche HB, Liu X, Morpurgo AF, Vandersypen LMK, Gate-induced insulating state in bilayer graphene devices. Nat. Mater. 2008; **7**: 151–157.

[79] Liu X, Fu L, Liu N et al., Segregation growth of graphene on CuNi alloy for precise layer control. J. Phys. Chem. 2011; **C 115**: 11976–11982.

[80] Takesaki Y, Kawahara K, Hibino H, Okada S, Tsuji M, and Ago H, Highly uniform bilayer graphene on epitaxial Cu–Ni(111) alloy. Chem. Mater. 2016; **28**: 4583–4592.

[81] Wang G, Chen D, Lu Z et al., Growth of homogeneous single-layer graphene on Ni-Ge binary substrate. Appl. Phys. Lett. 2014; **104**: 062103.

[82] Liu N, Fu L, Dai B et al., Universal segregation growth approach to wafer-size graphene from non-noble metals. Nano Lett. 2011; **11**: 297.

[83] Lin T, Huang F, Wan D et al., Self-regulating homogenous growth of high-quality graphene on Co–Cu composite substrate for layer control. Nanoscale. 2013; **5**: 5847.

[84] Rümmeli MH, Zeng M, Melkhanova S et al., Insights into the early growth of homogeneous single-layer graphene over Ni–Mo binary substrates. J. Chem. Mater. 2013; **25**: 3880.

[85] Wu T, Zhang X, Yuan Q et al., Fast growth of inch-sized single-crystalline graphene from a controlled single nucleus on Cu-Ni alloys. Nature Mater. 2016; **15**: 43–47.

[86] Papon R, Sharma KP, Mahyavanshi RD et al., CuNi binary alloy catalyst for growth of nitrogen-doped graphene by low pressure chemical vapor deposition. Phys. Stat. Sol. 2016; RRL **10**: 749–752.

[87] Geng D, Wu B, Guo Y et al., Uniform hexagonal graphene flakes and films grown on liquid copper surface. Proc. Natl. Acad. Sci. 2012; **109**: 7992.

[88] Wang J, Zeng MQ, Tan LF et al., High-mobility graphene on liquid p-block elements by ultra-low-loss CVD growth. Sci. Rep. 2013; **3**: 2670.

[89] Fujita JI, Miyazawa Y, Ueki R et al., Fabrication of large-area graphene using liquid gallium and its electrical properties. Jpn. J. Appl. Phys. 2010; **49**: 06GC01.

[90] Zeng M, Tan L, Wang J et al., Liquid metal: an innovative solution to uniform graphene films. Chem. Mater. 2014; **26**(12): 3637–3643.

[91] Bonaccorso F, Sun Z, Hasan T et al., Graphene photonics and optoelectronics. Nature Photonics 2010; **4**: 611–622.

[92] Li X, Zhu H, Wang K et al., Graphene-on-silicon Schottky junction solar cells. Adv. Mater. 2010; **22**: 2743–2748.

[93] Meng JH, Liu X, Zhang XW et al., Interface engineering for highly efficient graphene-on-silicon Schottky junction solar cells by introducing a hexagonal boron nitride interlayer, Nano Energy 2016; **28**: 44–50.

[94] Ayhan ME, Kalita G, Kondo M et al., Photoresponsivity of silver nanoparticles decorated graphene–silicon Schottky junction. RSC Adv. 2014; **4**: 26866–26871.

[95] Bartolomeo AD, Graphene Schottky diodes: an experimental review of the rectifying graphene/semiconductor heterojunction. Phys. Reports. 2016; **606**: 1–58.

[96] Chandramohan S, Ko KB, Yang JH et al., Performance evaluation of GaN light-emitting diodes using transferred graphene as current spreading layer. Jour. Appl. Phys. 2014; **115**: 054503.

[97] Wang L, Liu W, Zhang Y et al., Graphene-based transparent conductive electrodes for GaN-based light emitting diodes: challenges and countermeasures. Nano Energy. 2015; **12**: 419–436.

[98] Wang L, Zhang Y, Li X et al., Improved transport properties of graphene/GaN junctions in GaN-based vertical light emitting diodes by acid doping. RSC Adv. 2013; **3**: 3359–3364.

[99] Xu K, Xu C, Xie Y et al., Graphene GaN-based Schottky ultraviolet detectors. IEEE Trans. Electron Devices. 2015: **62**: 9.

[100] Li Q, Liu M, Zhang Y, Liu Z. Hexagonal boron nitride-graphene heterostructures: synthesis and interfacial properties. Small. 2016; **12**: 32–50.

[101] Britnell L, Gorbachev RV, Jalil R et al., Field-effect tunneling transistor based on vertical graphene heterostructures. Science. 2012; **335**: 947–950.

[102] Sharma S, Kalita G, Vishwakarma R et al., Opening of triangular hole in triangular-shaped chemical vapor deposited hexagonal boron nitride crystal. Sci. Rep. 2015; **5**: 10426.

[103] Sharma S, Sharma K, Rosmi MS et al., Morphology-controlled synthesis of hexagonal boron nitride crystals by chemical vapor deposition. Cryst. Growth Des. 2016; **16**(11): 6440–6445.

[104] Dean CR, Young AF, Meric I et al., Boron nitride substrates for high-quality graphene electronics. Nat. Nanotechnol. 2010; **5**: 722.

[105] Levendorf MP, Kim CJ, Brown L et al., Graphene and boron nitride lateral heterostructures for atomically thin circuitry. Nature 2012; **488**: 627.

[106] Ci L, Song L, Jin C et al., Atomic layers of hybridized boron nitride and graphene domains. Nat. Mater. 2010; **9**: 430–435.

[107] Gao Y, Zhang Y, Chen P et al., Toward single-layer uniform hexagonal boron nitride–graphene patchworks with zigzag linking edges. Nano Lett. 2013; **13**(7): 3439–3443.

[108] Gao T, Song X, Du H et al., Temperature-triggered chemical switching growth of in-plane and vertically stacked graphene-boron nitride heterostructures. Nat. Commun. 2015; **6**: 6835.

[109] Splendiani A, Sun L, Zhang Y et al., Emerging photoluminescence in monolayer MoS_2. Nano Lett. 2010; **10**: 1271.

[110] Mak KF, Lee C, Hone J, Shan J, Heinz TF, Atomically thin MoS_2: a new direct-gap semiconductor. Phys. Rev. Lett. 2010; **105**: 136805.

[111] Mak KF, He K, Shan J et al., Control of valley polarization in monolayer MoS2 by optical helicity. Nat. Nanotechnol. 2012; **7**: 494.

[112] Yu Y, Li C, Liu Y et al., Controlled scalable synthesis of uniform, high-quality monolayer and few-layer MoS_2 films. Sci. Rep. 2013; **3**: 1866.

[113] Park J, Lee W, Choi T et al., Layer-modulated synthesis of uniform tungsten disulfide nanosheet using gas-phase precursors. Nanoscale. 2015; **7**: 1308–1313.

[114] Roy K, Padmanabhan M, Goswami S et al., Graphene-MoS_2 hybrid structures for multifunctional photoresponsive memory devices. Nat. Nanotechnol. 2013; **8**: 826–830.

[115] Lv R, Robinson JA, Schaak RE, Sun D, Sun Y, Mallouk TE, Terrones M, Transition metal dichalcogenides and beyond: synthesis, properties, and applications of single- and few-layer nanosheets. Acc. Chem. Res. 2015; **48**: 56.

[116] Thangaraja A, Shinde SM, Kalita G et al., An effective approach to synthesize monolayer tungsten disulphide crystals using tungsten halide precursor. Appl. Phys. Lett. 2016; **108**(5): 4941393.

[117] Loan PTK, Zhang W, Lin CT et al., Graphene/MoS_2 heterostructures for ultrasensitive detection of DNA hybridisation. Adv. Mater. 2014; **26**: 4838 .

[118] Liu X, Balla I, Bergeron H et al., Rotationally commensurate growth of MoS_2 on epitaxial Graphene. ACS Nano. 2016; **10**(1): 1067–1075.

[119] Shi Y, Zhou W, Lu AY et al., Van der Waals epitaxy of MoS_2 layers using graphene as growth templates. Nano Lett. 2012; **12**(6): 2784–2791.

7

Graphene-Enhanced Optical Signal Processing

Jian Wang and Xiao Hu

Abstract

Graphene has emerged as an attractive material for a myriad of optoelectronic applications due to its variety of remarkable optical, electronic, thermal and mechanical properties. So far, the main focus has been on graphene based photonics and optoelectronics devices. Due to the linear band structure allowing interband optical transitions at all photon energies, graphene has remarkably large third-order optical susceptibility $\chi^{(3)}$, which is only weakly dependent on the wavelength in the near-infrared frequency range. Graphene possesses the properties of the enhancement four-wave mixing (FWM) of conversion efficiency. So, we believe that the potential applications of graphene also lies in nonlinear optical signal processing, where the combination of its unique large $\chi^{(3)}$ nonlinearities and dispersionless over the wavelength can be fully exploited. In this chapter, we give a brief overview of our recent progress in graphene-assisted nonlinear optical device which is graphene-coated optical fiber and graphene-silicon microring resonator and their applications, including degenerate FWM based tunable wavelength conversion of quadrature phase-shift keying (QPSK) signal, two-input optical computing, three-input high-base optical computing, graphene-silicon microring resonator enhanced nonlinear optical device for on-chip optical signal processing, and nonlinearity enhanced graphene-silicon microring for selective conversion of flexible grid multi-channel multi-level signal.

Keywords: graphene, optical nonlinearities, four-wave mixing (FWM), wavelength conversion, quadrature phase-shift keying (QPSK), optical computing, silicon microring resonators, graphene-silicon microring resonator (GSMR), all optical signal processing, on-chip all optical signal processing

1. Introduction

Graphene, a monolayer of carbon atoms arranged in a two-dimensional honeycomb lattice, is a basic building block for graphitic materials of all other dimensionalities. In 2004, a research team based in Manchester successfully isolated graphene by mechanical exfoliation [1]. Since then, many extraordinary properties have been reported, such as extremely high mobility of charge carrier 200,000 $cm^2V^{-1}s^{-1}$. Graphene possesses linear, massless band structure $E_{\pm}(p) = \pm V|p|$, where the upper (lower) sign corresponds to the electron (hole) band, p is the quasi-momentum, and $V \approx 10^6$ m/s is the Fermi velocity. The high mobility of charge carriers and zero bandgap of graphene can be employed as an ideal medium for high-frequency applications, such as radio-frequency switches [2].

The photonic properties of graphene are equally remarkable. On the one hand, many breakthroughs in research on graphene, including ultrafast photodetectors [3], broadband polarizers [4], and modulators [5], benefit from its unique band structure. The graphene electro-absorption modulator [5] is based on interband transitions, which can be tuned by applying drive voltage, correspondingly changing the Fermi energy (E_F) of graphene. Here, it has to emphasize that at the short wavelength range (that is, infrared and visible), the graphene optical absorptions are determined by interband transitions, whereas at the long wavelength range (that is, terahertz), they are dominated by intraband transitions. In the mid- to far-infrared and THz ranges, graphene exhibits a strong plasmonic response. Doped and patterned graphene can support localized plasmonic resonances, which significantly enhance the absorption [6]. On the other hand, the interband optical absorption in zero-gap graphene could be saturated readily from the visible region to the near-infrared region under strong excitation due to Pauli blocking [7]. Graphene could behave as a fast saturable absorber over a wide spectral range for the mode locking of fiber lasers. Since 2009, a wide variety of laser configurations and operational wavelengths [7–11] is based on graphene mode-locked lasers. Graphene has been suggested as a material that might have large $\chi^{(3)}$ nonlinearities, which is also due to its linear band structure allowing interband optical transitions at all photon energies. It has been indicated [12] that the nonlinear response of graphene is essentially dispersionless over the wavelength and much stronger compared to bulk semiconductors. It has been experimentally demonstrated that the nonlinear refractive index of graphene is as $n_2 \approx 10^{-7}$ cm^2W^{-1} [13] using the Z-scan technique. After that, optical bistability, self-induced regenerative oscillations, and four-wave mixing (FWM) have been consecutively observed in graphene-silicon hybrid optoelectronic devices [14]. FWM has also been demonstrated in graphene in various configurations, e.g., slow-light graphene-silicon photonic crystal waveguide [15], graphene optically deposited onto fiber ferrules [16], and graphene-coated microfiber [17, 18]. Moreover, FWM-based wavelength conversion of a 10-Gb/s non-return-to-zero (NRZ) signal with mechanically exfoliated graphene was first reported in Ref. [19]. Advanced optical modulation formats play an important role in enabling high-capacity optical transport networks [20] where wavelength conversion function is highly desired.

All-optical wavelength conversions have been studied by many previous works using semiconductor optical amplifiers (SOAs), highly nonlinear fibers (HNLFs), and periodically poled lithium niobates (PPLNs). Silicon-on-insulator (SOI) waveguides feature low cost, ultra-compact footprint and are compatible with standard metal-oxide-semiconductor (CMOS) technology, comparing to SOAs, HNLFs, and PPLNS. The nonlinear interaction will also be strongly enhanced in silicon waveguide due to its tight light confinement. To further enhance the nonlinear interaction in silicon waveguide devices, resonator structures such as microrings, microdisks, and photonic crystal nanocavities can be introduced. Among these integrated resonant structures, microring resonator has been attractive to researcher in the past several years mainly because of its small size and potential for telecom and datacom applications. Microring resonators have accelerated the demonstration of very low power continuous-wave (CW) nonlinear optics, and similar benefits are expected for its operation in processing high bandwidth optical signal. For applications to optical data signal processing, the main challenge is that the bandwidth of the microring must be large enough to contain all the spectral components of the optical signals. Additionally, all-optical wavelength conversions of advanced optical modulation formats have not been realized in integrated microring structures.

In this chapter, we go over our recent progress in graphene-assisted nonlinear optical device and their applications, including degenerate FWM-based tunable wavelength conversion of QPSK signal [21], two-input optical high-base hybrid doubling and subtraction functions [22], three-input high-base optical computing [23], graphene-silicon microring resonator-enhanced nonlinear optical device for on-chip optical signal processing [24], and nonlinearity enhanced graphene-silicon microring for selective conversion of flexible grid multi-channel multi-level signal.

2. Optical properties of graphene

The primary mechanism of optical absorption in graphene involves two processes: carrier intraband transitions and interband transitions [25–27]. For short wavelengths (that is, infrared and visible range), the graphene optical absorption is determined by interband transitions, whereas for long wavelengths (that is, terahertz range), it is dominated by intraband transitions. **Figure 1** pictorially depicts the basic mechanisms. Interband transitions refer to an exchange of charge carriers between the conduction and valence bands, the energy of a photon $\hbar\omega$ should be satisfying the relationship $\hbar\omega \geq 2E_F$ as shown in **Figure 1(a)**. For n-doped graphene, the optical photon ($\hbar\omega_2$) with energy less than $2E_F$ cannot be absorbed which is because of the electron states in resonance occupied the conduction band. For p-doped graphene, the optical photon ($\hbar\omega_2$) with energy less than $2E_F$ cannot be absorbed which is due to the unavailable electrons for the interband transition. For the low frequency THz range ($\hbar\omega \leq 2E_F$), the intraband absorption mechanism is shown in **Figure 1(b)**. And graphene behaves like a conductive film and its optical conductivity closely follows its electrical conductivity, which can be described by a simple Drude model.

Due to the linear band structure of graphene allowing interband optical transitions at all photon energies, graphene is considered to be a material that might have large $\chi^{(3)}$ nonlinearities. In Ref. [12], the principle of graphene-based degenerate FWM is proposed. Two continuous-wave (CW) pumps with frequencies ω_1 and ω_2 are combined together graphene-based device with high third-order nonlinearity ($\chi^{(3)}$) and mix together to generate a new coherent beam with frequency ω_{con} (converted idler), as depicted in **Figure 2**.

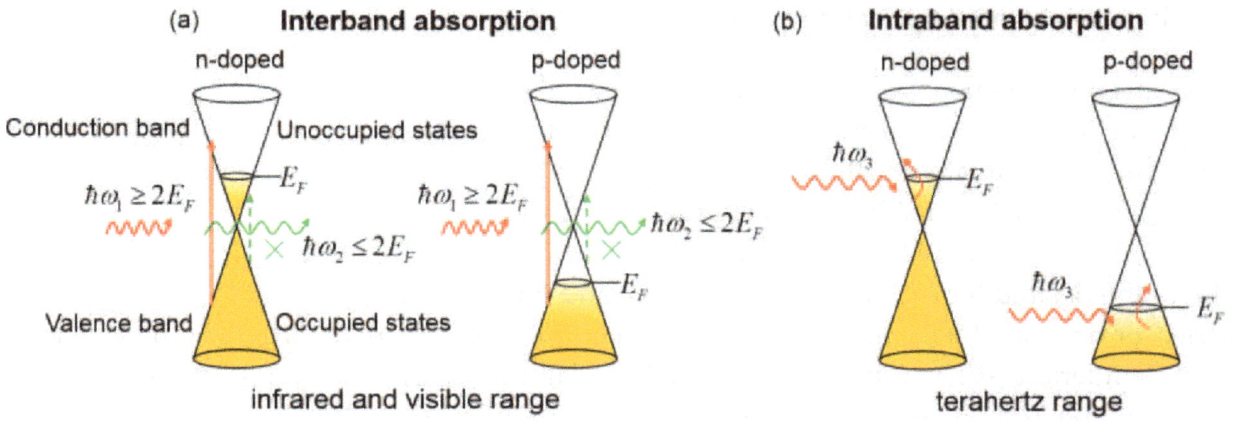

Figure 1. (a) Interband and (b) intraband absorption mechanisms in graphene.

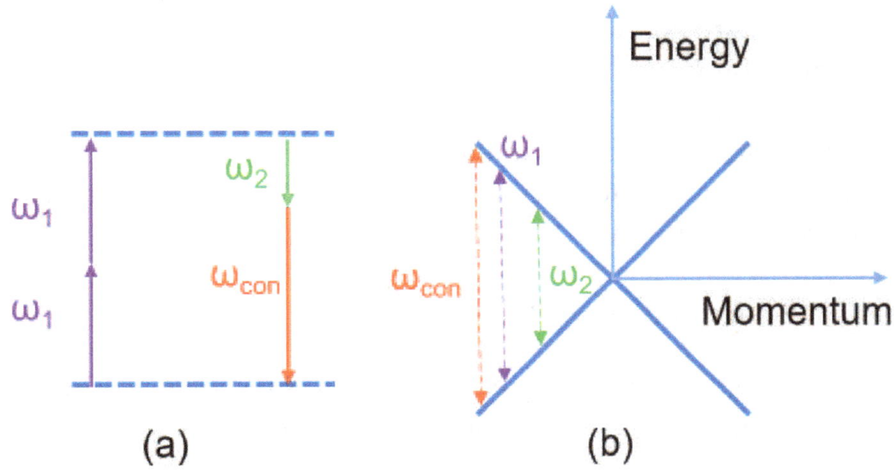

Figure 2. (a) Diagram of energy conservation in the degenerate FWM process. (b) Band structure of graphene with the three resonant photon energies (arrows) involved in degenerate FWM.

3. Fabrication and characterization of graphene-assisted nonlinear optical devices

There is a wide choice in terms of size, quality, and price for any particular graphene-based application: (1) for the liquid phase and thermal exfoliation method, the produced graphene or reduced graphene oxide flakes are suited for composite materials, conductive paints, and so on; (2) for the method of synthesis on SiC, the ultrahigh quality graphene with crystallites approaching hundreds of micrometers in size can be used for high-performance electronic devices; and (3) for chemical vapor deposition (CVD) method, the fabricated large-area uniform polycrystalline graphene films are ready for use in lower-performance active and nonactive devices. Here, considering the quality, size, and price of graphene, we choose the CVD method to grow graphene. Then, the prepared graphene was connected with single mode fiber and silicon microring resonator.

3.1. Graphene-coated optical fiber device

Monolayer graphene was grown by the chemical vapor deposition (CVD) method [28]. Graphene was primarily grown on Cu foils (25-μm thick with a purity of >99.99 wt% obtained from Alfa Aesar) in a hot wall furnace. The growth process can be briefly summarized as follows: (1) load the fused silica tube with the Cu foil, evacuate, back fill with hydrogen, heat to 1050°C and maintain a H_2(g) pressure of 42 mTorr under a 2.5 sccm flow; (2) stabilize the Cu film at the desired temperature, up to 1050°C, and introduce 40 sccm of CH_4(g) for a desired period of time at a total pressure of 450 mTorr; (3) after exposure to CH_4, cool the furnace to room temperature. Then, poly(methyl methacrylate) (PMMA) film was spin coated on the surface of the graphene-deposited Cu foil. Then the Cu foil was etched away with 1 M $FeCl_3$ solution. The resultant PMMA/graphene film (5 mm × 5 mm) was washed in deionized water several times and transferred to the Si/SiO_2 substrate or deionized water solution for next step. To fabricate the graphene-coated optical fiber device, the floating PMMA/graphene sheet was mechanically transferred onto the fiber pigtail cross-section and dried in a cabinet. After drying at room temperature for about 24 hours, the carbon atoms could be self-assembled onto the fiber end-facet, thanks to the strong viscosity of graphene. The PMMA layer can be removed by boiling acetone. By connecting this graphene-on-fiber component with another clean and dry FC/PC fiber connector, as shown in **Figure 3**, the nonlinear optical device was finally constructed and used to FWM-based wavelength conversion applications.

3.2. Graphene-silicon microring resonator (GSMR)

The fabrication process of the nonlinearity enhanced GSMR is shown in **Figure 4**. We fabricate the nonlinearity enhanced GSMR on a commercial SOI wafer with a 340-nm-thick silicon slab

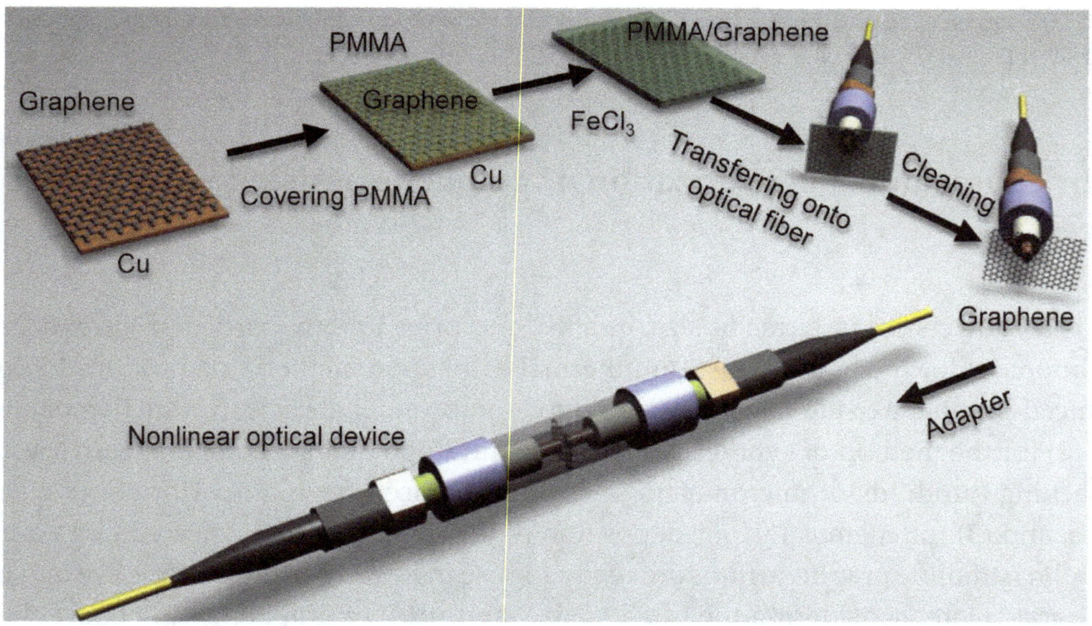

Figure 3. Fabrication process of the graphene-assisted nonlinear optical device.

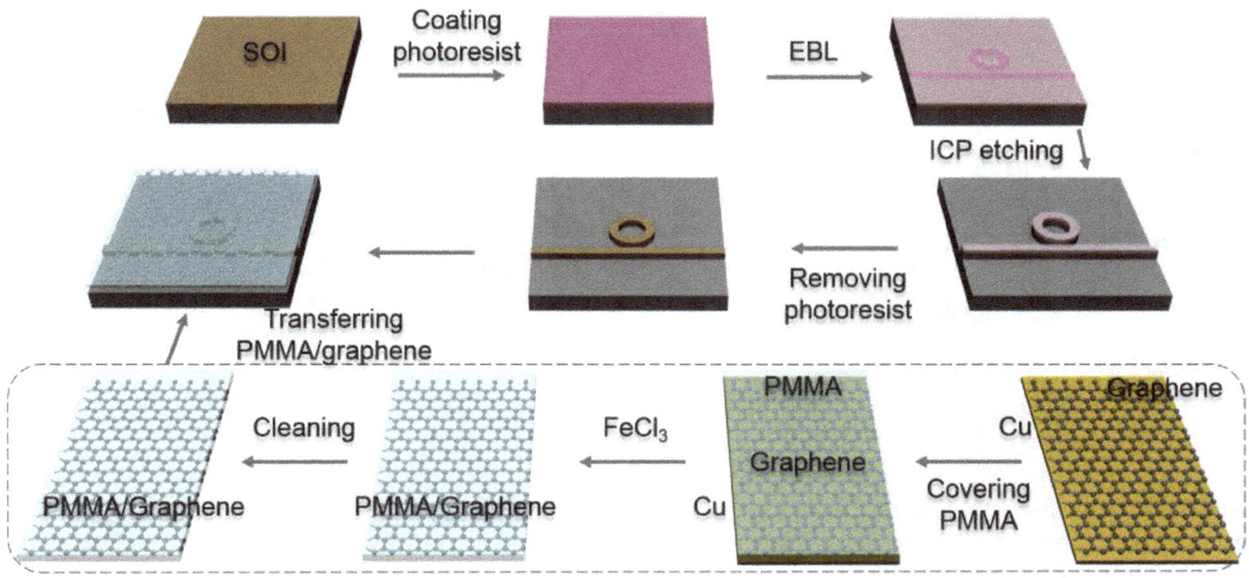

Figure 4. Fabrication process of the GSMR.

on the top of a 3-μm silica buffer layer. First, the device pattern is transferred to photoresist by E-beam lithography (EBL). The upper silicon layer is etched downward for 200 nm to form a ridge waveguide through induced coupled plasma (ICP) etching. After removing the photoresist, a silicon microring resonator was fabricated. Second, monolayer graphene was grown on a Cu foil (25-μm thick with a purity of >99.99 wt% obtained from Alfa Aesar) by the chemical vapor deposition method [28]. Poly(methyl methacrylate) (PMMA) film was next spin coated on the surface of the graphene-deposited Cu foil, and the Cu foil was etched away with 1 M $FeCl_3$ solution. The resultant PMMA/graphene film was then washed with deionized water several times. Finally, the floating PMMA/graphene sheet was mechanically transferred onto the top of silicon microring resonator.

Figure 5(a) illustrates the Raman spectrum of the transferred graphene on the silicon microring resonator. Compared to pristine graphene, the blue shifts of the positions of the G and 2D peaks are consistent to the nature of the p-doped graphene [29]. Additionally, the intensity ratio of the 2D to the G peak is about 1.8, significantly smaller than that of the pristine graphene (4–5), which is another evidence of the p-doped graphene [38]. The heavily p-doped graphene is particularly fabricated to achieve optical transparency in the infrared with negligible linear losses, which can be explained as follows: due to the p-doping of graphene, the Fermi level (E_F) is lower than half the photon energy (−νh/2, blue dashed line) and there are no electrons available for the interband transition [5] (**Figure 5(b)**) and intraband graphene absorption is near-absent in the infrared [30]. The device consists of a silicon microring resonator coupled to a straight waveguide with a gap of 150 nm. The waveguide is bidirectional tapered up to a width of 20 μm over a length of 600 μm to connect dual TE-polarized grating couplers. A scanning electron micrograph (SEM) of the fabricated GSMR resonator with partial part of the straight waveguide is shown in **Figure 5(c)**. It is fabricated by standard complementary metal-oxide-semiconductor (CMOS) processes on an SOI substrate with a 3 μm-thick buried oxide layer. The width and height of the ridge waveguide in

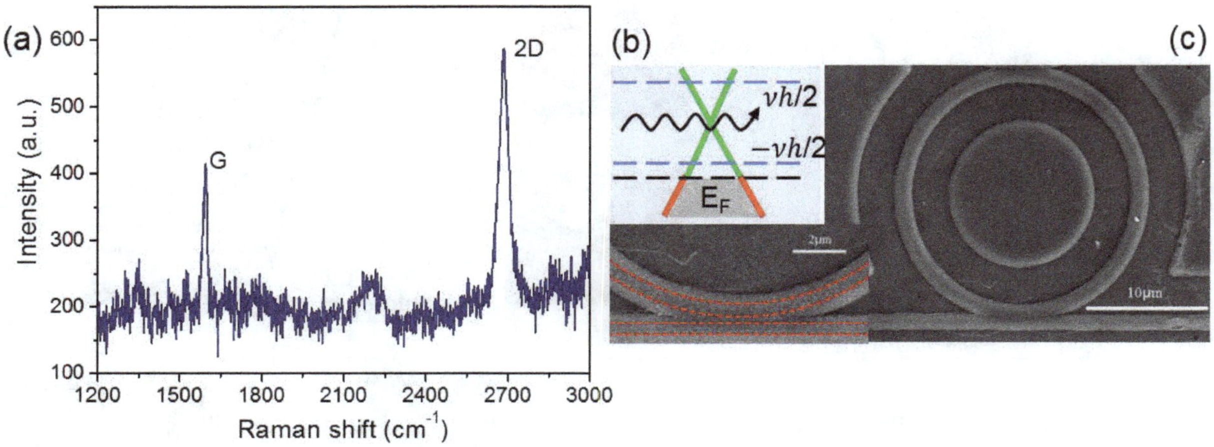

Figure 5. (a) Raman spectrum of the single-layer graphene sample. (b) The doped principle: the Fermi level (E_F) less than half the photon energy (−vh/2, blue dashed line), therefore, there are no electrons available for the interband transition. (c) SEM image of the GSGM resonator. The inset shows the detailed view of the coupling region. Red-dashed lines show the outline of the silicon waveguides.

the structure are 450 and 200 nm, respectively. The total insertion loss is about 10 dB at input wavelength of 1550 nm. Here, the input power is defined as the power in the straight waveguide coupled into the silicon microring resonator. We use standard polymer-based transfer method to cover a graphene sample on the top of the silicon waveguide and the detailed picture of the straight waveguide coupled with an arc region of the microring resonator is shown in the inset of **Figure 5(c)**. Due to the deposition of polymer on the GSMR, the gap is covered and the silicon waveguides before transfer are marked with red-dashed lines.

4. Graphene-enhanced nonlinear optical device for optical transport networks signal processing

It is well known that advanced optical modulation formats have become of great importance to enable high-capacity optical transport networks [20] where wavelength conversion function is highly desired. The nonlinear response of graphene is very high and essentially dispersionless over the wavelength. Such nonlinearity of graphene can be utilized to realize various nonlinear functional devices for telecommunication networks, such as, wavelength converters.

4.1. Degenerate FWM-based tunable wavelength conversion of QPSK signal

Figure 6 shows the experimental set-up for degenerate FWM-based wavelength conversion using a single-layer graphene grown by the CVD method [21]. The CW output from an external cavity laser (ECL1) serves as the signal light for the degenerate FWM and is modulated with QPSK signal at 10 Gbaud by a single-polarization optical I/Q modulator. An arbitrary waveform generator (AWG) running at a 10 GS/s sampling rate is used to produce the electrical signal. The modulated 10 Gbaud QPSK signal is then amplified by an erbium-doped

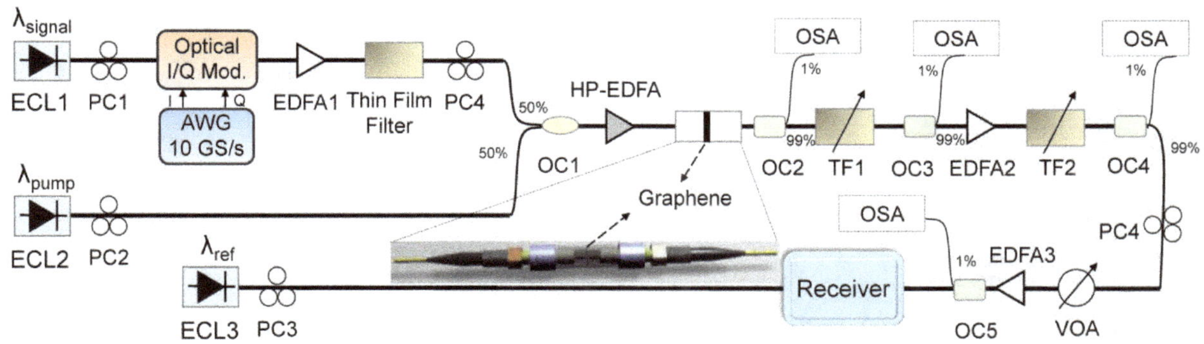

Figure 6. Experimental set-up for degenerate FWM-based wavelength conversion in graphene. Inset: "sandwiched structure" graphene sample used as a nonlinear optical device. ECL: external cavity laser; AWG: arbitrary waveform generator; TF: tunable filter; OC: optical coupler; PC: polarization controller; OSA: optical spectrum analyzer; and VOA, variable optical attenuator.

optical fiber amplifier (EDFA) followed by a thin film filter to suppress the amplified spontaneous emission (ASE) noise. Afterwards, the 10 Gbaud QPSK signal is combined with another CW light from ECL2, which serves as the pump light through a 3-dB coupler, amplified using a high-power EDFA (HP-EDFA), and launched into the single-layer graphene sample. The polarization states of the CW pump and QPSK signal are adjusted to achieve optimized conversion efficiency of degenerate FWM in graphene. The amplified CW pump and QPSK signal take part in the degenerate FWM process when passing through the single-layer graphene sample and a newly converted idler is generated. After the FWM wavelength conversion, the converted idler is selected using two tunable filters (TF1, TF2) for coherent detection. First, the converted idler is selected using TF1. Since the power level of the converted idler is relatively low, the selected converted idler is amplified by EDFA2. Second, in order to suppress the amplified spontaneous emission (ASE) noise originated from EDFA2, another TF2 is employed. Hence, the TF1 is used to select the converted idler, and TF2 is used to suppress the ASE noise. The CW output from ECL3 serves as a reference light for coherent detection. A variable optical attenuator (VOA) and a low noise EDFA (EDFA3) are employed to adjust the received signal-to-noise ratio (OSNR) for bit-error rate (BER) measurements.

To characterize the performance of QPSK wavelength conversion, we further measure the BER curve as a function of the received OSNR for back-to-back (B-to-B) signal and converted idler. **Figure 7** plots measured BER performance for tunable QPSK wavelength conversion with the converted idler generated at 1546.88, 1539.92, and 1557.90 nm, respectively. The power of HP-EDFA is estimated to be 31 dBm. The measured conversion efficiencies for converted idlers at 1546.88, 1539.92, and 1557.90 nm are −36.2, −48.2, and −39.8 dB, respectively. As shown in **Figure 7**, the observed OSNR penalty is around 1 dB at a BER of 1×10^{-3} (7% forward error correction (FEC) threshold) for QPSK wavelength conversion with the converted idler at 1546.88 nm. The received OSNR penalties of ~2.2 dB at a BER of 1×10^{-3} are observed for converted idlers at 1539.92 and 1557.90 nm. The increased OSNR penalty is mainly due to the reduced conversion efficiency for converted idlers at 1539.92 and 1557.90 nm. The right insets of **Figure 7** depict corresponding constellations of the B-to-B signals and converted idlers. The OSNR penalty is believed to be originated from the relatively low conversion efficiency (<−35 dB), which can be ascribed to the very limited interaction between the single-layer graphene and the propagating

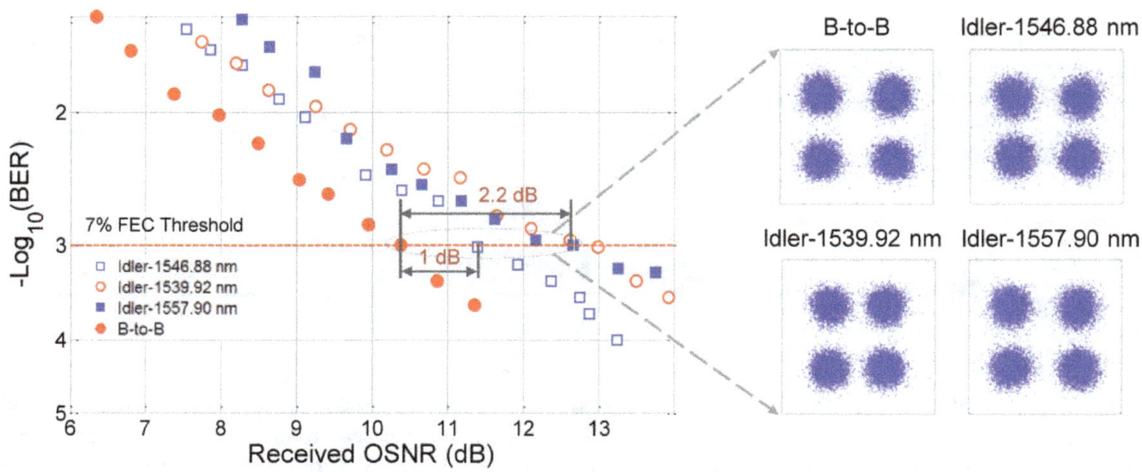

Figure 7. Measured BER versus received OSNR for wavelength conversion of QPSK signal. Insets show constellations of QPSK.

light only at the end face of optical fiber. With further improvement, one might enhance the conversion efficiency by mechanically transferring graphene sample grown by the CVD method onto the D-shaped fiber or microfiber to ensure more direct graphene-light interaction.

4.2. Two-input optical high-base hybrid doubling and subtraction functions

Figure 8 illustrates the concept and principle of two-input hybrid quaternary arithmetic functions [22]. From the constellation in the complex plane (**Figure 8(a)**), it is clear that one can use four-phase levels (π/4, 3π/4, 5π/4, 7π/4) of (D)QPSK to represent quaternary base numbers (0, 1, 2, 3). To implement two-input hybrid quaternary arithmetic functions, the aforementioned graphene-assisted nonlinear optical device is employed. Two-input quaternary numbers (A, B) are coupled into the nonlinear device, then two converted idlers (idler 1, idler 2) are simultaneously generated by two degenerate FWM processes. **Figure 13(b)** illustrates the degenerate FWM process. We derive the electrical field (E) and optical phase (φ) relationships of two degenerate FWM processes under the pump nondepletion approximation expressed as

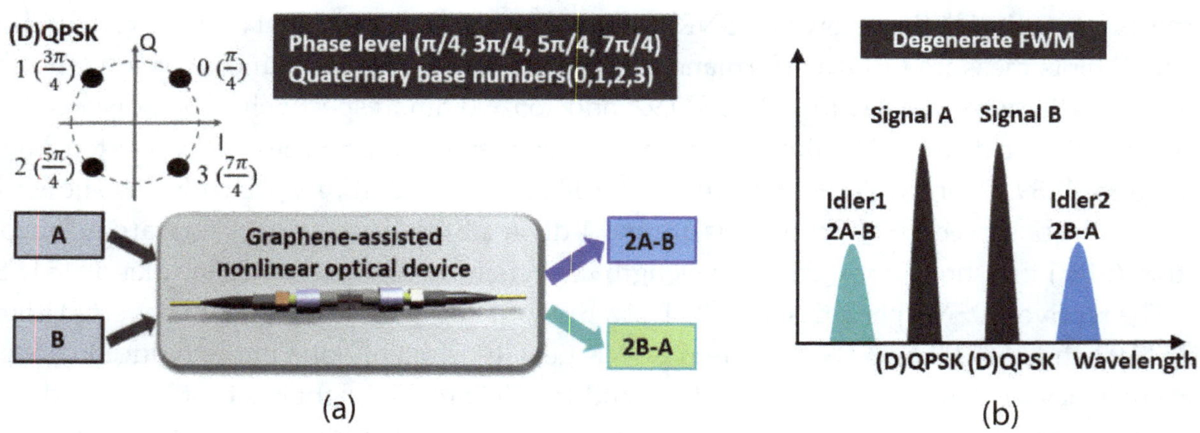

Figure 8. (a) Concept and (b) principle of hybrid quaternary arithmetic functions (2A-B, 2B-A) using degenerate FWM and (D)QPSK signals.

$$E_{i1} \propto E_A \cdot E_A \cdot E_B^*, \quad \varphi_{i1} = \varphi_A + \varphi_A - \varphi_B \tag{1}$$

$$E_{i2} \propto E_B \cdot E_B \cdot E_A^*, \quad \varphi_{i2} = \varphi_B + \varphi_B - \varphi_A \tag{2}$$

where the subscripts A, B, i1, i2 denote input signal A, signal B, converted idler 1, idler 2, respectively. Owing to the phase wrap characteristic with a periodicity of 2π, it is implied from the linear phase relationships in Eqs. (1) and (2) that idler 1 and idler 2 carry out modulo 4 operations of hybrid quaternary arithmetic functions of doubling and subtraction (2A-B, 2B-A).

The measured typical spectrum obtained after the CVD single-layer graphene-coated fiber device is depicted in **Figure 9**. Two 10-Gbaud NRZ-(D)QPSK signals at 1550.10 (A) and 1553.60 nm (B) are employed as two inputs. The power of two input signals (A, B) is set to about 32 dBm. The conversion efficiency is -36 dB. One can clearly see that two converted idlers are obtained by two degenerate FWM processes with idler 1 at 1546.60 nm (2A-B) and idler 2 at 1557.20 nm (2B-A). The resolution of the measured spectrum is set to 0.02 nm. The steps in the measured spectrum are actually the modulation sidebands of two NRZ-(D)QPSK carrying signals. As shown in **Figure 10**, in order to verify the hybrid quaternary arithmetic functions, we measure the phase of symbol sequence for two input signals and two converted idlers. By carefully comparing the quaternary base numbers for two input signals and two converted idlers, one can confirm the successful implementation of two-input hybrid quaternary arithmetic functions of 2A-B and 2B-A.

4.3. Three-input high-base optical computing

We also propose an approach to performing three-input optical addition and subtraction of quaternary base numbers using multiple nondegenerate FWM processes based on graphene-coated fiber device [23]. The concept and principle of three-input high-base optical computing are similar to **Figure 8**. In the experiment, the wavelengths of three-input signals A, B, and C are fixed at 1548.52, 1550.12, and 1552.52 nm, respectively. **Figure 11** depicts the measured typical optical spectrum obtained after the single-layer graphene-coated fiber device. One can clearly see that three converted idlers are generated by three nondegenerate FWM processes

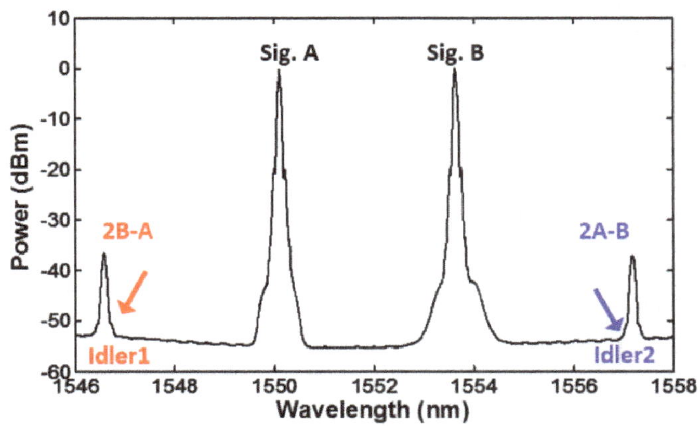

Figure 9. Measured spectrum for 10-Gbaud two-input hybrid quaternary arithmetic functions.

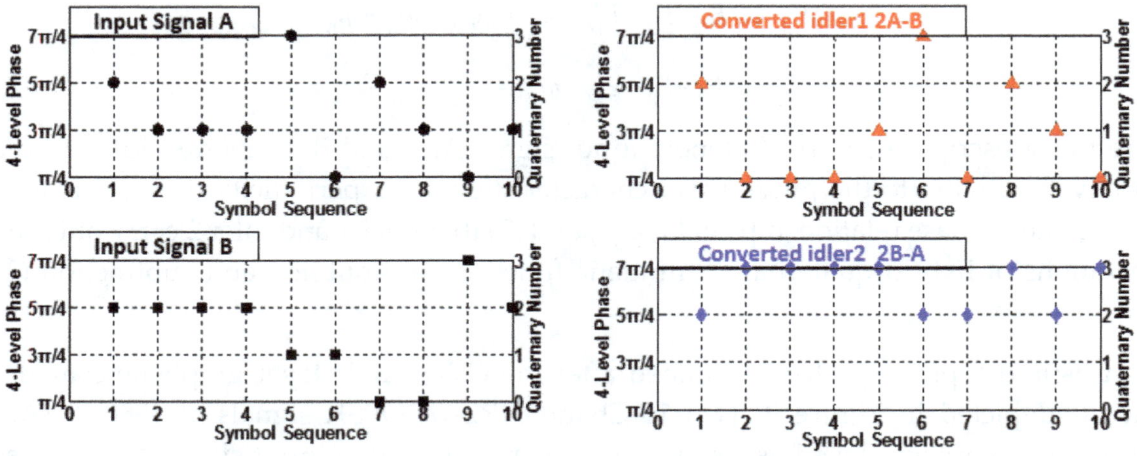

Figure 10. Measured phase of symbol sequence with coherent detection for 10-Gbaud two-input hybrid quaternary arithmetic functions.

with idler 1 at 1546.13 nm (A + B − C), idler 2 at 1550.92 nm (A + C − B), and idler 3 at 1554.13 nm (B + C − A), respectively. The power of HP-EDFA is estimated to be 31 dBm. The conversion efficiencies of three nondegenerate FWM processes are measured to be larger than −34 dB.

To characterize the performance of the proposed graphene-assisted modulo 4 functions of three-input high-base optical computing, we further measure the BER curves as a function of the received OSNR for back to back (B-to-B) signals and three converted idlers. **Figure 12** depicts the measured BER curves for 10-Gbaud modulo 4 operations of three-input quaternary hybrid addition and subtraction of A + B − C, A + C − B, and B + C − A. As shown in **Figure 12**, the observed OSNR penalties for modulo 4 operations of three-input quaternary hybrid addition and subtraction are accessed to be less than 7 dB at a BER of 2 × 10^{-3} (7% enhanced forward error correction (EFEC) threshold). The increased OSNR penalties might be mainly due to the

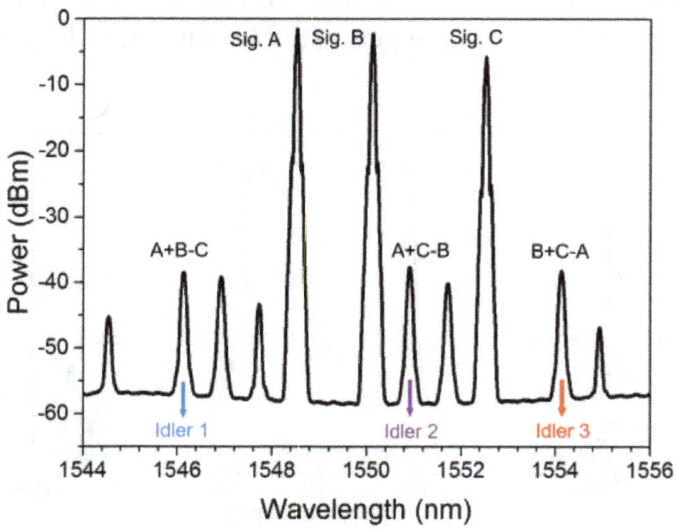

Figure 11. Measured spectrum for 10-Gbaud three-input quaternary hybrid addition and subtraction.

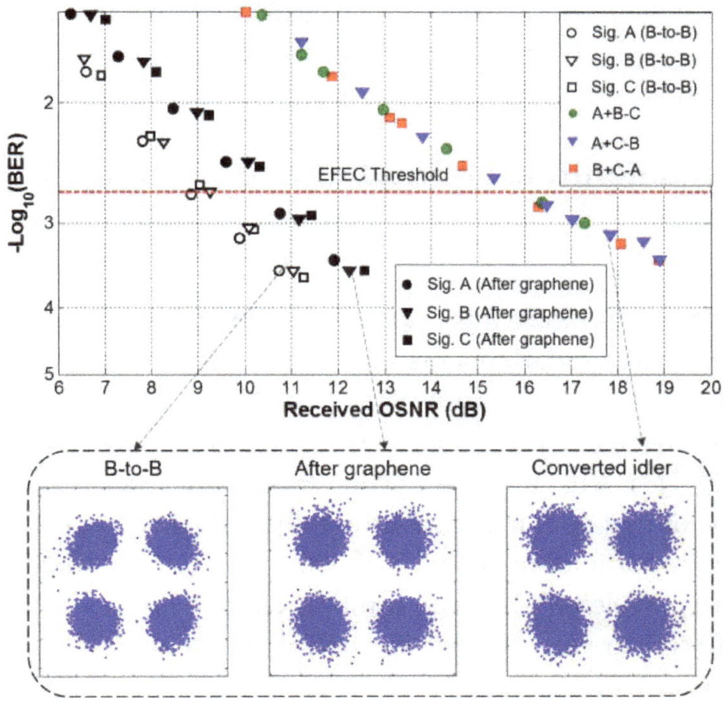

Figure 12. Measured BER curves for 10-Gbaud modulo 4 operations of three-input quaternary hybrid addition and subtraction of A + B − C, A − C − B, and B + C − A. Insets show constellations of (D)QPSK signals.

relatively low conversion efficiency for converted idlers and accumulated distortions transferred from three-input signals (A, B, and C). The insets in **Figure 12** depict the corresponding constellations of the B-to-B signals and converted idlers. The BER curves and constellations of three output signals (A, B, and C) after graphene are also shown in **Figure 12** for reference.

For the graphene-assisted modulo 4 functions of three-input high-base optical computing, we also study the performance tolerance to the relative time offset between three input signals. **Figure 13** depicts the BER performance as a function of the relative time offset between three signals (signal offset) under an OSNR of ∼17 dB. It is found that the BER is kept below EFEC

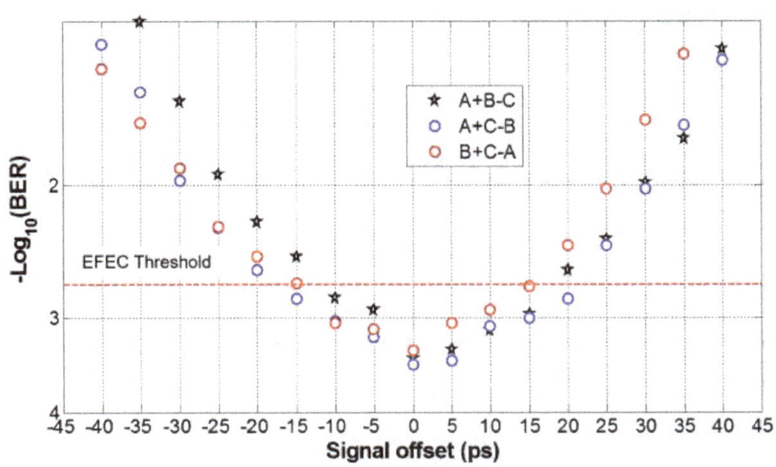

Figure 13. Measured BER performance versus signal offset.

threshold when the signal offset is within 15 ps. The obtained results shown in **Figure 13** indicate a favorable performance tolerance to the signal offset.

5. GSMR-enhanced nonlinear optical device for on-chip optical signal processing

Microring resonators have accelerated the demonstration of very low power continuous-wave (CW) nonlinear optics, and similar benefits are expected for its operation in processing high bandwidth optical signal. Actually, it is only very recently that the first demonstration of optical signal processing based a resonant cavity has been reported. Wavelength conversion at 2.5 Gb/s in a single microring [31] and 10 Gb/s in a silicon-cascaded microring resonator [32] have been demonstrated. On the other hand, it is well known that advanced optical modulation formats have become of great importance to enable high-capacity optically routed transport networks and design of modern wavelength-division multiplexed (WDM) fiber systems [20]. However, all-optical wavelength conversions of advanced optical modulation formats have not been realized in integrated ring structures.

5.1. GSMR-enhanced all-optical up and down wavelength conversion

Figure 14 shows the experimental set-up for degenerate FWM based up and down wavelength conversion using a GSMR [24].

In the experiment, the radius of the silicon microring resonator is 10 μm, and the corresponding free spectral range is around 10 nm. The grating coupler exhibits a 50-nm coupling range with 3-dB coupling loss and the central wavelength of the grating is 1550 nm. Based on the characterization of the fabricated GSMR, two neighboring resonant wavelengths of 1548 and 1558 nm are chosen as the pump and signal light for up wavelength conversion, and the converted idler

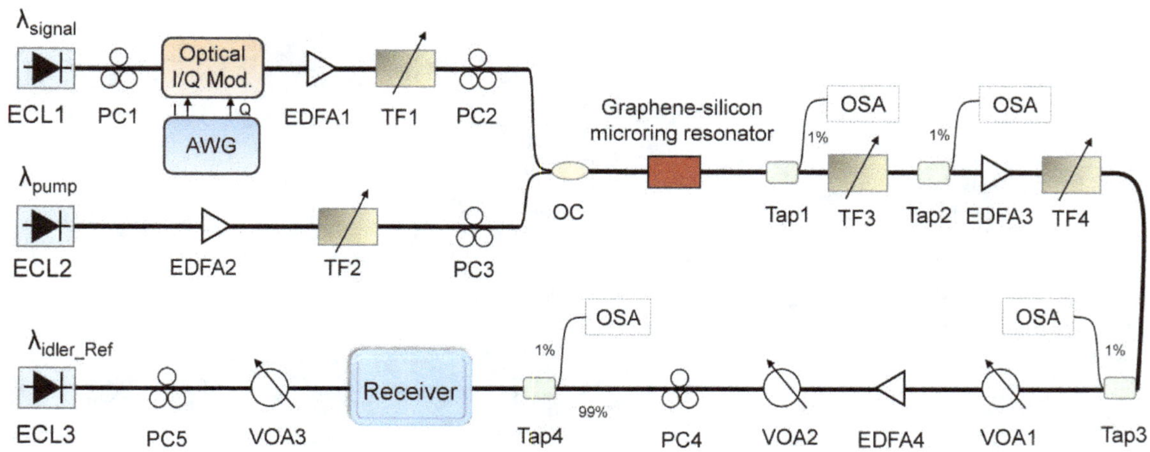

Figure 14. Experimental set-up for degenerate FWM-based up and down wavelength conversion in GSMR. ECL: external cavity laser; AWG: arbitrary waveform generator; EDFA: erbium-doped optical fiber amplifier; TF: tunable filter; OC: optical coupler; PC: polarization controller; OSA: optical spectrum analyzer; and VOA: variable optical attenuator.

wavelength is around 1538 nm. Similarly, for down wavelength conversion, the signal and pump light wavelengths are chosen as 1538 and 1548 nm, and the converted idler wavelength is around 1558 nm. **Figure 15(a)** and **(b)** shows the typical output degenerate FWM spectra obtained after the GSMR for up and down wavelength conversions of QPSK signal.

We define the conversion efficiency as the power ratio of converted idler to signal. **Figure 16(a)** and **(b)** plots the experimentally measured and fitted conversion efficiency as a function of input pump power for up and down wavelength conversion, respectively. The nonlinear Kerr coefficient increment caused by graphene is responsible for the enhanced FWM in the GSMR [33]. One can clearly see that the conversion efficiency increases with the pump power. The saturation of the conversion efficiency at relatively high pump power level results from the two-photon absorption and free carrier absorption in silicon. The device can be tuned by the thermo-optic effect. The resonant wavelength of GSMR as a function of temperature is depicted in **Figure 17**. When the temperature changes from 20 to 40°C, the resonant wavelength can be linearly tuned from 1556.80 to 1559.11 nm. With future improvement, the tuning range of the device can be also remarkably increased by using microheater structures for temperature tuning [34, 35].

To further characterize the performance of QPSK wavelength conversion, we measure the BER curves as a function of the received OSNR for back-to-back signals and up/down converted idler. **Figure 18** plots measured BER performance for QPSK wavelength conversion with the converted idlers generated at 1538.64 nm (up conversion) and 1558.15 nm (down conversion), respectively. The output powers of EDFA1 and EDFA2 are set to be around 25.1 and 25.3 dBm, respectively. The measured conversion efficiencies for converted idlers at 1538.64 nm (up conversion) and 1558.15 nm (down conversion) are −38.34 and −40.2 dB, respectively. The observed OSNR penalties for QPSK up and down wavelength conversion are less than 1.4 dB at a BER of 1×10^{-3} [7% forward error correction (FEC) threshold]. The insets of **Figure 18** depict the corresponding constellations of the back-to-back signals and converted idlers. We also evaluate the BER performance for up wavelength converted idler when the pump power increases from 9.3 to 15.3 dBm. As shown in **Figure 19**, the minimum penalty is less than 0.8 dB when the pump power is 13.3 dBm. The OSNR penalty is around 2 dB with a pump power 9.3 dBm.

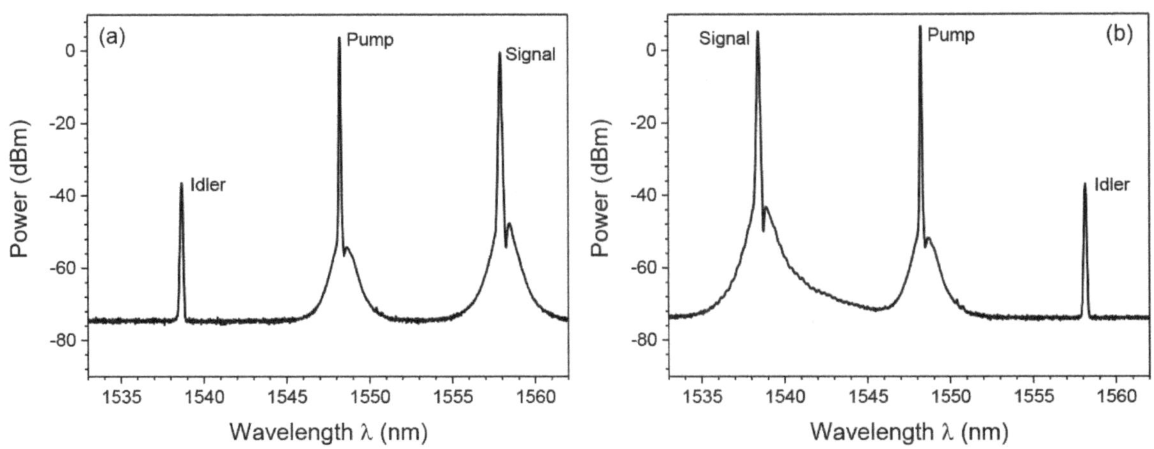

Figure 15. Measured degenerate FWM spectra for (a) up and (b) down wavelength conversion.

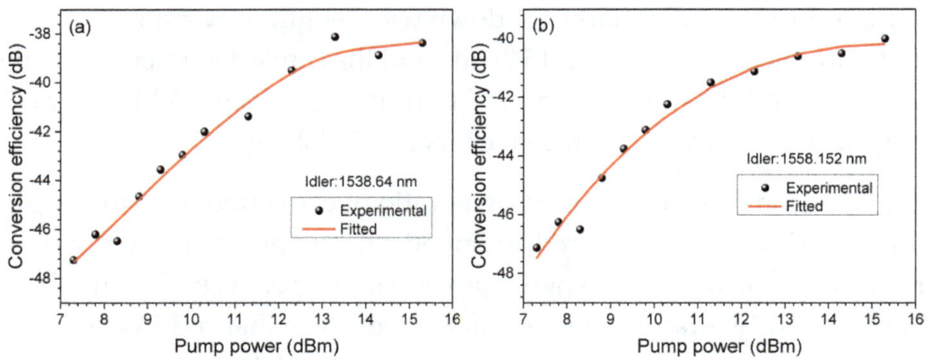

Figure 16. Experimentally measured and fitted conversion efficiency versus input pump power for (a) up and (b) down wavelength conversion.

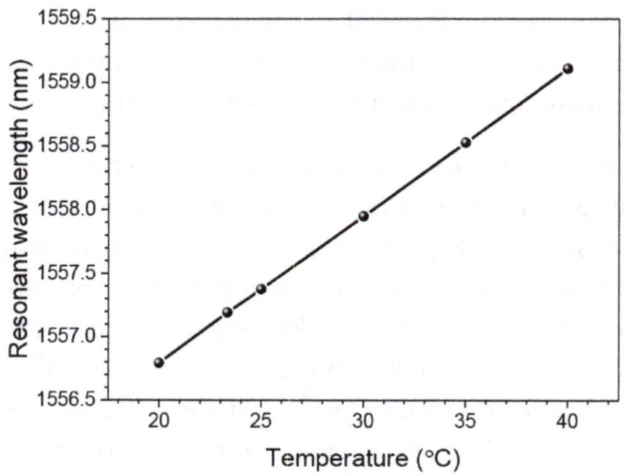

Figure 17. Measured resonant wavelength of GSMR versus temperature.

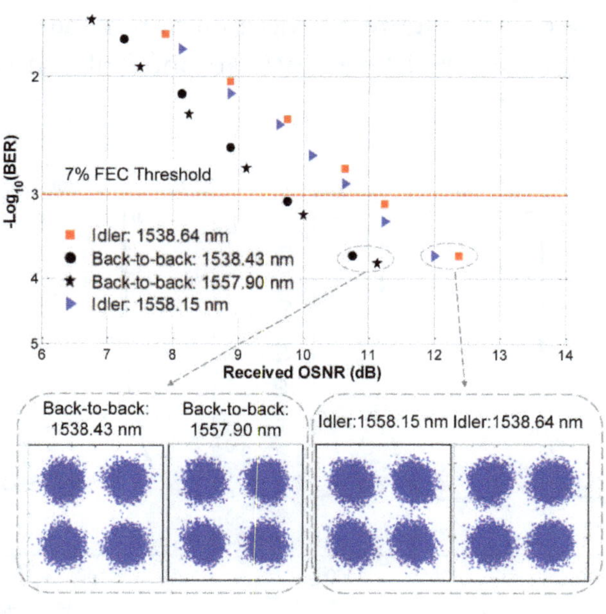

Figure 18. Measured BER versus received OSNR for up and down wavelength conversion of QPSK signal. Insets show constellations of QPSK.

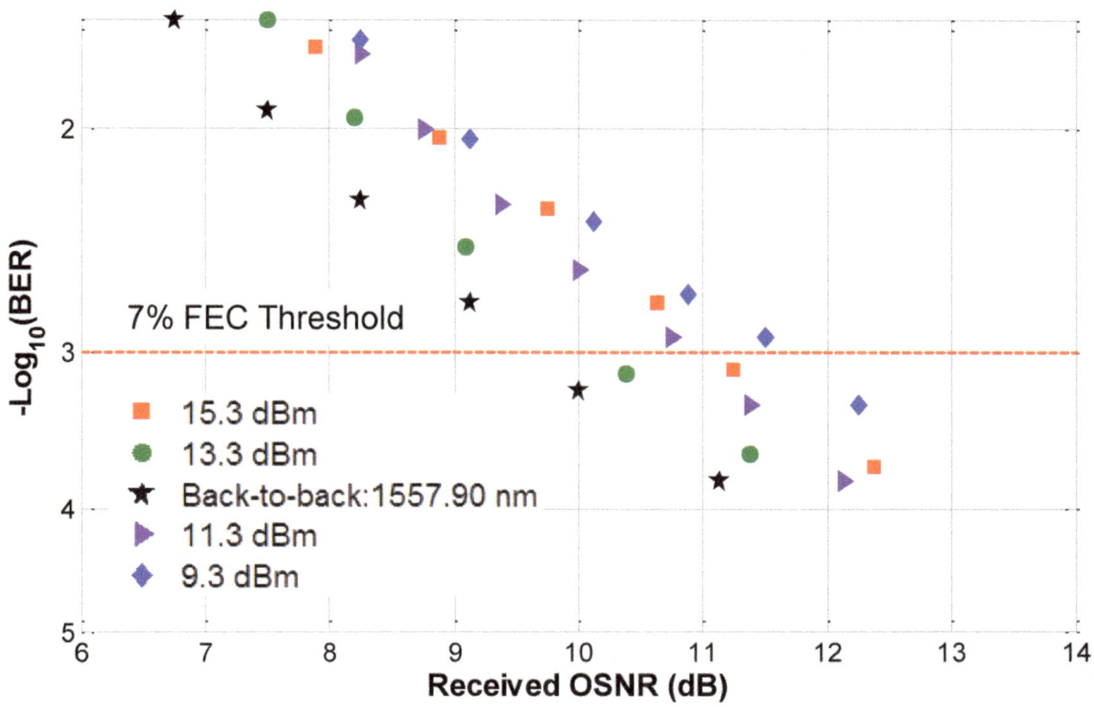

Figure 19. Measured BER versus received OSNR for up wavelength conversion of QPSK signal when pump power increases from 9.3 to 13.3 dBm.

The obtained results shown in **Figures 15–19** imply favorable performance achieved for up and down wavelength conversion of QPSK signal using the fabricated GSMR.

5.2. Nonlinearity enhanced graphene-silicon microring for selective conversion of flexible grid multi-channel multi-level signal

Figure 20 illustrates the concept and operation principle of the GSMR-enhanced FWM for selective conversion of flexible grid multichannel multilevel signal. As shown in **Figure 20(a)**, introducing a resonator structure and graphene to the silicon waveguide can enable enhanced nonlinearity together with channel-selective operation. As shown in **Figure 20(b)**, one continuous-wave (CW) pump and a four-channel flexible grid data-carrying signal are fed into the GSMR. When propagating along the silicon waveguide, pump photons are annihilated to create signal photons and newly converted idler photons by the FWM process. At the output of the silicon waveguide, the data information carried by the input signal is converted to the idlers. Since the resonant dip of the nonlinearity enhanced GSMR is aligned to one of the signals, only the data information of the selected signal light can be converted to the idler. Moreover, the selective conversion operation is transparent to the channel spacing of the multichannel multilevel signal, thus it is compatible with flexible grid optical network. As shown in **Figure 20(c)**, by tuning the pump wavelength and thermal tuning the resonant wavelength of the nonlinearity enhanced GSMR, the desired channel of the flexible grid multichannel multilevel signal can be selected.

We evaluate the system performance of the selective conversion of flexible grid multichannel 16-QAM signal using the nonlinearity enhanced GSMR. **Figure 21(a)–(d)** plot measured the BER performance when the channel spacing is 200, 100, 50, and 20 GHz, respectively. The

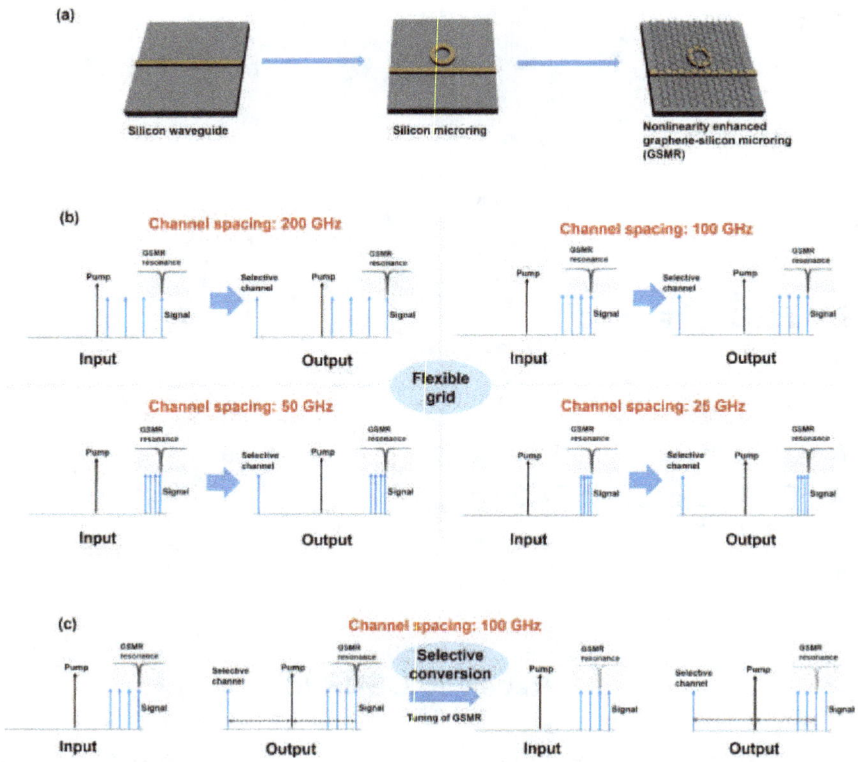

Figure 20. Concept and principle of GSMR-enhanced FWM for channel-selective wavelength conversion of flexible grid multi-channel signal. (a) From silicon waveguide to silicon microring and GSMR with enhanced nonlinearity. (b) Flexible grid. (c) Selective conversion.

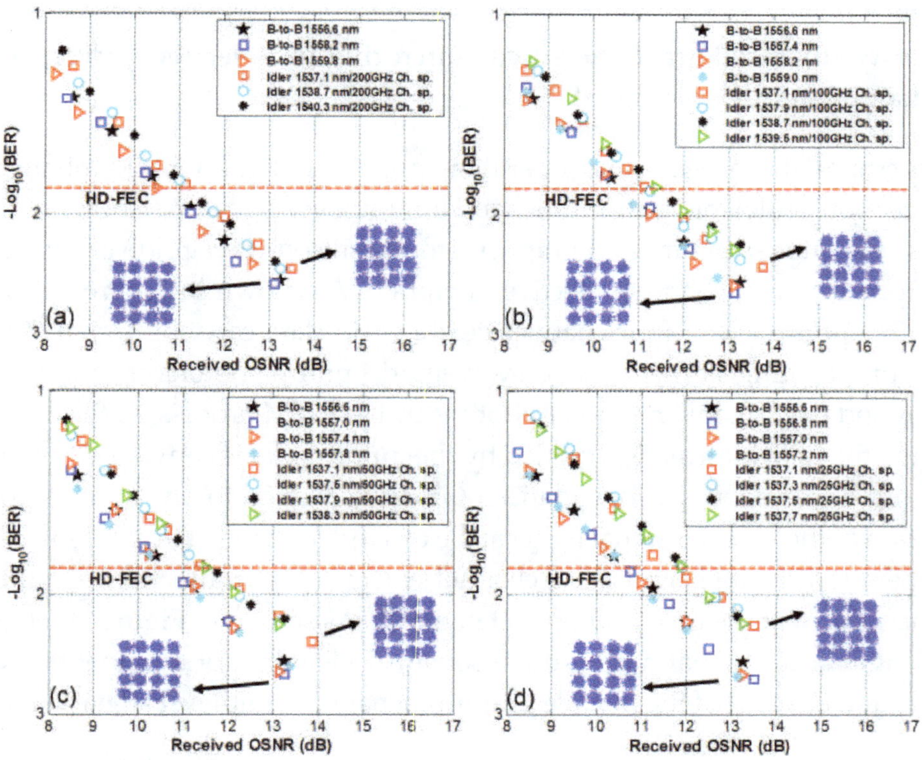

Figure 21. Measured BER curves and constellations of selective conversion of flexible grid multichannel 16-QAM signal in a nonlinearity enhanced GSMR. The channel spacing is (a) 200 GHz, (b) 100 GHz, (c) 50 GHz, and (d) 25 GHz, respectively.

optical signal-to-noise ratio (OSNR) penalties are all below 1 dB for 200, 100, and 50 GHz channel spacing. The OSNR penalties for 25-GHz channel spacing are about 1.2 dB. The insets of **Figure 21(a)–(d)** depict the corresponding constellations of the B-to-B signal and converted idlers. The clear constellations of converted idlers indicate favorable performance achieved for selective conversion of flexible grid multichannel multilevel signal in a nonlinearity enhanced GSMR.

6. Discussions and conclusions

Looking back on the previous works of graphene-assisted nonlinear optical signal processing, Hendry and co-workers experimentally demonstrated the graphene-based FWM for the first time [12]. They also indicated that graphene might have large $\chi^{(3)}$ nonlinearity due to its linear band structure allowing interband optical transitions at all photon energies. After that FWM was also observed in graphene-silicon hybrid optoelectronic devices [14] and graphene-coated microfiber [17, 18]. Moreover, FWM-based wavelength conversion of a 10-Gb/s NRZ signal with mechanically exfoliated graphene was first reported in Ref. [19].

In this chapter, we have reviewed our recent progress in graphene-assisted nonlinear optical device and their applications, including degenerate FWM-based tunable wavelength conversion of quadrature phase-shift keying (QPSK) signal, two-input optical high-base hybrid doubling and subtraction functions, three-input high-base optical computing, graphene-silicon microring resonator enhanced nonlinear optical device for on-chip optical signal processing, and nonlinearity enhanced graphene-silicon microring for selective conversion of flexible grid multichannel multilevel signal.

For FWM-based tunable wavelength conversion of QPSK signal, the total effective nonlinear Kerr coefficient of the graphene-assisted nonlinear optical device is actually the combined contributions from the graphene and the device material (e.g. silica in graphene-coated fiber). It has to emphasize that spectrally efficient advanced modulation formats have been widely used in optical fiber transmission systems. The advanced optical modulation formats play an important role in enabling high-capacity optical transport networks where wavelength conversion function is highly desired. Therefore, exploring wavelength conversion of advanced modulation formats based on FWM in graphene is very interesting and meaningful.

For graphene-assisted optical computing (two-input and three input), the innovative schemes to achieve hybrid quaternary arithmetic functions of doubling and subtraction using optical nonlinearities in graphene and (D)QPSK signals are presented. We experimentally demonstrate 10-Gbaud quaternary arithmetic functions of 2A-B and 2B-A by exploiting degenerate FWM in graphene. The power penalties of converted idlers at a BER of 2×10^{-3} are measured to be about 7.4 dB for 2A-B and 7.0 dB for 2B-A. With future improvement, graphene-assisted nonlinear optical devices might be employed to facilitate more optical signal processing applications.

For, three-input high-base optical computing, an innovative scheme to perform graphene-assisted modulo 4 functions of three-input high-base optical computing is presented. By exploiting multiple nondegenerate FWM processes in a single-layer graphene-coated fiber device and adopting (D)QPSK signals, we experimentally demonstrate 10-Gbaud modulo 4 operations of three-input quaternary hybrid addition and subtraction of A+ B − C, A + C − B, and B + C − A, respectively. Additionally, in order to verify the enhancement of graphene-coated fiber device, we measure the output spectrum without graphene for reference under the same experimental conditions. Moreover, we repeat the experiment by adding extra 2 and 5 m single-mode fibers in the set-up and get almost the same experimental results. The nondegenerate FWM processes in graphene-assisted nonlinear optical devices (e.g., graphene coated fiber device) are enhanced by comparing the conversion efficiency between the two cases with and without graphene.

For, GSMR-enhanced all-optical up and down wavelength conversion, using the graphene-assisted nonlinear optical device, we experimentally demonstrate up and down wavelength conversion of a 10-Gbaud quadrature phase-shift keying (QPSK) signal by exploiting degenerate four-wave mixing (FWM) progress in the fabricated GSMR. To enhance the nonlinear interactions in silicon waveguide devices, resonator structures such as microring resonators can be introduced. Graphene can be also utilized to enhance the nonlinear interactions in silicon waveguide devices. So, in order to further enhance the nonlinearity of silicon microring resonator, the GSMR is fabricated. It is expected that the GSMR might find more interesting on-chip optical signal-processing applications.

Actually, there are different platforms for nonlinear optical signal processing, e.g., silica in fiber, silicon, and graphene. **Table 1** summarizes a brief comparison among silica in fiber, silicon, and graphene. The Kerr coefficients of silica in fiber, silicon, and graphene are $\sim 10^{-20}$, $\sim 10^{-18}$, and $\sim 10^{-11}$ m^2/W, respectively. Nonlinear optical signal processing based on silica in fiber has lower power loss. Although silica in fiber is compatible with optical fiber transmission systems, its third-order nonlinearity is lower and the desired fiber is longer (i.e., larger footprint). Silicon has higher $\chi^{(3)}$ nonlinearity and the compactness of silicon photonic device is suitable for chip-scale optical signal processing functions. However, silicon photonic device is not compatible with optical fiber transmission systems. In contrast, graphene has even larger $\chi^{(3)}$ nonlinearity and the fabricated graphene-assisted nonlinear optical fiber device with the graphene placed within the connector of two fibers is fully compatible with existing optical fiber transmission systems. The $\chi^{(3)}$ nonlinearity of graphene is several orders of magnitude larger than silica in fiber and silicon, which is due to the unique linear band structure of π-electrons [12, 36, 37]. The graphene-assisted nonlinear optical fiber device is compact. The combined effective nonlinearity of graphene-assisted nonlinear optical fiber device is increased and the graphene enhances the FWM process. However, the measured conversion efficiency of graphene-assisted nonlinear optical fiber device in the experiment is ~33 dB, which is lower than highly nonlinear fiber. It is noted that practically fabricated graphene is not perfect and any imperfections during the fabrication of graphene can break the band structure and degrades the dispersionless and $\chi^{(3)}$ nonlinearity properties. Additionally, the FWM conversion efficiency is also dependent on the number of graphene layers. Previous work demonstrated that the nonlinear response of graphene was sensitive to the number of graphene layers [12]. It is expected that for a few graphene layers the nonlinearity increases in proportion to the number of layers, suggesting the tremendous potential of graphene as a

Platforms	Kerr coefficient	Loss	Compactness	Compatibility*
Silica in fiber [38]	$\sim 10^{-20}$ m^2/W	Low	Large footprint	Fully compatible
Silicon [39]	$\sim 10^{-18}$ m^2/W	High	Compact	Incompatible
Graphene-coated fiber [13]	$\sim 10^{-11}$ m^2/W	Low	Compact	Fully compatible

*Compatibility of silica in fiber, silicon photonic device, and graphene-assisted nonlinear optical fiber devices with existing optical fiber transmission systems.

Table 1. Comparison among silica, silicon, and graphene.

platform for efficient nonlinear optical signal processing. In this scenario, it is possible to further enhance the FWM response by appropriately increasing the number of graphene layers employed in the experiment.

As mentioned above, a silicon photonic device is not compatible with optical fiber transmission systems. However, silicon has higher $\chi^{(3)}$ nonlinearity and the compactness of the silicon photonic device is suitable for chip-scale optical signal processing functions. Recently, silicon waveguides used as wavelength converters have shown some superior performance compared with SOA, HNLFs, and PPLN. The advantages of silicon waveguide-based wavelength converters are as follows: (1) in silicon waveguides, high-contrast index leads to tight light confinement and the nonlinear effects are greatly enhanced; (2) silicon waveguides also feature broad bandwidth, high speed, low cost, low power consumption, and complementary metal-oxide-semiconductor (CMOS) compatibility. However, the conversion efficiency of FWM in the silicon waveguides is fundamentally hampered by the large nonlinear losses of silicon, which is caused by two-photon absorption (TPA) and free carrier absorption (FCA) in silicon [40]. Therefore, the saturation of the conversion efficiency is consequence of the nonlinear absorption increment when the intensity of the input light reaches a certain level at the near-infrared region. The incorporation of reverse biased p-i-n junctions in the silicon waveguides can remove the free carrier and fasten the lifetime of the free carrier. In such a way, the conversion efficiency of FWM is increased [41]. However, the external electric modulation gives rise to the power consumption which is not available for large-scale photonic integration. Graphene has a third-order nonlinear susceptibility which is several orders of magnitude larger than that of silicon. When the monolayer graphene is transferred on the silicon waveguides, the nonlinear optical performances of the device are enhanced owing to the evanescently coupling between the silicon waveguide and graphene over a distance of hundreds of micrometers [14, 15]. Therefore, combination of the giant nonlinearity of graphene and the strong electromagnetic field confinement of silicon waveguides may be an interesting way for all optical signal processing.

Acknowledgements

This work was supported by the National Natural Science Foundation of China (NSFC) under grants 61222502, 11574001 and 11274131, the National Basic Research Program of China (973 Program) under grant 2014CB340004, the Program for New Century Excellent

Talents in University (NCET-11-0182), the Wuhan Science and Technology Plan Project under grant 2014070404010201, and the seed project of Wuhan National Laboratory for Optoelectronics (WNLO). The authors thank the engineer in the Center of Micro-Fabrication and Characterization (CMFC) of WNLO for the support in the fabrication of graphene-assisted nonlinear optical device and the facility support of the Center for Nanoscale Characterization and Devices of WNLO.

Author details

Jian Wang* and Xiao Hu

*Address all correspondence to: jwang@hust.edu.cn

Wuhan National Laboratory for Optoelectronics, School of Optical and Electronic Information, Huazhong University of Science and Technology, Wuhan, Hubei, China

References

[1] K. S. Novoselov, A. K. Geim, S. V. Morozov, D. Jiang, Y. Zhang, S. A. Dubonos, and A. A. Firsov, "Electric field effect in atomically thin carbon films." Science 306, 666–669 (2004).

[2] K. Kim, J. Y. Choi, T. Kim, S. H. Cho, and H. J. Chung, "A role for graphene in silicon-based semiconductor devices." Nature 479, 338–344 (2011).

[3] F. Xia, T. Mueller, Y. M. Lin, A. Valdes-Garcia, and P. Avouris, "Ultrafast graphene photodetector." Nat. Nanotechnol. 4, 839–843 (2009).

[4] Q. Bao, H. Zhang, B. Wang, Z. Ni, C. H. Y. X. Lim, Y. Wang, and K. P. Loh, "Broadband graphene polarizer." Nat. Photonics 5, 411–415 (2011).

[5] M. Liu, X. Yin, E. Ulin-Avila, B. Geng, T. Zentgraf, L. Ju, and X. Zhang, "A graphene-based broadband optical modulator." Nature 474, 64–67 (2011).

[6] A. Grigorenko, M. Polini, and K. Novoselov, "Graphene plasmonics." Nat. Photonics 6, 749–758 (2012).

[7] Q. Bao, H. Zhang, Y. Wang, Z. Ni, Y. Yan, Z. X. Shen, K.P. Loh, and D. Y. Tang, "Atomic-layer graphene as a saturable absorber for ultrafast pulsed lasers." Adv. Fun. Mat. 19, 3077–3083 (2009).

[8] Z. Sun, T. Hasan, F. Torrisi, D. Popa, G. Privitera, F. Wang, F. Bonaccorso, D. M. Basko, and A. C. Ferrari, "Graphene mode-locked ultrafast laser." ACS Nano 4, 803–810 (2010).

[9] Q. Bao, H. Zhang, Z. Ni, Y. Wang, L. Polavarapu, Z. Shen, Q. Hua, D. Tang, and K. P. Loh, "Monolayer graphene as a saturable absorber in a mode-locked laser." Nano Res. 4, 297–307 (2011).

[10] Y. Meng, S. Zhang, X. Li, H. Li, J. Du, and Y. Hao, "Multiple-soliton dynamic patterns in a graphene mode-locked fiber laser." Opt. Express 20, 6685–6692 (2012).

[11] N. Tolstik, E. Sorokin, and I. T. Sorokina, "Graphene mode-locked Cr: ZnS laser with 41 fs pulse duration." Opt. Express 22, 5564–5571 (2014).

[12] E. Hendry, P. J. Hale, J. Moger, A. K. Savchenko, and S. A. Mikhailov, "Coherent nonlinear optical response of graphene." Phy. Rev. Lett. 105, 097401 (2010).

[13] H. Zhang, S. Virally, Q. Bao, K.P. Loh, S. Massar, N. Godbout, and P. Kockaert, "Z-scan measurement of the nonlinear refractive index of graphene." Opt. Lett. 37, 1856–1858 (2012).

[14] T. Gu, N. Petrone, J. F. McMillan, A. van der Zande, M. Yu, G. Q. Lo, D. L. Kwong, J. Hone, and C. W. Wong, "Regenerative oscillation and four-wave mixing in graphene optoelectronics." Nat. Photonics, 6, 554–559 (2012).

[15] H. Zhou, T. Gu, J. F. McMillan, N. Petrone, A. van der Zande, J. C. Hone, and S. Zhou, "Enhanced four-wave mixing in graphene-silicon slow-light photonic crystal waveguides." Appl. Phys. Lett. 105, 091111 (2014).

[16] B. Xu, A. Martinez, K. Fuse, and S. Yamashita, "Generation of four wave mixing in graphene and carbon nanotubes optically deposited onto fiber ferrules." OSA. CLEO: Science and Innovations, CMAA6, 2011.

[17] Y. Wu, B. Yao, Y. Cheng, Y. Rao, Y. Gong, X. Zhou, B. Wu, and K. S. Chiang, "Four-wave mixing in a microfiber attached onto a graphene film." IEEE Photonics Techno. Lett. 26, 249–252 (2014).

[18] Y. Wu, B. C. Yao, Q. Y. Feng, X. L. Cao, X. Y. Zhou, Y. J. Rao, and K. S. Chiang, "Generation of cascaded four-wave-mixing with graphene-coated microfiber." Pho. Res. 3, A64–A68 (2015).

[19] B. Xu, A. Martinez, and S. Yamashita, "Mechanically exfoliated graphene for four-wave-mixing-based wavelength conversion." IEEE Photonics Techno. Lett. 20, 1792–1794 (2012).

[20] P. J. Winzer, and R J. Essiambre, "Advanced modulation formats for high-capacity optical transport networks." J. Lightwave Techno. 24, 4711–4728 (2006).

[21] X. Hu, M. Zeng, A. Wang, L. Zhu, L. Fu, and J. Wang, "Graphene-assisted nonlinear optical device for four-wave mixing based tunable wavelength conversion of QPSK signal." Opt. Express 23, 26158–26167 (2015).

[22] A. Wang, X. Hu, L. Zhu, M. Zeng, L. Fu, and J. Wang, "Experimental demonstration on two-input optical high-base hybrid doubling and subtraction functions in graphene." Opt. Express 23, 31728–31735 (2015).

[23] X. Hu, A. Wang, M. Zeng, Y. Long, L. Zhu, L. Fu, and J. Wang, "Graphene-assisted multiple-input high-base optical computing." Sci. Rep. 6, 1–5 (2016).

[24] X. Hu, Y. Long, M. Ji, A. Wang, L. Zhu, Z. Ruan, Y. Wang, and J. Wang, "Graphene-silicon microring resonator enhanced all-optical up and down wavelength conversion of QPSK signal." Opt. Express, 24, 7168–7177 (2016).

[25] F. Bonaccorso, Z. Sun, T. Hasan, and A. C. Ferrari, "Graphene photonics and optoelectronics." Nat. Photonics 4, 611–622 (2010).

[26] Q. Bao, and K. P. Loh, "Graphene photonics, plasmonics, and broadband optoelectronic devices." ACS Nano 6, 3677–3694 (2012).

[27] B. Sensale-Rodriguez, R. Yan, M. M. Kelly, T. Fang, K. Tahy, W. S. Hwang, and H. G. Xing, "Broadband graphene terahertz modulators enabled by intraband transitions." Nat. Commun. 3, 1–7 (2012).

[28] X. Li, W. Cai, J. An, S. Kim, J. Nah, D. Yang, and S. K. Banerjee, "Large-area synthesis of high-quality and uniform graphene films on copper foils." Science. 324, 1312–1314 (2009).

[29] M. Kalbac, A. Reina-Cecco, H. Farhat, J. Kong, L. Kavan, and M. S. Dresselhaus, "The influence of strong electron and hole doping on the Raman intensity of chemical vapor-deposition graphene." ACS Nano 4, 6055–6063 (2010).

[30] M. Jablan, H. Buljan, and M. Soljačić, "Plasmonics in graphene at infrared frequencies." Phys. Rev. B 80, 245435 (2009).

[31] A. Pasquazi, R. Ahmad, M. Rochette, M. Lamont, B. E. Little, S. T. Chu, and D. J. Moss, "All-optical wavelength conversion in an integrated ring resonator." Opt. Express. 18, 3858–3863 (2010).

[32] F. Morichetti, A. Canciamilla, C. Ferrari, A. Samarelli, M. Sorel, and A. Melloni, "Travelling-wave resonant four-wave mixing breaks the limits of cavity-enhanced all-optical wavelength conversion." Nat. Commun. 2, 296 (2011).

[33] M. Ji, H. Cai, L. Deng, Y. Huang, Q. Huang, J. Xia, and Y. Wang, "Enhanced parametric frequency conversion in a compact silicon-graphene microring resonator." Opt. Express. 23, 18679–18685 (2015).

[34] X. Wang, J. A. Martinez, M. S. Nawrocka, and R. R. Panepucci, "Compact thermally tunable silicon wavelength switch: modeling and characterization." IEEE Photonics Technol. Lett. 20, 936–938 (2008).

[35] F. Gan, T. Barwicz, M. A. Popovic, M. S. Dahlem, C. W. Holzwarth, P. T. Rakich, and F. X. Kartner, "Maximizing the thermos-optic tuning range of silicon photonic structures." Photon. in Switching 19, 67–68 (2007).

[36] A. K. Geim, and K. S. Novoselov, "The rise of graphene." Nat. Mat. 6, 183–191 (2007).

[37] K. S. Novoselov, A. K. Geim, S. V. Morozov, D. Jiang, M. I. Katsnelson, I. V. Grigorieva, S. V. Dubonos, and A. A. Firsov, "Two-dimensional gas of massless Dirac fermions in graphene." Nature 438, 197–200 (2005).

[38] L. Prigent and J.-P. Hamaide, "Measurement of fiber nonlinear Kerr coefficient by four-wave mixing." Photonics Techno. Lett. IEEE 5, 1092–1095 (1993).

[39] M. Dinu, F. Quochi, and H. Garcia, "Third-order nonlinearities in silicon at telecom wavelengths." Appl. Phys. Lett. 82, 2954–2956, 2003.

[40] T. Uesugi, B. S. Song, T. Asano, and S. Noda, "Investigation of optical nonlinearities in an ultra-high-Q Si nanocavity in a two-dimensional photonic crystal slab," Opt. Express 14, 377–386 (2006).

[41] A. Gajda, L. Zimmermann, M. Jazayerifar, G. Winzer, H. Tian, R. Elschner, T. Richter, C. Schubert, B. Tillack, and K. Petermann, "Highly efficient CW parametric conversion at 1550 nm in SOI waveguides by reverse biased p-i-n junction," Opt. Express 20, 13100–13107 (2012).

8

Surface-Modified Graphene for Mid-Infrared Detection

Mehrdad Siahsar, Mahboubeh Dolatyari,
Ali Rostami and Ghasem Rostami

Abstract

In this chapter, morphology variation and electronic structure in a surface-modified graphene are demonstrated by both calculation and experimental results. The results indicate that the band structure and morphology of modified graphene sheets are altered because of changing in the type of hybridization of carbon atoms in the graphene sheet. Accordingly, the band gap of graphene can be tuned by surface modification using organic molecules. Then, modified graphene is used for fabrication of infrared detectors. The properties of unmodified graphene photodetectors were also measured so as to compare with modified graphene photodetectors. The results demonstrate that modification of graphene using organic ligands improved the detection parameters such as fast response time, electrical stability and low dark current. Moreover, the sensitivity of photodetectors based on modified graphene was significantly improved.

Keywords: fabrication, photodetector, organic ligand, modified, unmodified, IR, graphene

1. Introduction

Graphene is a two-dimensional sp^2-bonded carbon atom on a honeycomb lattice [1]. The particular arrangement of carbon atoms in graphene leads to a novel energy dispersion relation, mean root of the appealing electronic characteristic, which corresponds to massless fermions [2–4]. In spite of the fact that graphene has much favourable properties, which have revolutionized the miscellaneous aspects of science and technology, it could not be an applicable material for special purposes because of its shortcomings that have to be overcome [5]. Although, in the last decade, the unique properties of graphene, namely electronic, mechanical, optical and chemical properties, and its possible applications have been widely investigated, the absence of a gapped semiconductor with properties like that of graphene is strongly felt, especially in the graphene-based photodetectors which do not enjoy fast response time stemming from zero band gap of graphene [6–18]. In addition, another salient weakness of graphene's application appears in the designing of transistor-based digital circuits [19]. Transistor made by graphene, in the off-state, has a current due to minimum conductivity of graphene [19]. Accordingly, the power consumption of fabricated circuits is considerable. The engineering of band structure of graphene sheet has been extensively studied. Band gap about 0.5 eV can be formed in nanoribbons with the width about 10 nm [20–22]. In this method, controling the value of band gap which is strongly affected by the structure of nano ribbons edges is difficult. Moreover, bilayer graphene can be used for this purpose, which is synthesized with difficulty [23, 24]. One can manipulate electrical and magnetic properties of graphene by introducing different defect state in graphene [25, 26]. Because all atoms are placed on the surface of graphene, density of electrons is high on it, so active is the graphene sheet that it can easily react with the gases in the surrounding atmosphere. Therefore, the chemical instability of graphene leads to use vacuum condition in order to obtain repetitive results.

Some ideas have been investigated so as to overcome the mentioned weaknesses. By studying the electron transport properties of graphene-like two-dimensional materials, one can reach to this result that using heterostructures of graphene and other two-dimensional materials have significant impact on the application of graphene [5, 27–32]. Instead of two-dimensional materials, zero- and one-dimensional organic or inorganic materials can be used in the above-mentioned heterostructures [33–36]. A focal point in this research is modification of the surface of the graphene sheet via zero-dimensional organic ligands. In order to realize the process of modification of graphene sheet, it is important to know that by adding hydrogen to the surface of graphene sheet, which is highly conducting, it could be converted into a novel material known as graphane, two-dimensional hydrocarbon, which is an insulator with direct band gap of 3.7 eV [37]. The alteration in the percentage of hydrogenation of graphene can lead to the creation of various band gap. In other words, the engineering of band gap synthesized can be possible by hydrogenation from zero in the graphene to 3.7 eV in the fully hydrogenated graphene [37]. This achievement should broaden the spectrum of photonics application of graphene.

To modify surface of graphene sheet, we introduce trap states and a band gap in graphene by means of organic ligands, which interact with graphene for fabricating sensitive infrared detector operated at 3–5 μm [38–40]. To achieve this, our focus is oriented to the analysis of altered structural arrangement of the organic ligands/graphene with strong interaction between them. Our analysis show that the geometry of synthesized material is affected by the adsorption of organic ligands on graphene layer, for the carbon atoms of graphene strongly interact with the hydrogen atoms of organic ligands. Not only was the structural analysis of the synthesized materials operated, but calculation of the band structure and electron density difference were also calculated to confirm the claim made about modified form of graphene layer.

X-ray diffraction (XRD), atomic force microscope (AFM), scanning electron microscope (SEM), transmission electron microscope (TEM) and electrical resistivity analysis are used to investigate the structure and morphology of the graphene sheets, which were modified by different concentration of various organic ligands. Finally, we used the synthesized materials as an active layer of metal-graphene-metal (MGM) photodetectors. The experimental investigations show that the detection parameter of fabricated photodetectors improved significantly, for example, temporal time of fabricated photodetectors in our reported work is up to 1000 times faster than the photodetector reported by Yu and co-workers [41].

2. Experimental

2.1. Synthesis and modification

In existance of all fabrication methods of graphene oxide, we prefer to use the following method because the synthesis route plays an important role on the chemical, optical and electrical properties of graphene oxide [42]. Graphite oxide (GO) was synthesized by natural graphite by Hummers' method [43]. A colloidal suspension of synthesized graphene oxide in purified water was prepared by sonication of GO in water (3 mg/mL) for 3 h. Hydrazine monohydrate (1 mL for 3 mg of GO, 98%, Aldrich) was subsequently added to the suspension, in order to remove oxygen components by the hydrazine reduction [38–40]. Additional stirring in an oil bath held at 80°C for 12 h yielded a black precipitation of reduced graphene oxide powder. The obtained materials were centrifuged by water for several times. The synthesized material was dried at 100°C and used for further characterizations and treatment. Organic ligands—thiosemicarbazide, thiophene-2-arbaldehyde, thiophene-2-carboxylic acid and pyridine-4-carboxylic acid—were utilized to modify graphene surface. To prepare modifying organic ligands, 0.5 mg of thiophene-2-arbaldehyde, thiophene-2-carboxylic acid and pyridine-4-carboxylic acid was separately solved in 50 mL water. Moreover, 5 and 10 mg of thiosemicarbazide were independently solved in 7 mL ethanol. Each of the prepared suspension solely was added to a suspension, which consisted of 27 mg of obtained graphene and 30 mL of deionized water. All of the obtained suspensions were stirred for 24 h at room temperature. After purification, the obtained materials were individually added to 30 mL deionized water and dispersed until a thin sheet was formed on top of the colloidal suspension [38–40].

2.2. Fabrication of photodetector

The sheet was transferred on the interdigitated Copper (Cu) contact, which is deposited on a fibre-glass substrate having fingers with a length of 0.5 cm, a width of 150 μm and a pitch of 150 μm. The thickness of the Cu layer was 500 nm [38–40].

2.3. Characterization

In this chapter, all structural characterization and measurement were performed by the following devices:

The crystal structure of modified and unmodified graphene was characterized by powder X-ray diffraction (PXRD) on a Siemens D500 using Cu-kα radiation (λ = 1.541 Å). Ultrasound radiation was performed using UP400s Germany (0.3 cm diameter Ti horn, 200 W, 24 kHz).

The surface morphology of synthesized materials was obtained on a Dual-scope C26 scanning probe and microscope DME atomic force microscope (AFM). The morphology of products was studied via a Tescan model MIRA3 field-emission scanning electron microscope with an accelerating voltage of 10 kV. TEM images were obtained on a Zeiss-EM10C-80KV transmission electron microscope with an accelerating voltage of 80 kV [38–40].

2.4. Computational details

Both the electronic band structure and density of states (DOS) of graphene were calculated by DFT. Calculations were carried out with the CASTEP code using local density approximation (LDA) and the non-local gradient-corrected exchange-correlation functional as parameterized by the Perdew-Zunger scheme (CA-PZ), which uses a plane wave basis set for the valance electrons and ultra-soft pseudo potential for the core electrons. The number of plane waves included in the basis was determined by cut-off energy (Ec) of 300.0 eV. The summation over the Brillouin zone was operated with a *k*-point sampling using a Monkhorst-Pack grid with parameters of 3 × 3 × 1 [40].

3. Results

3.1. Sensitivity

Achilles heel of the current generation of graphene-based photodetectors is their very poor sensitivity. This issue is addressed by adding organic ligands, which introduce band gap. Adjusting the created band gap leads to reduce the dark current, a parameter which plays an absolutely important role in determining sensitivity. The effect of almost all of the organic ligands, which are used to modify graphene sheet, is clear in the improvement of the sensitivity of fabricated photodetectors. A schematic of fabricated photodetector is represented in **Figure 1**.

As shown in **Table 1**, the measured sensitivity, which is calculated by $(R_d - R_i)/R_i$ where R_d and R_i are electrical resistance under dark and infrared light illumination condition, for graphene-based photodetector is 0.5 [38]. Whereas, the sensitivity of photodetectors based on graphene fabricated by Hwang and co-workers, in the highest applied voltage (0.1 V), is 0.5, which is in good agreement with the sensitivity of our fabricated unmodified graphene-based detector [17, 40]. The sensitivity has been boosted by using of back gate. Because of that the applied voltage excites the surface plasmon polariton. Therefore, one outcome of this excitation is enhancement of the sensitivity up to two orders of magnitude [40, 44]. This means the sensitivity can reach to 1. In this case, the sensitivity of the detector is three times less than our modified detector (this can be adjusted by kind of applied organic ligands). In addition, another method to improve sensitivity has been introduced as graphene/silicon-heterostructure [29]. In this condition, the highest sensitivity cannot reach to 1. Using quantum dots is the other way to achieve this goal. Quantum dots (which decorate the graphene sheet) as a zero-dimensional materials significantly affect the sensitivity. This factor can be changed with regard to the types of quantum dots [35]. That nanorods as one-dimensional materials have significant effect on improving the sensitivity of fabricated detectors is an undeniable effect [33].

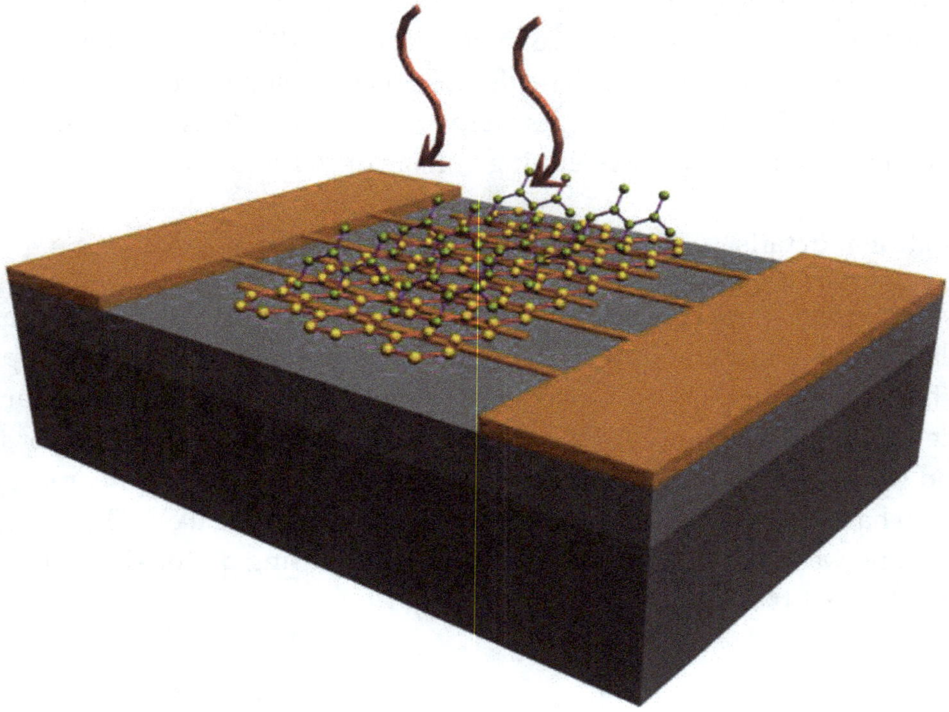

Figure 1. Scheme of fabricated photodetector [40].

Modification of graphene surface by organic ligands such as hydrazine, thiophene-2-carboxylic acid and low-dose thiosemicarbazide increases the sensitivity up to about 1. The sensitivity for photodetector modified by pyridine-4-carboxylic acid is 1.5, which is three times better. Finally, the best ligand in order to improve the sensitivity is high-dose thiosemicarbazide, which causes 500% improvement in the sensitivity compared with unmodified graphene [38–40].

Absorption spectrum of thiophene-2-caboxylic acid, thiophene-2-carbaldehyde and pyridine-4-carboxylic acid has been reported in reference [39]. One possible justification for the

Organic ligand	Sensitivity	Responsivity (A/W)	Rise time (ms)
Unmodified graphene	0.5	10	50
Hydrazine	0.9	14	35
Thiophene-2-carbaldehyde	0.092	0.1	40
Thiophene-2-carboxylic acid	0.9	20	---
Pyridine-4-carboxylic acid	1.5	2.5	38
High-dose thiosemicarbazide	3	9	18
Low-dose thiosemicarbazide	1	10	40

Table 1. Sensitivity, responsivity and rise time of fabricated detectors.

considerable value of photoresponsivity of fabricated photodetector, which was modified with thiophene-2-carboxylic acid is that mentioned ligand has strong absorption in 3–5 μm (2000–3300 cm^{-1}) range [39]. Not only does graphene sheet absorb the IR light, tiophene-2-carboxylic acid ligand absorbs the IR light as well. Consequently, the responsivity of considered detector modified with thiophene-2-carboxylic acid is higher than other fabricated photodetectors [40]. Besides, the devices discussed were stable at room temperature for many days and represented some characteristics during repeated measurement [45].

3.2. I-V characteristic

Figure 2 shows the bias dependence of the I-V characteristic of modified and unmodified graphene-based fabricated photodetectors, which were recorded between 0 and 4 V with voltage steps of 0.1 V. As shown in **Figure 2**, the photo-responsivity ascends with increasing the bias voltage. As stated in **Table 1**, the photo-responsivity value of 10 A/W is observed for detector fabricated by unmodified graphene [39]. Thiosemicarbazide organic ligands do not have remarkable effect on photo-responsivity [40]. The responsivity of fabricated detector is augemented 40% (14A/W) by using hydrazine ligands [38]. By applying thiophene-2-carboxylic acid ligand, one can significantly raise the parameter of photodetector up to 20 A/W [39]. On the other hand, some organic ligands can decrease the mentioned parameter up to 0.1 A/W [39].

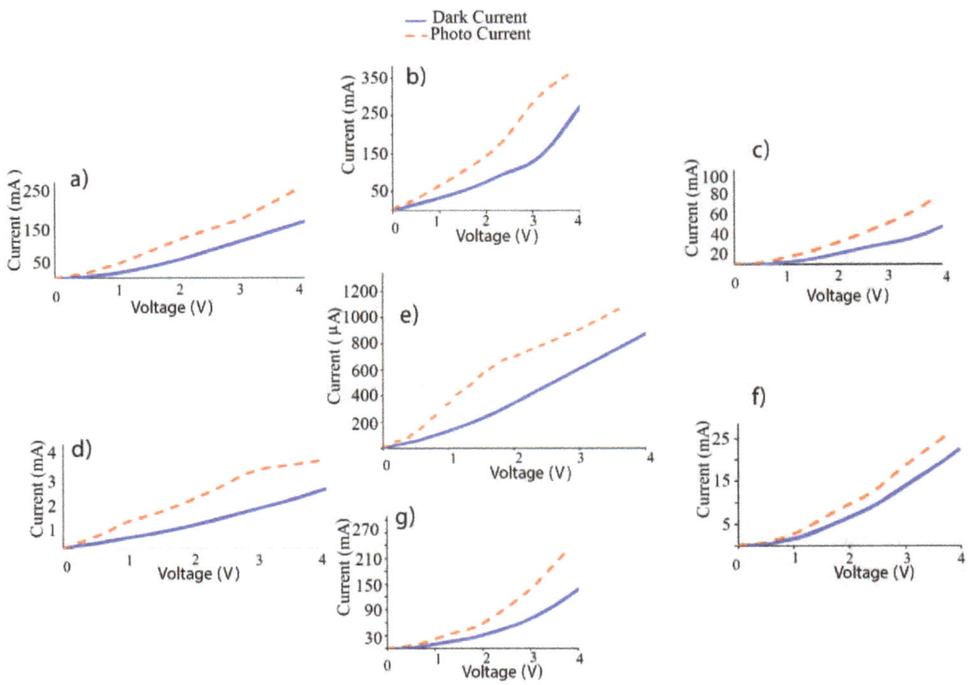

Figure 2. Current-voltage characteristics of fabricated detector based on (a) unmodified grapheme, (b) modified by hydrazine, (c) modified by mercapto acetic acid, (d) modified by low dose thiosemicarbazide, (e) modified by high-dose thiosemicarbazide, (f) modified by thiophene-2-carbaldehyde and (g) modified by thiophene-2-carboxylic acid under illumination of 3–5 μm infrared light [38–40].

3.3. Response time

Figure 3 shows the response time of photodetectors under illumination of IR lamp. The rise time of photodetector based on unmodified graphene is 50 ms [39]. Except fabricated detector based on graphene modified with thiophene-2-carboxylic acid which was too slow to record rise time, all suggested photodetectors based on modified graphene have similar rise time in comparison with photodetectors based on unmodified graphene [38–40]. Among all mentioned photodetectors, photodetector fabricated by graphene which was modified by high-dose thiosemicarbazide is the fastest with the rise time of 18 ms [40]. **Figure 3** shows the response time of investigated photodetectors. All measurements were performed under the illumination of IR lamp with power of 0.1 W in the wavelength range 3–5 μm at room temperature [38–40].

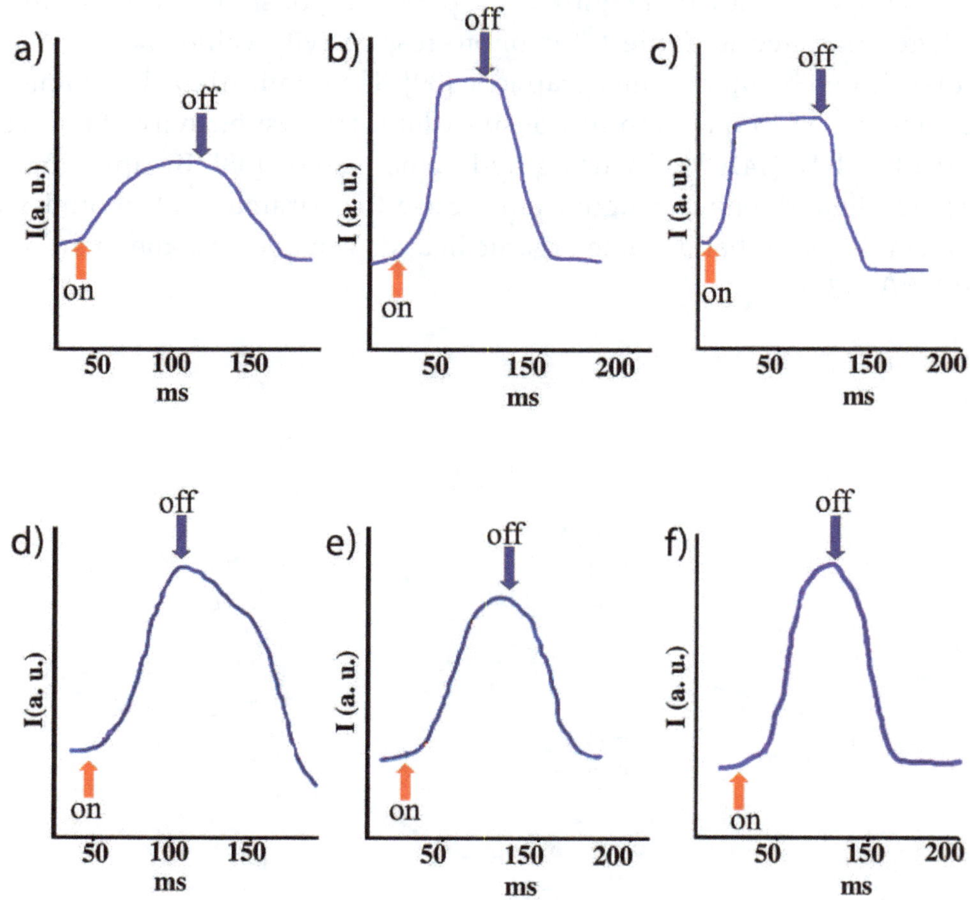

Figure 3. The response time of fabricated detector based on (a) unmodified graphene, (b) modified by low-dose thiosemicarbazide, (c) modified by high-dose thiosemicarbazide, (d) modified by thiophene-2-carbaldehyde, (e) modified by pyridine-4-carboxylic acid and (f) modified by hydrazine [38–40].

3.4. SEM

Figure 4 shows the SEM images of unmodified and modified graphene sheets. It is obvious, from the **Figure 4**, that the graphene sheets are bended in the course of modification process [40]. This phenomenon is in accordance with the calculated result [40]. We attribute the bending process of the graphene sheets to breaking of the translational symmetry of C-C sp^2 bond after the formation of C-H sp^3 bonds [45]. As shown in **Figure 5**, when hydrogen atoms attach to graphene, carbon atoms move out of plane. Therefore, the graphene sheets bend [45]. One can control the rate of bending of sheets not only by type but also by applying different dose of the organic ligands. As illustrated in **Figure 4**, the synthesized unmodified graphene exhibits almost long and uniform layer. The graphene modified by pyridine-4-carboxylic acid consists of nanosheets with their sizes in about micrometres. However, modification of the graphene using thiophene-2-carboxylic acid leads rolling of micro-sized sheets and forms the flake-like nanorods. One salient example of controlling of bending process via changing the dose of organic ligands is applying different dose of thiosemicarbazide so as to modify graphene sheet [40]. It is obvious that when the dose of thiosemicarbazide is low, sheets will be bended and nanobelts were synthesized. Sheets will be rolled completely, whenever high dose of thiosemicarbazide is applied [40].

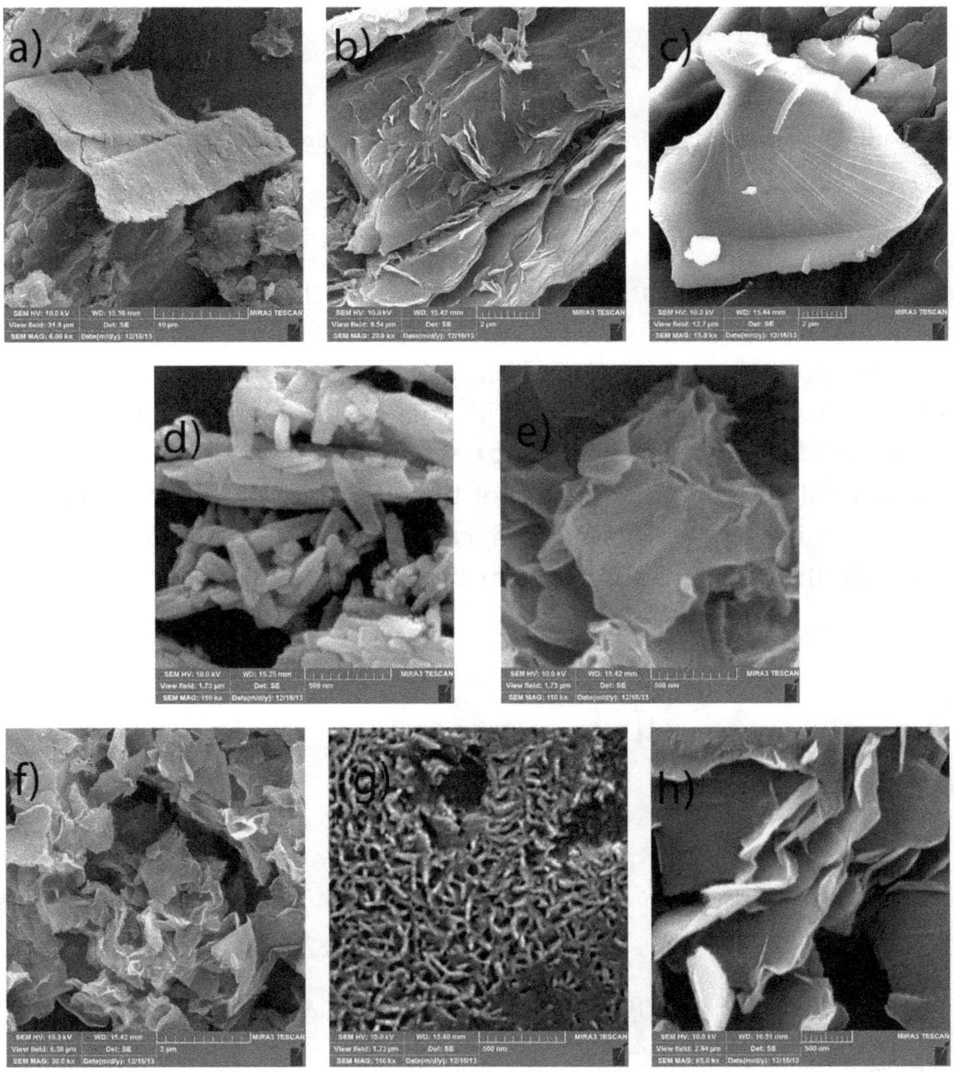

Figure 4. SEM images of (a) graphene and graphene modified by (b) pyridine-4-carboxylic acid, (c) thiophene-2-carbaldehyde, (d) thiophene-2-carboxylic acid, (e) hydrazine, (f) low-dose thiosemicarbazide, (g) high-dose thiosemicarbazide and (h) mercapto acetic acid [38–40].

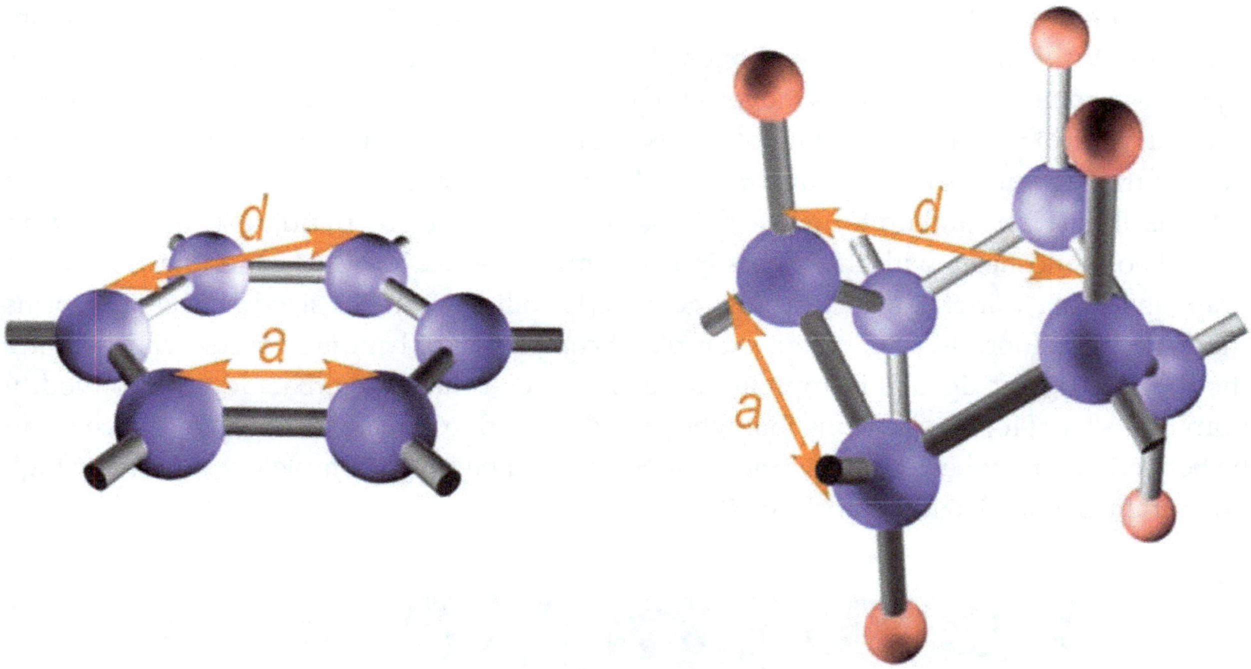

Figure 5. Scheme of out-of-plan distortion of graphene by hydrogenation process [45].

3.5. TEM

TEM images, which are shown in **Figure 6**, confirm this claim that the process of modification of graphene bended and finally rolled the sheets by revealing the detailed sub-structured information of the synthesized materials. **Figure 6(a)** shows unmodified graphene, which is almost uniform with some wrinkles. The TEM images of the graphene sheet modified low-dose thiosemicarbazide shown in **Figure 6 (b)**, which represent more wrinkles and bending in the sheet. **Figure 6(c)** is the TEM images of the high-dose-modified graphene sheet with thiosemicarbazide. It is obvious that thiosemicarbazide ligand completely rolled the sheet, and nanotubes with the diameter about 20 nm are achieved [40].

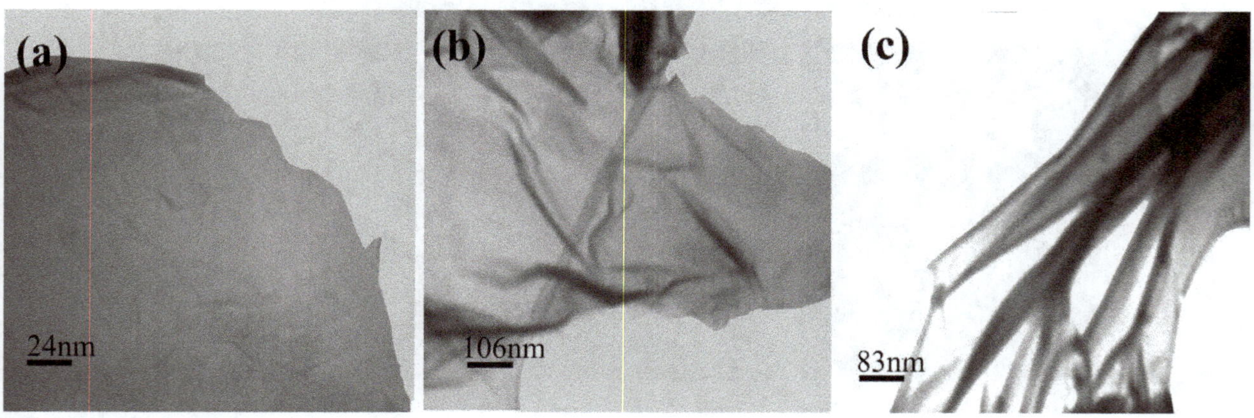

Figure 6. TEM image of (a) unmodified graphene, (b) modified by low-dose thiosemicarbazide and (c) modified by high-dose thiosemicarbazide [40].

3.6. AFM

Figure 7 shows the atomic force microscopy (AFM) topology images of the samples that confirm SEM and TEM results. The AFM images indicate that the organic ligands have well attached to the graphene sheet. With comparing of unmodified and modified graphene, one can conclude that the surface morphology of the graphene sheet become more and more uneven by applying organic ligands and using high concentration of them. Comparison of unmodified graphene with the reduced graphene oxide using high-dose hydrazine shows the height of wrinkles decreases [38]. This attributes to the removing of moisture trapped in the wrinkles in the course of chemical reduction of graphene oxide [40]. The surface of the nanorods, the result of modification of graphene with tiophene-2-carboxylic acid, which was seen in SEM images (see **Figure 4**), was also shown in AFM images [39].

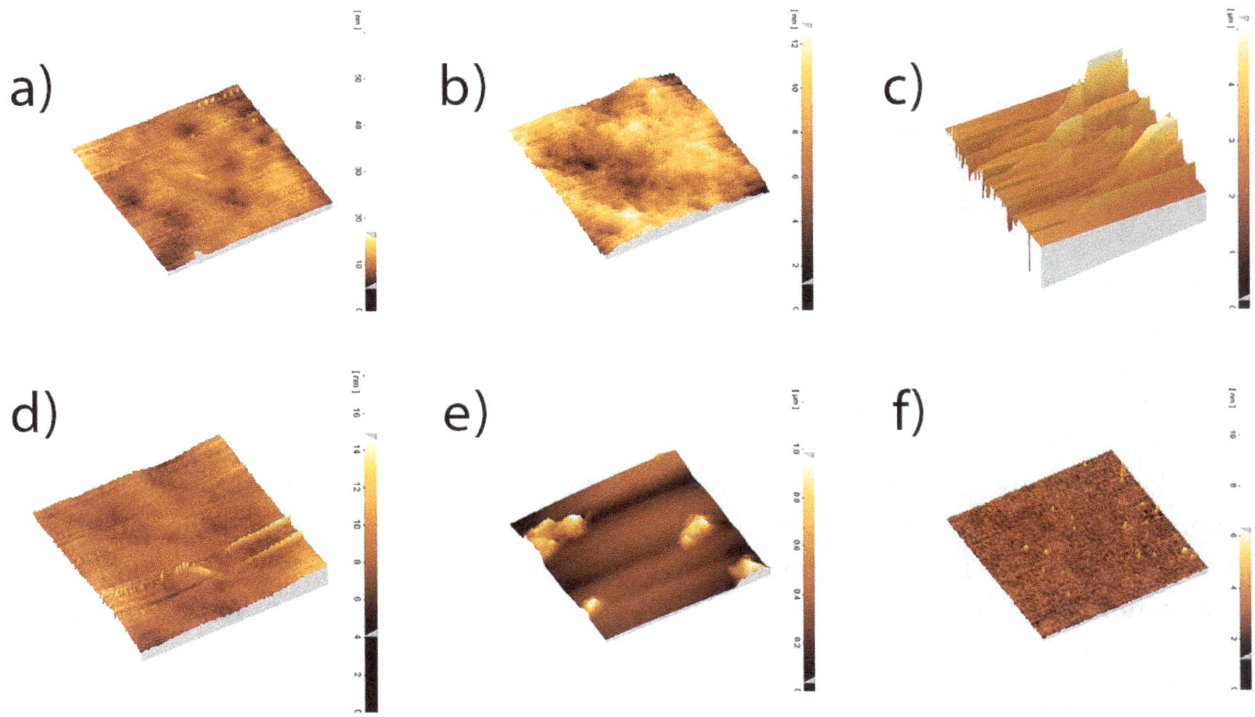

Figure 7. AFM image of (a) graphene (b) modified by low-dose thiosemicarbazide, (c) modified by high-dose thio semi carbazide, (d) modified by pyridine-4-carboxylic acid, (e) thiophene-2-carboxylic acid and (f) thiophene-2-carbaldehyde [38–40].

3.7. XRD

The powder X-ray diffraction (PXRD) patterns of the graphene oxide, unmodified and modified graphene are demonstrated in **Figure 8**. As illustrated in this figure, graphene oxide exhibits a strong and partly broad peak at $2\theta = 12.02°$ with an interlayer distance of 0.77 nm. Graphite flakes exhibit a strong and sharp peak at about $2\theta = 26°$, which indicate a higher order structure with an interlayer distance of 0.34 nm along the (002) orientation [46]. The comparison of XRD pattern of graphite with the graphene oxide reveals that the interlayer spacing of GO, which is about 0.77 nm, is higher than the interlayer space of the graphene

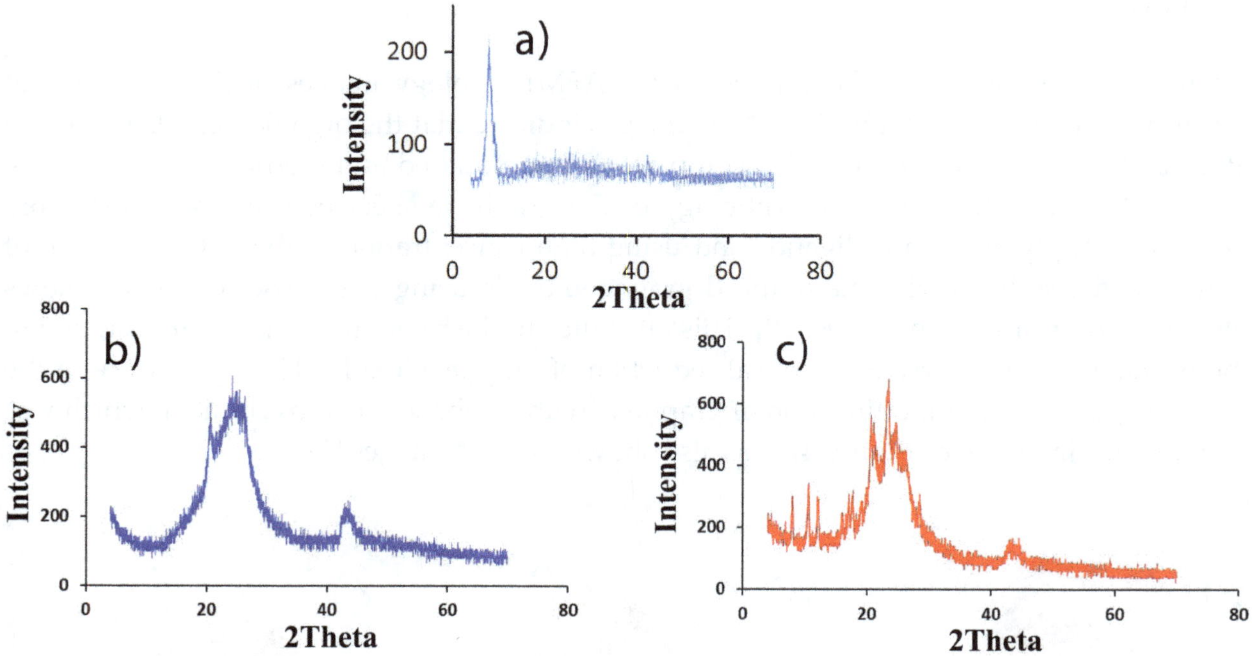

Figure 8. XRD patterns of (a) graphene oxide, (b) graphene and (c) graphene modified by thiosemicarbazide [38–40].

flakes because of the existence of oxygenated functional groups and intercalated water molecules in graphene oxide which are the result of chemical oxidation. In other words, graphene oxide is the graphene densely covered with hydroxyl and other groups [45]. By removing these groups from the surface of graphene sheet, graphene oxide XRD peak at $2\theta = 12°$ is diminished. This disappearance confirms the formation of graphene. The short range order in stacked stacks of graphene leads to the broadening of the characteristic diffraction peak of graphite. The XRD peak of graphene occurred at $2\theta = 24°$, which corresponds to an interlayer distance of 0.37 nm, is slightly shifted to the lower angles. The reason for this movement in XRD pattern stems from the remaining of trivial amount of residual oxygenated functional group or other structural defects [47, 48].

The XRD pattern of the modified graphene with thiosemicarbazide (see **Figure 8**) not only does show the slight shift to the left, but also indicates the new peaks appeared at lower angles. One possible rationale for this change is the creation of the new local thiosemicarbazide-graphene phase, which in turn could be an evidence for a structural chemical reaction between the free electrons of graphene and electrons of thiosemicarbazide to form a chemical bond, which results in the formation of new crystalline phase [40].

4. Simulation

Modifying surface of synthesized graphene with organic ligands dramatically increases the photosensitivity and significantly decreases the response time. In addition, shallow defects, an outgrowth of modification process, in electronic structure of graphene cause the improvement in responsivity compared to reported results [41]. Modification of graphene sheet with ligands is occurred when the surface atoms of graphene interact with hydrogen atoms of organic ligands [40]. The formation of the defect electron trapping centres is the outcome of the surface modification (see **Figures 9–12**), which leads to band gap opening in electronic surface of hybrid material.

Figure 9. Scheme of graphane, fully hydrogenated graphene [37].

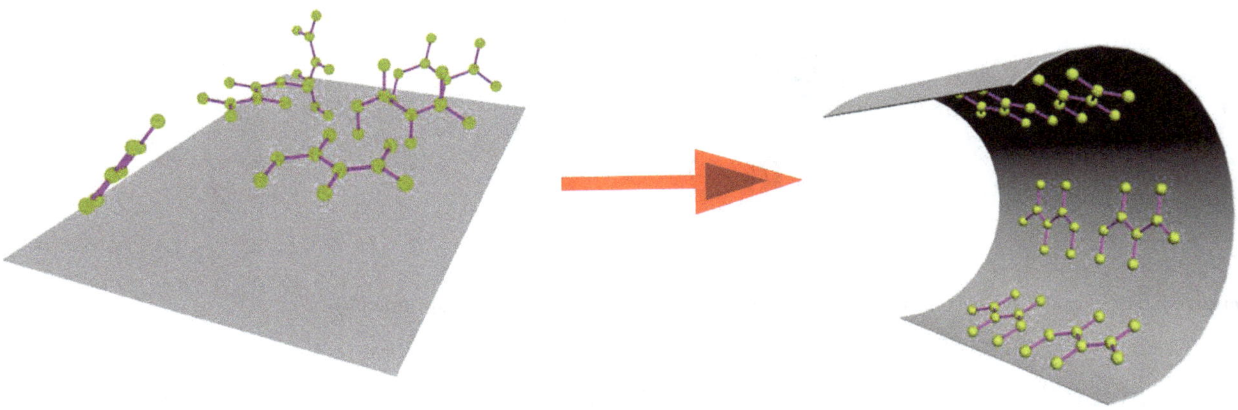

Figure 10. Schematic of rolling of graphene [40].

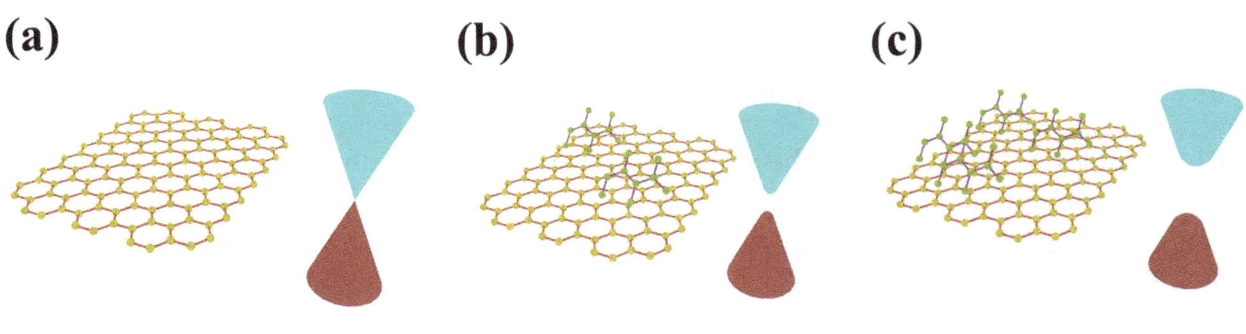

Figure 11. Schematic representation of band gap opening (a) unmodified graphene, (b) modified by low-dose thiosemicarbazide and (c) modified by high-dose thiosemicarbazide [41].

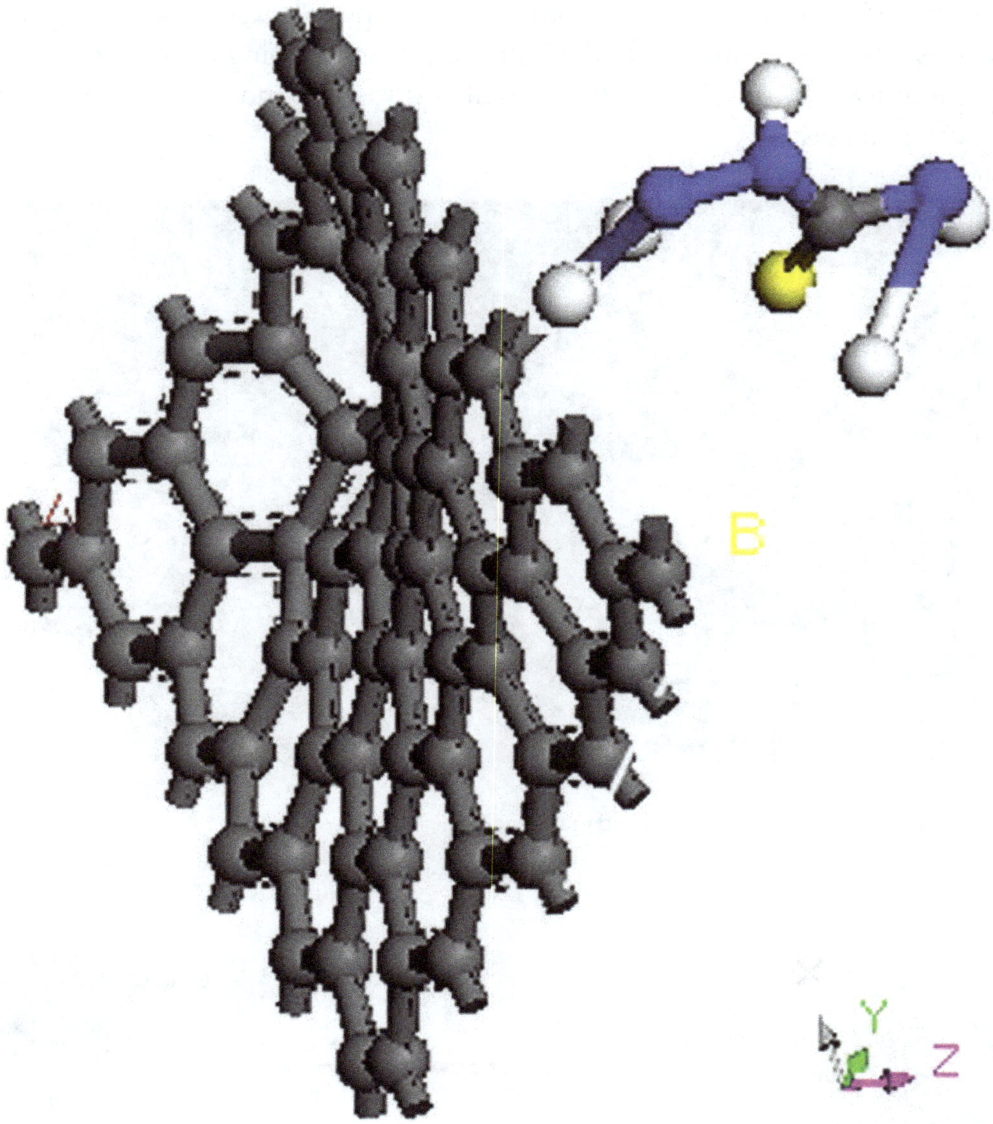

Figure 12. Schematic representation of band gap opening in the modified graphene [40].

DFT computations are executed to optimize the configuration of thiosemicarbazide adsorbed on graphene sheet [40]. The out-of-plane dislocation of graphene sheet, which is bonded to a thiosemicarbazide, is shown in **Figure 12**. As shown in this figure, the atomic dislocation of the graphene sheet is to be upward to hydrogen atoms of the organic ligand. The thiosemicarbazide is adhered to the graphene sheet via N-H aromatic π electron hydrogen bonding in which the thiosemicarbazide is a polar molecule [40].

DFT calculations indicate that the organic ligand can remarkably change the electronic state of graphene. Band length is changed due to the fact that hybridization of graphene change from sp^2 to sp^3 by attaching the hydrogen atoms to graphene sheet, which in turn leads to form the band gap. **Figure 13** shows the synthesized material behaves as a semiconductor with the band gap of 0.19 eV [40]. The significant part of the enhancement of the photo-sensitivity of graphene-based photodetector is a band gap creation in the band diagram of it.

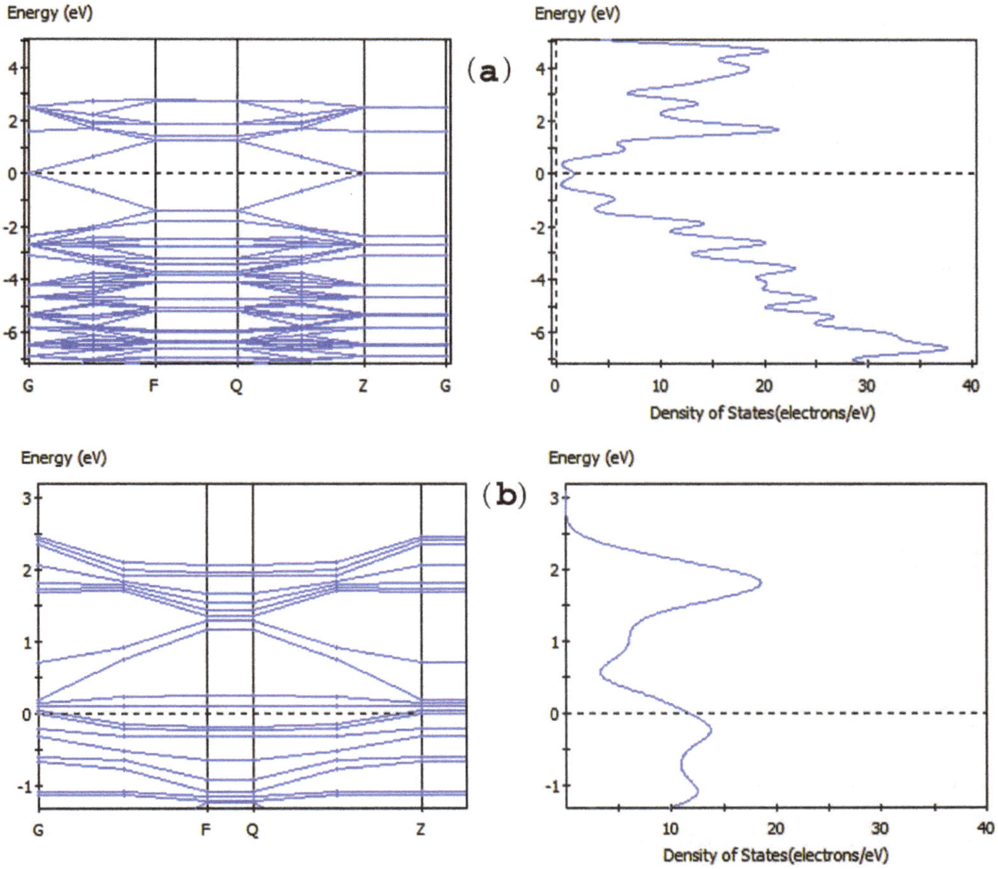

Figure 13. Calculated band structure and density of states for (a) graphene and (b) graphene modified by thiosemicarbazide [40].

5. Conclusion

In this work, graphene was efficiently prepared. The synthesized graphene was modified using different organic ligands. The chemical and structural properties of synthesized materials were investigated. Moreover, metal-graphene-metal (MGM) mid-IR photodetectors, in which the synthesized, modified and modified graphene used as an active layer, were fabricated and analysed at room temperature. The organic ligand used to modify graphene sheet can create and tune the band gap. Opening of the band gap overcomes the main weakness of graphene-based photodetectors such as high dark current, low sensitivity, repeatability and the effect of surrounding gases. Also, fast response time of fabricated photodetector reported that graphene-based photodetectors are the other benefits of our proposed structure.

Acronyms

SEM Scanning electron microscope

AFM Atomic force microscope

PXRD Powder X-ray diffraction

TEM Transmission electron microscope

MGM Metal-graphene-metal

IR Infrared

DOS Density of states

LDA Local density approximation

Ec Cut-off energy

GO Graphite oxide

Author details

Mehrdad Siahsar[1], Mahboubeh Dolatyari[2*], Ali Rostami[1,2] and Ghasem Rostami[2]

*Address all correspondence to: m.dolatyari@gmail.com

1 OIC Research Group, University of Tabriz, Tabriz, Iran

2 SP-EPT Labs, ASEPE Company, Industrial Park of Advanced Technologies, Tabriz, Iran

References

[1] K. S. Novoselov, A. K. Geim, S. V. Morozor, D. Jiang, Y. Zhang, S. V. Dubonos, I. V. Grigorieva, and A. A. Firsov. Electric Field Effect in Atomically Thin Carbon Films. Science. 2004;**306**:666–669.

[2] A. C. Ferrari, J. C. Meyer, V. Scardaci, C. Casiraghi, M. Lazzeri, F. Mauri, S. Piscanec, D. Jiang, K. S. Novoselov, and S. Roth. Raman Spectrum of Graphene and Graphene Layers. Phys. Rev. Lett. 2006;**97**(18):187401-1–187401-4.

[3] C. Popovici, C. S. Fischer, and L. von Smekal. Fermi Velocity Renormalization and Dynamical Gap Generation in Graphene. Phys. Rev. B. 2013;**88**(20):205429-1–205429-9.

[4] K. S. Novoselov, A. K. Geim, S. V. Morozov, D. Jiang, M. I. Katsnelson, I. V. Grigorieva, S. V. Dubonos, and A. A. Firsov. Two-Dimensional Gas of Massless Dirac Fermions in Graphene. Nature. 2005;**438**:197–200.

[5] Y. Liu, N. O. Weiss, X. Duan, H. Cheng, Y. Huang, and X. Duan. Van der Waals Heterostructures and Devices. Nat. Rev. Mater. 2016;**1**(9):16042.

[6] A. Varykhalov, J. Sãnchez-Barriga, A. Shikin, C. Biswas, E. Vescovo, A. Rybkin, D. Marchenko, and O. Rader. Electronic and Magnetic Properties of Quasifreestanding Graphene on Ni. Phys. Rev. Lett. 2008;**101**(15):157601-1–157601-4.

[7] Y. Wang, S. W. Tong, X. F. Xu, B. Özyilmaz, and K. P. Loh. Interface Engineering of Layer-by-Layer Stacked Graphene Anodes for High-Performance Organic Solar Cells. Adv. Mater. 2011;**23**(13):1514–1518.

[8] C. Soldano, A. Stefani, V. Biondo, L. Basiricò, G. Turatti, G. Generali, L. Ortolani, V. Morandi, G. P. Veronese, R. Rizzoli, R. Capelli, and M. Muccini. ITO-Free Organic Light-Emitting Transistors with Graphene Gate Electrode. ACS Photonics. 2014;**1**(10):1082–1088.

[9] D. Schall, D. Neumaier, M. Mohsin, B. Chmielak, J. Bolten, C. Porschatis, A. Prinzen, C. Matheisen, W. Kuebart, B. Junginger, W. Templ, A. L. Giesecke, and H. Kurz. 50 GBit/s Photodetectors Based on Wafer-Scale Graphene for Integrated Silicon Photonic Communication Systems. ACS Photonics. 2014;**1**(9):781–784.

[10] A. K. Geim and K. S. Novoselov. The Rise of Graphene. Nat. Mater. 2007;**6**(3):183–191.

[11] Y. Zhang, T. Liu, B. Meng, X. Li, G. Liang, X. Hu, and Q. J. Wang. Broadband High Photoresponse from Pure Monolayer Graphene Photodetector. Nat. Commun. 2013;**4**: 1811-1–1811-11.

[12] R. Zhang and R. Cheung. Mechanical Properties and Applications of Two-Dimensional Materials. In: Dr. Pramoda Nayak , editor. Two-dimensional Materials - Synthesis, Characterization and Potential Applications. Rijeka, Crotia: InTech; 2016. p. 219–246. DOI: 10.5772/64017

[13] A. M. Alexeev, M. D. Barnes, V. K. Nagareddy, M. F. Craciun, and C. D. Wright. A Simple Process for the Fabrication of Large-Area CVD Graphene Based Devices via Selective in Situ Functionalization and Patterning. 2D Mater. 2016;**4**(1):011010.

[14] Y. J. Kim, Y. Kim, K. Novoselov, and B. H. Hong. Engineering Electrical Properties of Graphene: Chemical Approaches. 2D Mater. 2015;**2**(4):042001.

[15] R. Vargas-Bernal. State-of-the-Art Electronic Devices Based on Graphene. In: Dr. Abhijit Kar, editor. Nanoelectronics and Materials Development. Rijeka, Crotia: InTech; 2016. p. 1–21. DOI: 10.5772/64320

[16] Y. D. Kim and M.-H. Bae. Light Emission from Graphene. In: Dr. Adrián Silva, editor. Advances in Carbon Nanostructures. Rijeka, Crotia: InTech; 2016. p. 83-100. DOI: 10.5772/64051

[17] G. Hwang, J. C. Acosta, E. Vela, S. Haliyo, and S. Régnier. Graphene as thin film infrared optoelectronic sensor. In: International Symposium on Optomechatronic Technology (ISOT); 2009; Istanbul, Turkey. IEEE; 2009. p. 169-174.

[18] G. Koley, A. Singh, and A. Uddin. Graphene-Based Sensors: Current Status and Future Trends. In: M. Aliofkhazraei, N. Ali, W. I. Milne, C. S. Ozkan, S. Mitura, and J. L. Gervasoni, editors. Graphene Science Handbook. Florida, USA: CRC Press; 2016. p. 211–234.

[19] Y.-W. Tan, Y. Zhang, K. Bolotin, Y. Zhao, S. Adam, E. H. Hwang, S. Das Sarma, H. L. Stormer, and P. Kim. Measurement of Scattering Rate and Minimum Conductivity in Graphene. Phys. Rev. Lett. 2007;**99**(24):246803-1–246803-4.

[20] M. Y. Han, B. Özyilmaz, Y. Zhang, and P. Kim. Energy Band-Gap Engineering of Graphene Nanoribbons. Phys. Rev. Lett. 2007;**98**(20):206805-1–206805-4.

[21] G. Guan, J. Lu, and H. Jiang. Preparation, Characterization, and Physical Properties of Graphene Nanosheets and Films Obtained from Low-Temperature Expandable Graphite. J. Mater. Sci. 2016;**51**(2):926–936.

[22] A. Celis, M. N. Nair, A. Taleb-Ibrahimi, E. H. Conrad, C. Berger, W. A. de Heer, and A. Tejeda. Graphene Nanoribbons: Fabrication, Properties and Devices. J. Phys. D: Appl. Phys. 2016;**49**(14):143001.

[23] Y. Zhang, T. Tang, C. Girit, Z. Hao, M. C. Martin, A. Zettl, M. F. Crommie, Y. R. Shen, and F. Wang. Direct Observation of a Widely Tunable Bandgap in Bilayer Graphene. Nature. 2009;**459**(7248):820–823.

[24] S. K. Jain, V. Juričić, and G. T. Barkema. Structure of Twisted and Buckled Bilayer Graphene. 2D Mater. 2016;**4**(1):015018.

[25] J. Zhou, M. M. Wu, X. Zhou, and Q. Sun. Tuning Electronic and Magnetic Properties of Graphene by Surface Modification. Appl. Phys. Lett. 2009;**95**(10):103108-1–103108-3.

[26] A. Nazari, R. Faez, and H. Shamloo. Modeling Comparison of Graphene Nanoribbon Field Effect Transistors with Single Vacancy Defect. Superlattices Microstruct. 2016;**97**:28–45.

[27] H. Li, Y. Zhou, and J. Dong. First-Principles Study of the Electron Transport Properties of Graphene-Like 2D Materials. In: A. Kar, editor. Nanoelectronics and Materials Development. InTech; 2016; p. 117–139. DOI: 10.5772/64109

[28] D. De Fazio, I. Goykhman, M. Bruna, A. Eiden, S. Milana, D. Yoon, U. Sassi, M. Barbone, D. Dumcenco, K. Marinov, and A. Kis. High Responsivity, Large-Area Graphene/MoS2 Flexible Photodetectors. ACS Nano. 2016;**10**(9):8252–8262.

[29] X. Wang, Z. Cheng, K. Xu, H. K. Tsang, and J. B. Xu. High-Responsivity Graphene/Silicon-Heterostructure Waveguide Photodetectors. Nat. Photonics. 2013;**7**(11):888–891.

[30] N. Jain, and B. Yu. Graphene-Enabled Heterostructures: Role in Future-Generation Carbon Electronics. In: M. Aliofkhazraei, N. Ali, W. I. Milne, C. S. Ozkan, S. Mitura, and J. L. Gervasoni, editors. Graphene Science Handbook. Florida, USA: CRC Press; 2016. p. 423–434.

[31] K. S. Novoselov, A. Mishchenko, A. Carvalho, and A. H. Castro Neto. 2D materials and van der Waals heterostructures. Science. 2016; **353**(6298): aac9439.

[32] D. Jariwala, T. J. Marks, and M. C. Hersam. Mixed-dimensional van der Waals heterostructures. Nat. Mater. 2017; 16: 170–181.

[33] H. Talebi, M. Dolatyari, G. Rostami, A. Manzuri, M. Mahmudi, and A. Rostami. Fabrication of Fast Mid-Infrared Range Photodetector Based on Hybrid Graphene-PbSe Nanorods. Appl. Opt. 2015;**54**(20):6386–6390.

[34] J. J. Navarro, S. Leret, F. Calleja, D. Stradi, A. Black, R. Bernardo-Gavito, M. Garnica, D. Granados, A. L. Vázquez de Parga, E. M. Pérez, and R. Miranda. Organic Covalent Patterning of Nanostructured Graphene with Selectivity at the Atomic Level. Nano Lett. 2016;**16**(1):355–361.

[35] G. Konstantatos, M. Badioli, L. Gaudreau, J. Osmond, M. Bernechea, F. P. Garcia de Arquer, F. Gatti, F. H. L. Koppens. Hybrid Graphene-Quantum Dot Phototransistors with Ultrahigh Gain. Nat. Nanotechnol. 2012;**7**:363–368.

[36] S. Mitra, S. Banerjee, A. Datta, and D. Chakravorty. A Brief Review on Graphene/Inorganic Nanostructure Composites: Materials for the Future. Indian J. Phys. 2016;**90**(9):1019–1032.

[37] J. O. Sofo, A. S. Chaudhari, and G. D. Barber. Graphane: A Two-Dimensional Hydrocarbon. Phys. Rev. B. 2007;**75**(15):153401.

[38] F. Jabbarzadeh, M. Siahsar, M. Dolatyari, G. Rostami, and A. Rostami. Modification of Graphene Oxide for Applying as Mid-Infrared Photodetector. Appl. Phys. B. 2015;**120**:637–643.

[39] F. Jabbarzadeh, M. Siahsar, M. Dolatyari, G. Rostami, and A. Rostami. Fabrication of New Mid-Infrared Photodetectors Based on Graphene Modified by Organic Molecules. IEEE Sens. J. 2015;**15**:2795–2800.

[40] M. Siahsar, F. Jabbarzadeh, M. Dolatyari, G. Rostami, and A. Rostami. Fabrication of High Sensitive and Fast Response MIR Photodetector Based on a New Hybrid Graphene Structure. Sens. Actuators A. 2016;**238**:150–157.

[41] W. J. Yu, Z. Li, H. Zhou, Y. Chen, Y. Wang, Y. Huang, and X. Duan. Vertically Stacked Multi-Heterostructures of Layered Materials for Logic Transistors and Complementary Inverters. Nat. Mater. 2012;**12**(3):246–252.

[42] Y. J. Kim, Y. H. Kahng, N. Kim, J. H. Lee, Y. H. Hwang, S. M. Lee, S. M. Choi, W. B. Kim, and K. Lee. Impact of Synthesis Routes on the Chemical, Optical, and Electrical Properties of Graphene Oxides and Its Derivatives. Curr. Appl. Phys. 2015;**15**(11):1435–1444.

[43] S. Bykkam, K. V. Rao, C. H. S. Chakra, and T. Thunugunta. Synthesis and Characterization of Graphene Oxide and Its Antimicrobial Activity Against Klebseilla and Staphylococus. Int. J. Adv. Biotechnol. Res. 2013;**4**(1):142–146.

[44] Y. Liu, R. Cheng, L. Liao, H. Zhou, J. Bai, G. Liu, L. Liu, Y. Huang, and X. Duan. Plasmon Resonance Enhanced Multicolour Photodetection by Graphene. Nat. Commun. 2011;**2**:579.

[45] D. C. Elias, R. R. Nair, T. M. G. Mohiuddin, S. V. Morozov, P. Blake, M. P. Halsall, A. C. Ferrari, D. W. Boukhvalov, M. I. Katsnelson, A. K. Geim, and K. S. Novoselov. Control of Graphene's Properties by Reversible Hydrogenation: Evidence for Graphane. Science. 2009;**323**(5914):610–613.

[46] S. Thakur and N. Karak. Green Reduction of Graphene Oxide by Aqueous Phytoextracts. Carbon. 2012;**50**(14):5331–5339.

[47] S. Park, J. An, I. Jung, R. D. Piner, S. J. An, X. Li, A. Velamakanni, and R. S. Ruoff. Colloidal Suspensions of Highly Reduced Graphene Oxide in a Wide Variety of Organic Solvents. Nano Lett. 2009;**9**(4):1593–1597.

[48] D. Li, M. B. Müller, S. Gilje, R. B. Kaner, and G. G. Wallace. Processable Aqueous Dispersions of Graphene Nanosheets. Nat. Nanotechnol. 2008;**3**:101–105.

9

Solution-Processed Graphene-Based Transparent Conductive Electrodes as Ideal ITO Alternatives for Organic Solar Cells

Minas M. Stylianakis, Dimitrios Konios,
Konstantinos Petridis and Emmanuel Kymakis

Abstract

The isolation of free-standing graphene in 2004 was the spark for a new scientific revolution in the field of optoelectronics. Due to its extraordinary optoelectronic and mechanical properties, graphene is the next wonder material that could act as an ideal low-cost alternative material for the effective replacement of the expensive conventional materials used in organic optoelectronic applications. Indeed, the enhanced electrical conductivity of graphene combined with its high transparency in visible and near-infrared spectra, enabled graphene to be an ideal low-cost indium tin oxide (ITO) alternative in organic solar cells (OSCs). The prospects and future research trend in graphene-based TCE are also discussed. On the other hand, solution-processed graphene combines the unique optoelectrical properties of graphene with large area deposition and flexible substrates making it compatible with printing and coating technologies, such as roll-to-roll, inkjet, gravure, and flexographic printing manufacturing methods. This chapter provides an overview of the most recent research progress in the application of solution-processed graphene-based films as transparent conductive electrodes (TCEs) in OSCs. (a) Chemically converted graphene (CCG), (b) thermally and photochemically reduced graphene oxide, (c) composite reduced graphene oxide-carbon nanotubes, and (d) reduced graphene oxide mesh films have demonstrated their applicability in OSCs as transparent, conductive electrodes.

Keywords: organic solar cells, transparent electrodes, graphene, reduction

1. Introduction

Organic solar cells (OSCs) based on two-dimensional (2D) nanomaterials, including graphene, transition metal dichalcogenides (TMDs), and Xenes (silicene, germanene, stanene, etc.) have experienced immense interest as possible candidates for clean energy generation offer the benefit of low-cost, light-weight, large-area, high mechanical flexibility, and low temperature/ high throughput manufacturing processes [1–3] offering important advantages over silicon technology. A key task toward the implementation of this technology is the adaptability and/ or the development of the materials used for their fabrication to further optimize their overall performance [4–12].

Transparent conductive electrodes (TCEs) with an outstanding combination of high electrical conductivity and good optical transparency have a crucial role in OSCs, as the first layer of the device that the incident solar electromagnetic radiation should transmit in order to be absorbed by the solar cell's active medium [13–15]. Conventionally, the most common materials are indium tin oxide (ITO) and fluorine tin oxide (FTO), which are widely used as the transparent electrode in many optoelectronic devices because of their good combination of high transparency and low resistance [16–18]. However, these metal oxides exhibit several issues due to the high cost, resulting mostly from the indium scarcity [19], their brittleness [20], the device degradation due to indium diffusion into the photoactive layer [21], and the requirement for high cost coating methods [20]. Significant effort toward the search of alternatives has therefore been motivated [15, 22–24] and several alternative materials have been proposed, including metallic nanowires (NWs), nanostructured carbon, and conductive polymers, among others [24, 25]. Therefore, developing new materials combining most desirable properties for transparent electrodes will contribute to satisfy the increasing demand for low-cost solution-processed flexible devices. Due to its exceptional properties, graphene is highly attractive and is believed to be the next wonder material for optoelectronics; thus, triggering its application as a transparent electrode for flexible energy-harvesting devices [26].

Graphene, a two-dimensional single-atom-thick (0.34 nm) carbon honeycomb, corresponds to the interlayer spacing of graphite [27]. It displays unique thermal conductivity ($\approx 5.0 \times 103$ W mK^{-1}) [28], superior mechanical strength (remarkable flexibility elastic modulus ~1 TPa), and outstanding chemical stability, especially high electron mobility (>15,000 cm^2 Vs^{-1}) [29], which overcomes the intrinsic performance limitations of traditional transparent electrode materials, as well as low sheet resistance (35 Ω sq^{-1} at 90% optical transmittance) [30], high optical transmittance (≈97.7% for single-layer graphene) [31], and good piezoresistive sensitivity [32].

Graphene production includes several methods. Depending on its quality, the properties of the synthesized graphene may differ. More specifically, the grain size, the shape, the thickness as well as the presence of defects influence graphene mechanical properties. From a highly ordered pyrolytic graphite (HOPG) through a mechanical exfoliation method ("scotch-tape"), originally graphene was produced [33]. Since then, the number of methods for graphene production has significantly increased. Some of them include liquid phase exfoliation (LPE) [34], chemical vapor deposition (CVD) [35], preparation from organic materials using solvothermal method [36], chemical cross-linking of polycyclic aromatic hydrocarbons [37], preparation from SiC through thermal decomposition [38], and carbon nanotube unzipping [39]. Although CVD method produces the less defective graphene films, it exhibits technical issues relating with the deposition of graphene films on flexible substrates making this method incompatible with roll-to-roll (R2R) mass production processes. On the contrary, the fact that graphene oxide (GO) can be produced in large quantities from low-cost graphite powder and its property to give stable dispersions in various solvents, make GO an ideal candidate for solution-processed graphene production [40]. On top of that, it can be the precursor for conductive inks production, adding to the printing electronics technology [27]. However, the introduction of oxygen functional groups on the GO lattice, disrupts the sp^2 conjugation system of the hexagonal graphene lattice, making GO an insulator. By removing the oxygen containing groups, GO can be partially reduced to conductive graphene-like sheets. The reduction process can be performed by chemical [41], thermal [42], or photochemical treatment [43], aiming to yield reduced graphene oxide (rGO) with similar properties to graphene [41].

This chapter covers the latest advances in solution-processed graphene-based thin films as the anode TCE in OSCs, substituting the conventional ITO. We highlight the latest advances on thermally, chemically and photochemically produced conductive graphene-based nanomaterials, as well as on graphene-based nanocomposites TCE films. In addition, we summarize some promising routes for the graphene-based TCEs treatment that advance their optoelectrical properties tailoring and achieve a balance between the sheet resistance (R_s) and transmittance of the solution-processed graphene-based TCEs. ITO replacement in graphene-based TCEs, establishes the era of lightweight, low cost, extended lifetime, and stability, as well as flexible and stretchable OSCs.

2. Applications of graphene-based TCE in OSCs

Numerous studies have aimed to investigate the key role of graphene-based materials as the anode (positive electrode) in OSCs. A TCE with efficient carrier transport and thus, high charge collection efficiency is required in OSC devices as outlined in **Figure 1**. Properties including high transparency (>80%), low sheet resistance ($R_s < 100$/sq), suitable work function (4.5–5.2 eV) as well as low cost and compatibility with R2R fabrication processes characterize the ideal TCE material. Despite the unique optoelectrical properties ($T > 90\%$, $R_s = 10–20\ \Omega$/sq) of ITO and its ideal work function, the abovementioned ITO disadvantages have intensified the research on graphene-based TCEs in OSCs. In this context, solution-processed graphene-based films meeting the standards for low electrical resistance, high optical transparency and tunable WF must be developed.

2.1. Chemically treated rGO transparent conductive electrodes

In 2008, Wu et al. compared the electrical properties of rGO films prepared according to two different reduction methods: chemical and thermal treatment [44]. As pristine material, graphene oxide (GO) prepared by Hummers' method was used. The conductivity properties of GO reduced by vacuum annealing at 1100°C, by a combination of hydrazine (N_2H_4) treatment,

Figure 1. The graphene-based TCE as the anode in a typical flexible OSC.

and Ar annealing at 400°C were explored. The results displayed that the vacuum annealing reduction resulted in rGO films with slightly better transparency and conductivity compared to the films reduced using the chemical and thermal treatment combination. In both cases, the film surfaces were free from spikes that can cause short circuit in optoelectronic devices. This is an advantage of graphene-based electrodes over carbon nanotube or metal nanowire mesh electrodes, which require a thick, spin-coated polymer buffer layer in order to prevent shorts, generated by rough surfaces (spikes). The transmittance and R_s with respect to the rGO film thickness with the two different fabrication methods are depicted in **Figure 2**. In a general term, for <20 nm film thickness, the optical transmittance was 480%, while the R_s varied from 5 to 1 MΩ sq^{-1}. Bilayer small molecule OSCs were fabricated on rGO/quartz and on ITO/glass substrates. The fabricated cells of the structure rGO/copper phthalocyanine (CuPc)/fullerene (C_{60})/bathocuproine (BCP)/Ag achieved a power conversion efficiency (PCE) of ~0.4%, whereas the ITO-based devices exhibited a PCE of ~0.84%. The observed PCE difference is attributed to the higher R_s of graphene-based electrodes compared to the ITO electrodes.

Yun et al. investigated the potential use of a solution-processed rGO thin film as transparent electrodes and characterized the effect of rGO films by tuning the film thickness and annealing treatment on the cell performances in OSCs [45]. Using a p-toluenesulfonyl hydrazide (p-$TosNHNH_2$) reducing agent, GO reduction took place. rGO dispersions with high concentrations and thin film processability were prepared as a result of the hydrazone groups attached to the edge and basal plane of graphene. In this way, issues created using other film preparation techniques (i.e., transfer problems during vacuum filtration) or high temperature reduction methods could be overwhelmed [46, 47]. They also evaluated the morphological and optical properties of the solution-processed rGO thin films by tuning the fabrication conditions, such

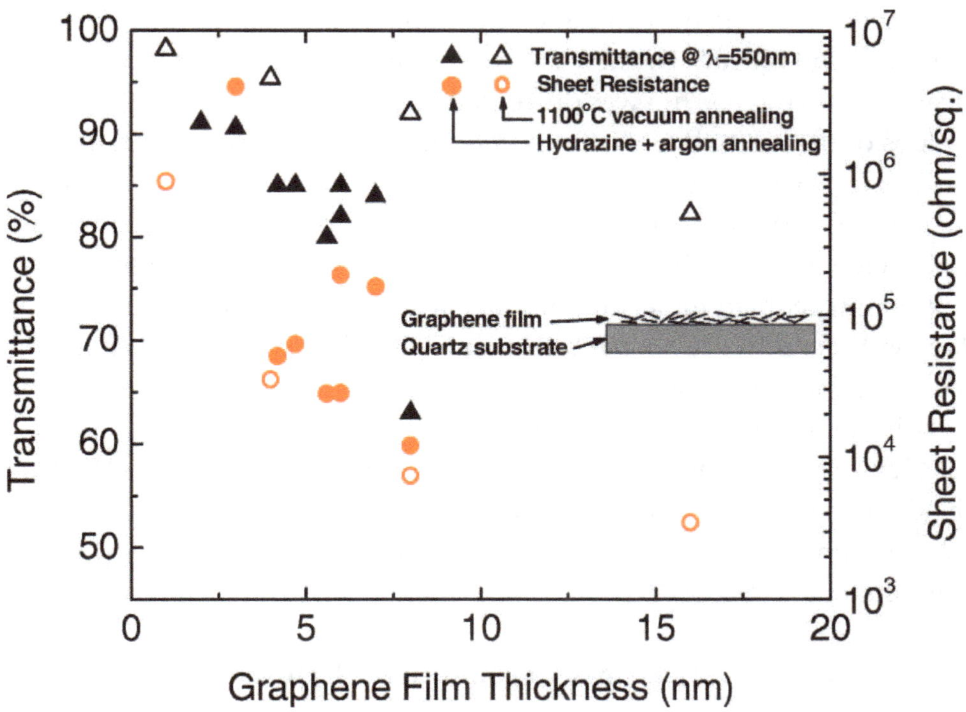

Figure 2. Transmittance at $\lambda = 550$ nm (triangles) and sheet resistance (circles) as a function of the graphene film thickness for both reduction methods: vacuum annealing at 1100°C (open symbols) and hydrazine treatment and Ar annealing at 400°C (filled symbols). Reprinted with permission from Wu et al. [44].

as spin-coating cycles and annealing temperatures. All rGO thin films fabricated through a spinning method showed a highly uniform morphology with thickness controllability. In addition, sheet resistance was efficiently decreased from ~103 to 10 kΩ sq^{-1} by the trade-off between the coating cycles and the annealing conditions. Finally, the effects of rGO-based electrodes on the OSC performances by controlling the number of coating cycles were studied. OSCs performance was improved upon decreasing the sheet resistance, for 200°C as the annealing temperature and seven times coated rGO film, achieving a PCE of 0.33%.

In 2015, Moaven et al. reported the synthesis of an rGO/Ag nanocomposite electrode appropriate for OSCs [48]. The reported technique is applicable to flexible substrates by spin coating of an aqueous solution of rGO/Ag nanocomposite on polyethylene terephthalate (PET) substrate in ambient conditions. The optical and electrical properties were determined by tuning the Ag concentration of the rGO/Ag nanocomposite and the electrode thickness. Flexible electrodes were prepared with R_s as low as 83 kΩ sq^{-1} and with a transmittance of 47%. Finally, OSCs onto PET, based on rGO/Ag anode electrode in different ratios were fabricated. The highest PCE observed was 0.24% onto PET flexible substrates which was improved by 25% compared to the conventional ITO/glass (0.18%).

Huang et al. in another article, demonstrated the preparation of highly conductive and transparent graphene-based electrodes with tunable WFs by mixing single walled carbon nanotubes (SWCNTs) with chemically reduced GO [49]. More specifically, dry powders of GO and SWCNTs were directly dispersed in anhydrous hydrazine. The yielded composite was coated in different thicknesses (1–5 layers) and the optical and electrical properties of the fabricated films were characterized. The transmittance was directly affected by the number of spin-cast layers, as presented in **Figure 3**. By increasing the number of spin-cast layers from one to five, the optical transmittance at 550 nm was decreased from 88.8 to 58.7%. On the other hand, sheet resistance values of the rGO-SWCNT films were conversely affected upon the increase of the number of spin-cast layers reaching R_s values as low as 254 Ω sq^{-1}. Furthermore, OSCs were fabricated using rGO-SWCNT films as the anode TCE. Through doping with alkali carbonates (Li, Na, Cs, etc.), it is possible to tune the WF of the solution-processed rGO-SWCNTs to match the lowest unoccupied molecular orbital (LUMO) of $PC_{61}BM$. In this way, an improved ohmic contact with the active layer can be achieved, leading to increased charge injection and better device performance. On top of that, SWCNTs can act as conductive percolation paths that short circuit the rGO sheets. Inverted OSCs with structure of rGO-SWCNT/PEDOT:PSS/P3HT:$PC_{61}BM$/V_2O_5/Al were fabricated to demonstrate the applicability of WF tuning of rGO-SWCNT. The devices based on P3HT:$PC_{61}BM$, incorporating the four-layer rGO-SWCNT film exhibited a maximum PCE of 1.27%. Additionally, an excellent flexibility even under bending angles of more than 60° was observed.

2.2. Thermally annealing rGO transparent conductive electrodes

The first application of solution-processed thermally reduced GO TCEs in OSC devices was performed by Yin et al. in 2010 [50]. In more detail, GO films were firstly spin-coated on SiO_2/Si substrates, followed by reduction through thermal annealing at 1000°C in the presence of Ar/H^2. Polymethylmethacrylate (PMMA) (~300 nm thick) was subsequently coated to be used as an intermediate transfer substrate before the final transfer of rGO films onto polyethylene terephthalate (PET) substrates. Flexible poly(3-hexylthiophene) (P3HT):phenyl-C_{61}-butyric acid

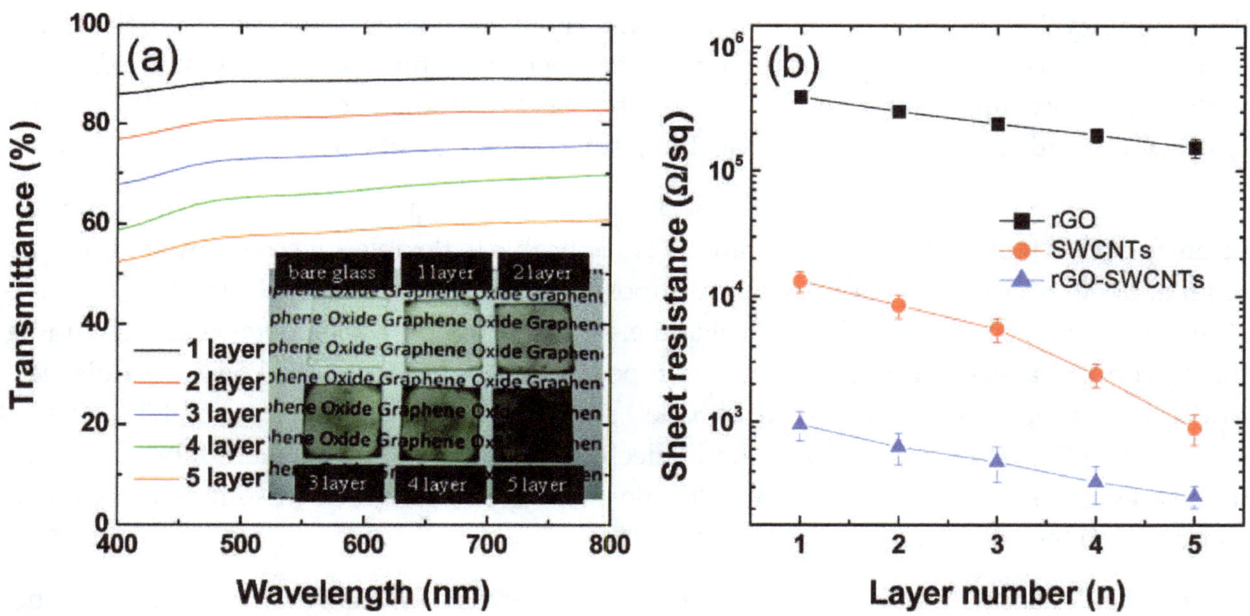

Figure 3. (a) rGO-SWCNT films transmission spectra featuring different numbers of deposited layers. Inset: photograph of the samples tested. (b) rGO, SWCNTs, and the hybrid rGO-SWCNT sheet resistances as a function of the layers' number. A significant decrease of the rGO sheet resistance is demonstrated after adhering to the SWCNTs. Reprinted with permission from Huang et al. [49].

methyl ester ($PC_{61}BM$)–based OSCs were fabricated on the rGO-coated PET substrates, and the electrical characteristics of the OSCs were investigated with respect to the thickness of the rGO electrode (**Figure 4**).

Higher device performance dependence on the rGO film R_s was observed when the rGO films optical transmittance exceeded 65%, while the light transmission efficiency dominated the OSCs performance in lower optical transmittance (<65%). The optimum power conversion efficiency (PCE) was achieved for device 3 (PCE = 0.78%), slightly lower than the one reported for CVD-graphene/PET-based photovoltaic devices (1.18%) [15]. This results from the lower R_s of the CVD grown graphene compared to the thermally derived rGO film.

rGO films were also investigated regarding their flexibility as displayed in **Figure 5**. BHJ OSC electrical properties incorporating rGO films as electrodes demonstrated excellent tolerance under high bending conditions and multiple number of bending cycles. The critical point beyond which the fabricated OSCs with rGO/PET films exhibited degradation due to bending depends on the rGO film thickness. In more detail, thicker rGO electrodes led to higher stability, when the OSCs were subjected to tensile stress. After various bending cycles, an electrical degradation of rGO-based OSCs was observed, which was analogous to the increased R_s of the rGO/PET films. In particular, after 1600 bending cycles, the R_s of the rGO/PET electrode increased from 16 to 18 kΩ sq^{-1} and 3.2 to 3.5 kΩ sq^{-1}, respectively, for the 4 and 16 nm thick rGO films.

In another study, Geng et al. reported a simple method for preparing graphene TCEs using seriatim a chemical and a thermal reduction method [51]. More specifically, a converted graphene (CCG) suspension obtained via controlled chemical reduction of exfoliated GO in the absence of dispersants was used as the pristine material and then, upon thermal annealing for 15 min at three different temperatures (200, 400, and 800°C) under vacuum in a furnace tube, the sp^2 carbon networks of the graphene sheets were recovered, with the resulting CCG

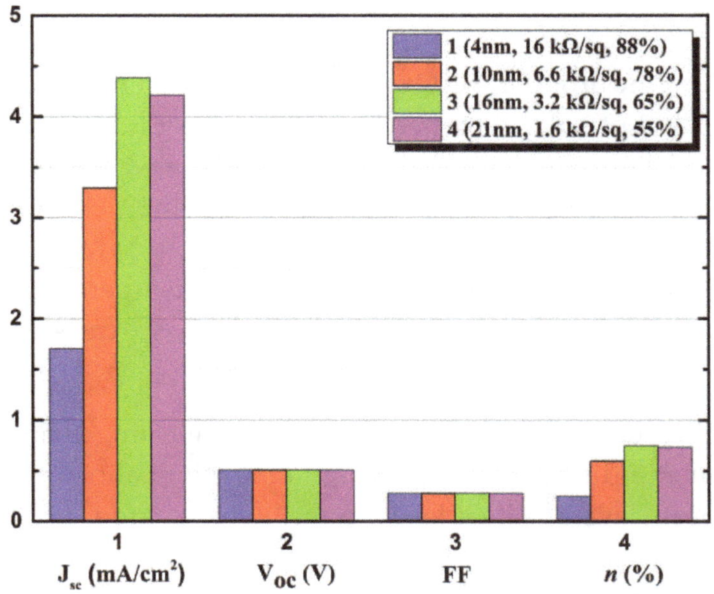

Figure 4. The effect of the rGO transparent films thickness in the performance of P3HT:$PC_{61}BM$ BHJ OSCs. Reprinted with permission from Petridis et al. [26]. Reproduced by permission of The Royal Society of Chemistry.

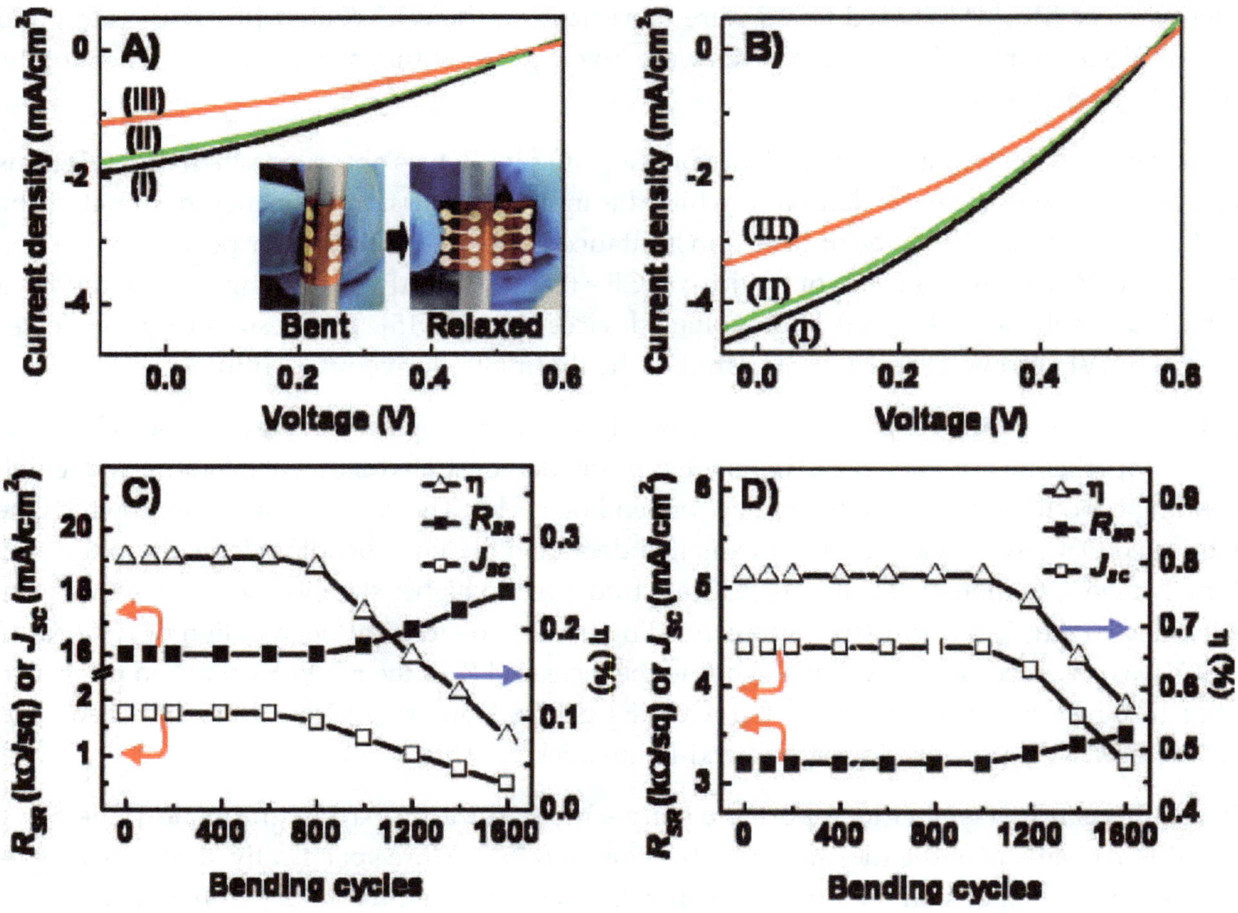

Figure 5. Current density-voltage (*J-V*) curves the various devices tested: (A) for device 1 after applying (i) 400, (ii) 800, and (iii) 1200 cycles of bending, and (B) for device 3 after (I) 800, (II) 1200, and (III) 1600 cycles of bending. (C) and (D) The short circuit current density, overall power conversion efficiency (n), and sheet resistance R_{SR} for devices 1 and 3, respectively, as a function of the number of bending cycles. Reprinted with permission from Yin et al. [50].

films exhibiting an R_s of the order of 103 $\Omega\ sq^{-1}$ at 50% transparency (at 550 nm). Owing to the apparently greater extent of sp^2 carbon networks restoration during the two-step reduction of the CCG films, each thermally annealed CCG film exhibited lower R_s than the thermally annealed GO as depicted in **Figure 6**. The transparency of the CCG films decreases linearly with respect to the volume of the CCG suspension used to prepare the vacuum-filtered film, as also displayed in **Figure 6**. By increasing the annealing temperature, the transparency of the CCG films decreased due to thermal improvement of the sp^2 carbon networks in the CCG sheets.

The recession of the R_s observed in **Figure 6**, as the annealing temperature increases, is attributed to the smaller induced distance owing to the elimination of the functional groups between the CCG sheets layers, which leads to facilitated charge carrier transfer across the CCG sheets. It is observed that interlayer distance in CCG-A800 films is reduced to 0.354 nm, approaching the value of bulk graphite. The transparency reduced from this point, while the R_s was reduced linearly to the order of $10^3\ \Omega\ sq^{-1}$, demonstrating that reduction impacted successfully and uniformly from the outermost to the inner layers [51].

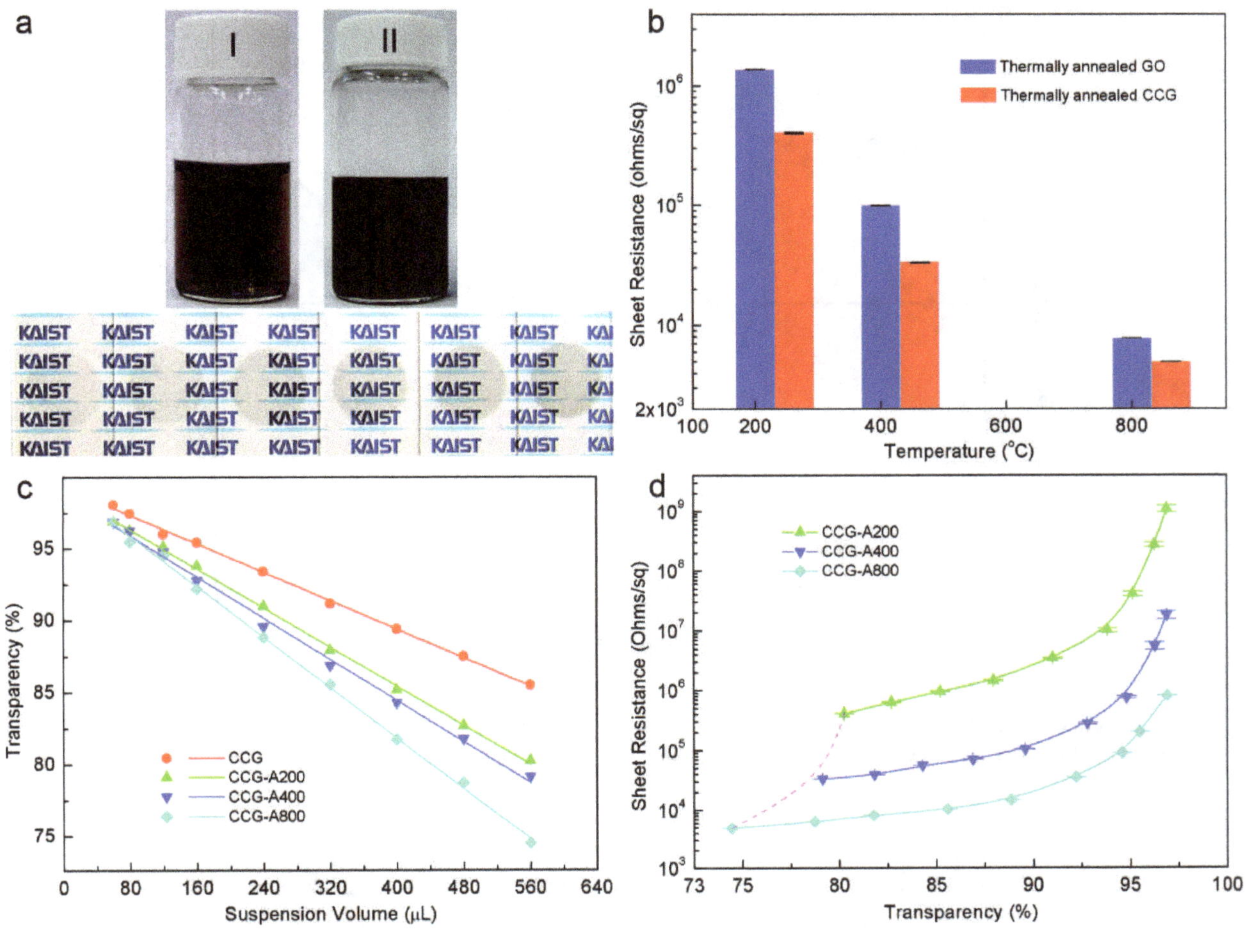

Figure 6. (a) A GO suspension with concentration of 0.1 mg/mL (I), the resultant CCG suspension (II), and the CCG films produced using 160, 240, 320, 400, 480, and 560 μL of a 10 mg/L CCG suspension. (b) GO and CCG films R_s annealed at 200, 400, and 800°C. (c) Transparency of CCG, CCG-A200, CCG-A400, CCG-A800 films with respect to the volume (μL) of the 10 mg/L CCG suspension used for film preparation, and (d) R_s values of CCG-A200, CCG-A400, CCG-A800 films as a function of film transparency. Reprinted with permission from Geng et al. [51].

An important conductivity improvement was induced due to the structural changes in the CCG resulted from the thermal reduction. In particular, sp^2 carbon network recover was crucial for increasing the charge carrier transport in individual CCG layers, while the interlayer distance was decreased to a level close to the value of bulk graphite; thereby, improving the charge carrier transport across the CCG layers. As a proof of concept, CCG films were used as TCEs in P3HT:PC_{61}BM-based OSCs achieving a maximum PCE of 1.01%, which is ~50% decreased compared to the PCE value in ITO-based devices, as presented in **Figure 7**.

2.3. Photochemically treated rGO transparent conductive electrodes

For the first time, Kymakis et al. presented a facile, laser-assisted technique to prepare transparent and highly conductive graphene-based films on top of flexible substrates, produced by spin-casting [52]. The experimental setup was realized with (1) a Ti:Sapphire pulsed laser source, (2) a 10 mm diameter lens, and (3) a high precision X-Y computer controlled translation stage where the film was placed and translated across the focused laser beam (**Figure 8**).

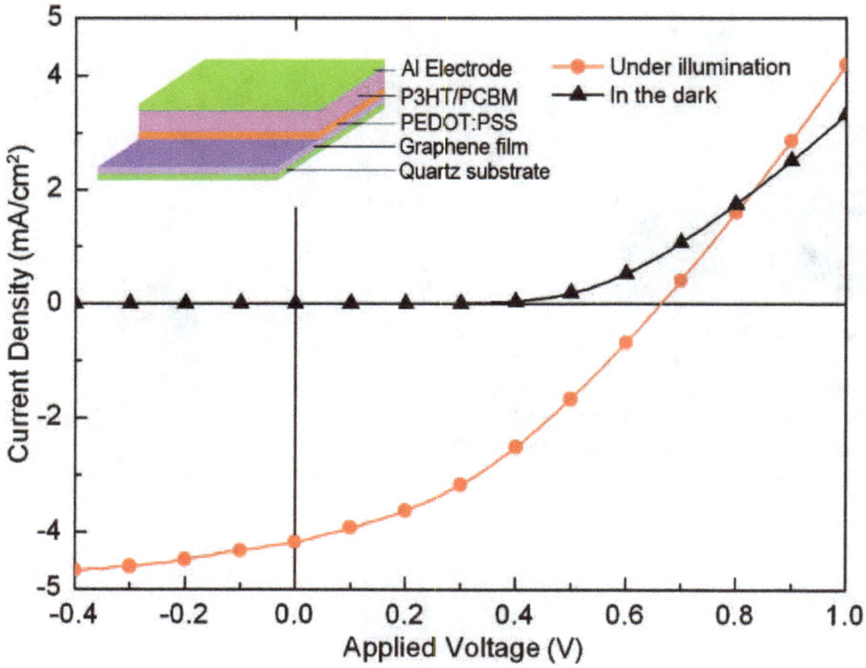

Figure 7. *J-V* curves for P3HT:PC61BM-based OSCs incorporating the CCG-A800 film as TCE. The inset shows the architecture of the solar cell device. Reprinted with permission from Geng et al. [51].

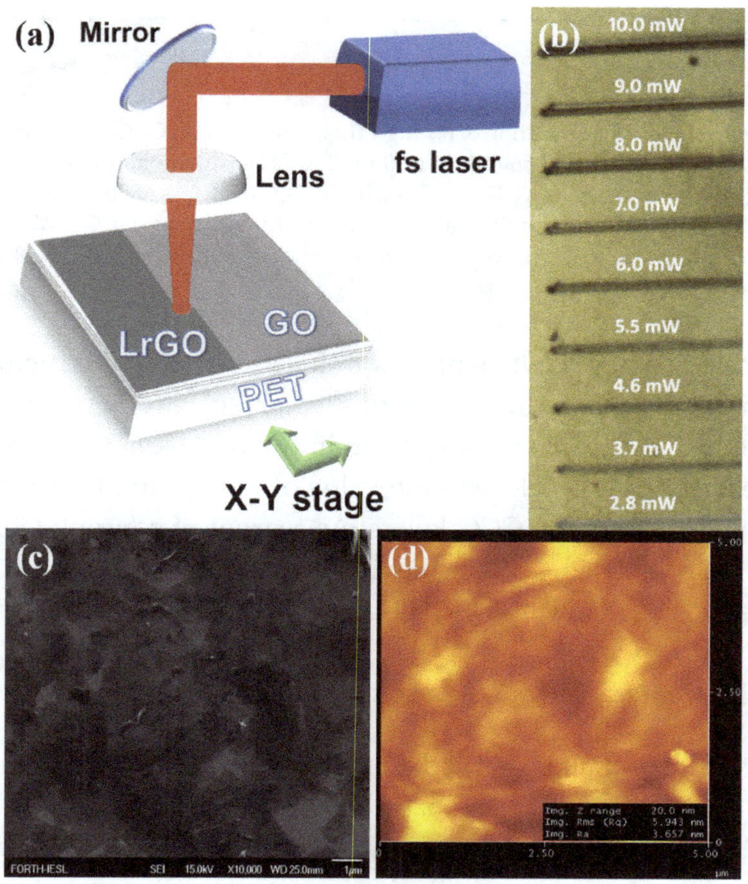

Figure 8. (a) Laser-based GO reduction experimental setup. (b) Scan lines obtained upon irradiation with 100 fs pulses at different fluences indicating the gradual color change due to reduction. (c) SEM and (d) AFM images of LrGO films on PET. Reprinted with permission from Kymakis et al. [52].

During the reduction process, the fs laser beam irradiated the as-spun GO layers, while it was translated across the layers. The reduction of GO can be obvious. Scan lines obtained upon irradiation with 100 fs pulses at different fluencies indicating the gradual color change due to the reduction are also presented in **Figure 8**. The brownish color of the as cast film was gradually turned into black, which is a strong indication that GO is rapidly reduced via the laser treatment in air, without using any reducing chemical agent. Scanning electron microscopy (SEM) and atomic force microscopy (AFM) analysis demonstrated that no or minor ablation effects occur during the reduction process.

The scope behind the production of solution cast laser reduced GO (LrGO) electrodes is to prepare highly flexible OSCs that can be utilized for compact R2R solar modules. In this context, the LrGO films were used to fabricate flexible OSC devices (**Figure 9**) in order to determine their photovoltaic characteristics and identify the combination of transparency and R_s that provides the best performance. Compared to the thermally reduced GO-based

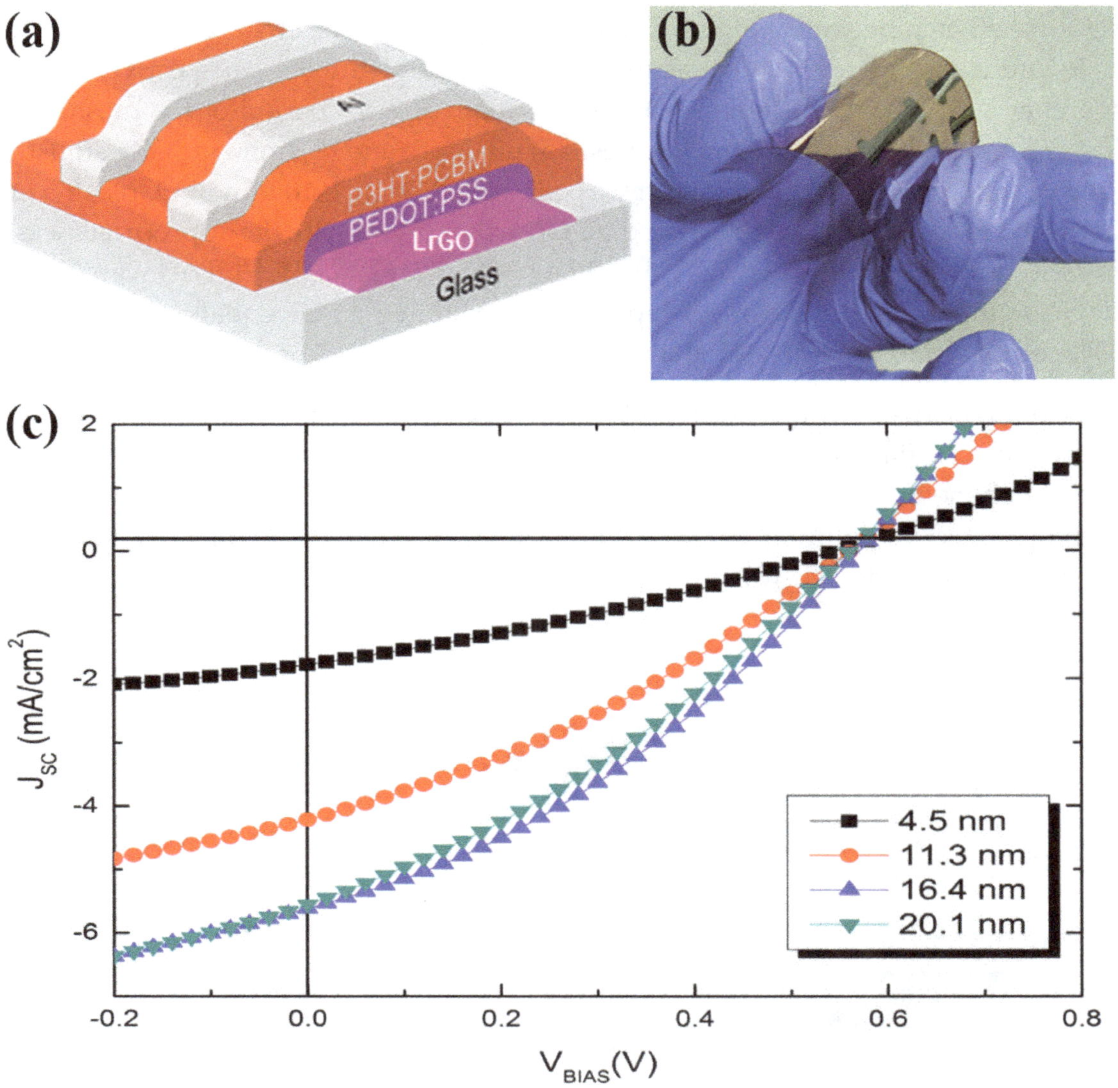

Figure 9. (a) Schematic and (b) picture of the flexible PET/rGO/PEDOT:PSS/P3HT:PC_{61}BM/Al photovoltaic devices fabricated and (c) The illuminated current density-voltage (*J-V*) curves of the solar cells with various LrGO film thicknesses. Reprinted with permission from Kymakis et al. [52].

OSC devices (PCE ~ 0.78%) [50], LrGO-based OSCs exhibited 41% efficiency improvement for a 16.4 nm thick film having an R_s of 1.6 kΩ sq^{-1} and 70% transparency. It is clear that the higher efficiency is attributed to the LrGO film, revealing the superiority of the laser ablation method, in contrast to the chemical method.

Since rGO TCEs presents low performance in terms of conductivity and transmittance (R_s ~1 kΩ sq^{-1}, 70% transmittance) compared to the ITO TCEs (R_s ~15 Ω sq^{-1}, 90% transmittance), new efforts are performed to improve graphene-based TCE performances. One very promising technique is the employment of a mesh structure with periodic lines on the rGO film [53], with OSC devices based on graphene mesh electrodes (GMEs) already exhibiting performances comparable to those using conventional ITO TCEs [54, 55]. Therefore, it is possible to control the films R_s and transparency by tailoring the grid width, spacing, and thickness [56].

In the same year, Zhang et al. demonstrated the preparation of graphene mesh electrodes (GMEs) by using the standard industrial photolithography and O_2 plasma etching process as illustrated in **Figure 10** [57]. The R_s and the transparency of the graphene TCEs before the mask-based etching were 150 Ω sq^{-1} and ~8%, while after the etching technique, the electrodes R_s and transparency were measured to be 750 Ω sq^{-1} and 65%, respectively. The GME transparency was directly designated by (a) pit depth and (b) period of the mesh, while its conductance was mainly affected by the mesh (a) pit depth and (b) linewidth. By adjusting the duration of the O_2 process, the pit depth was controlled, with 4–10 min of O_2 plasma etching time required for high transparent mesh pits. To further investigate the appropriateness of the GMEs, OSC devices of the structure GME/PEDOT:PSS/P3HT:PC_{61}BM/LiF/Al were fabricated. The devices employing optimum pit depth GMEs exhibited a PCE equal to 2.04% decreased by ~33% compared to the conventional ITO-based OSCs (PCE ~3%).

In 2015, Konios et al. presented a different laser-assisted method to directly pattern a mesh on the surface of rGO films deposited onto flexible substrates, for the production of large area reduced graphene oxide micromesh (rGOMM) electrodes [58]. By using this technique, the use of complicated photolithographic [59], ion beam [60], chemical etching [61], template [62] and O_2 plasma methods [52] was overtaken. The proposed technique is a one-step method,

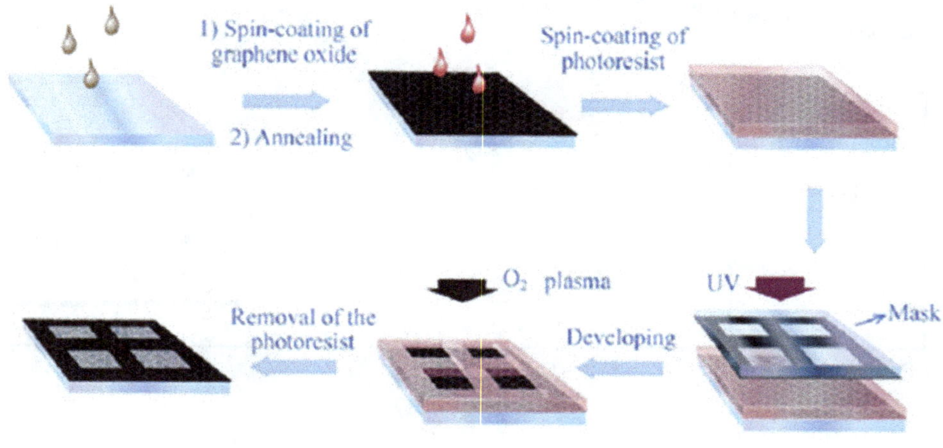

Figure 10. Illustration of the fabrication steps for the preparation of GMEs. Reprinted with permission from Zhang et al. [57].

since no TEM grid patterning mask [63], and no transfer step are needed. In addition, the method does not use any photo-resistive material [64] or pre-patterned elastomeric stamps [65], while the application of a fs laser pulses allows the patterning of micro size holes on top of any flexible low cost material. In addition, this technique can be easily controlled over the entire illuminated area induces minimum thermal damage in the surrounding layers and more importantly is compatible to R2R production processes.

The electrode transparency can be accurately controlled with a small R_s increase, successfully handling with the existing trade-off between transparency and electrical conductivity. The rGO thin films optoelectrical properties of depended on the interplay between the periodicity and the geometrical characteristics of the mesh pattern structure. The rGO mesh electrode experimental setup and SEM images of the laser-induced patterns are presented in **Figure 11**. The proposed technique main advantage is that it permits fine-tuning of the optoelectrical properties via variation of the irradiation dose (energy, number of pulses) and/or the periodicity and thus the neck width of the mesh. A significant improvement of the transmittance by ~65% was performed after laser treatment. As a proof of concept, reduced GO micromesh (rGOMM) were employed as the anode TCE in flexible devices based on poly[N-90-heptadecanyl-2,7-carbazole-alt-5,5-(40,70-di-2-thienyl20,10,30-benzothiadiazo-le)] (PCDTBT):[6, 6]-phenyl C71 butyric acid methyl ester ($PC_{71}BM$) and compared with those deposited on ITO (**Figure 12**).

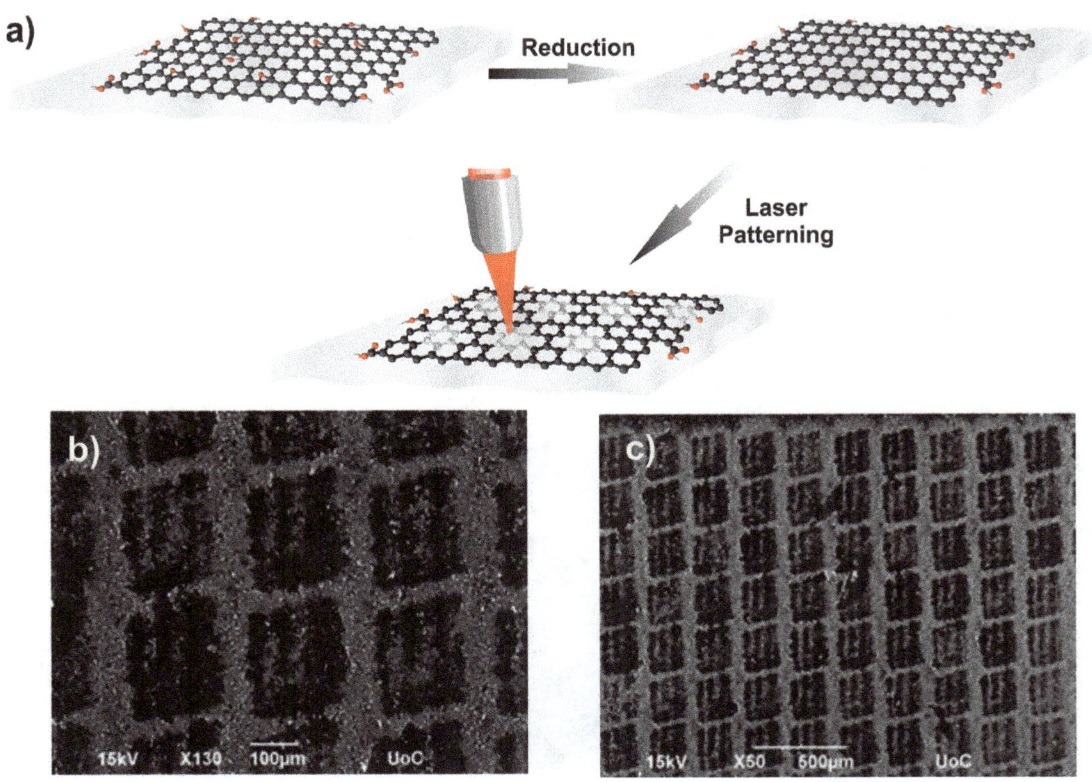

Figure 11. (a) Illustration of the laser-induced patterns on the rGO films and (b, c) SEM images of the laser-induced patterns: the darker spots correspond to the laser-processed areas where the mesh lines correspond to the lighter colored paths. Reprinted with permission from Konios et al. [58].

The optimum photovoltaic parameters for the rGOMM-based devices were extracted for transparency ~59.1% and R_s ~565 Ω sq^{-1}, with the resulting PCE of 3.05%, the highest reported so far for flexible OSC devices incorporating solution-processed graphene-based electrodes. Another important issue investigated in this study was the determination of the effectiveness of the proposed method when tested in large area photovoltaic cells, as presented in **Figure 12**. This is very important for upscaling from lab solar cells to solar modules. In this context, stress tests demonstrated that the photovoltaic performance deterioration for both rGOMM and ITO electrodes tested was almost the same; for 135 mm^2 active area PCE reduction measured was 63.2% for ITO and 64.9% for rGOMM, respectively. Therefore, the proposed method can be effectively applied when upscaling to large area photovoltaic cells or solar modules without compromising the photovoltaic efficiency compared with the widely commercialized ITO transparent electrode.

During the last 5 years, an intensive research effort was conducted in the field of solution-processed graphene-based TCEs (**Figure 13**). The optoelectrical properties of graphene-based TCEs in OSCs are presented in **Table 1**. The increase of the PCE of OSC devices incorporating graphene-based TCEs has been improved from 0.13 to 3.7% for rigid devices, while the progress on the flexible graphene-based devices was also incredible with an efficiency improvement from 0.78 to 3.05%.

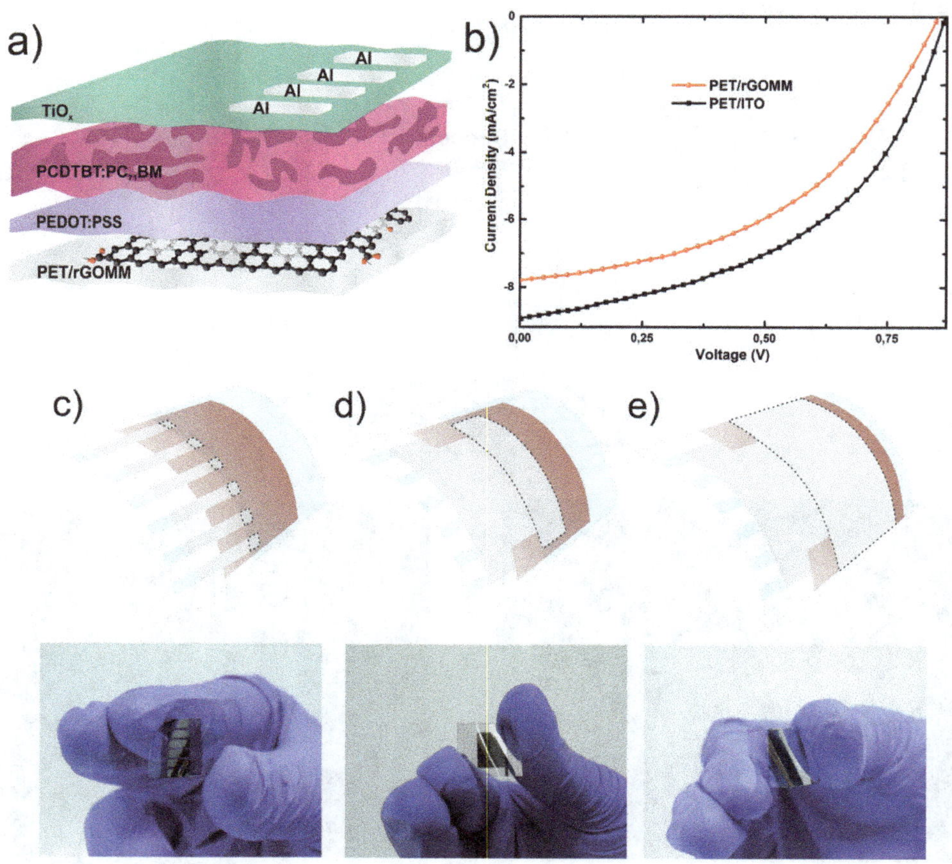

Figure 12. (a) The architecture of the BHJ OSC device with the laser-induced rGOMM as TCE. (b) Photovoltaic performance of the OSCs with rGOMM (spheres) and ITO (squares) as TCE. Schematic illustration and photographs of (c) 4 mm^2, (d) 50 mm^2 and (e) 135 mm^2 active areas devices tested. Reprinted with permission from Konios et al. [58].

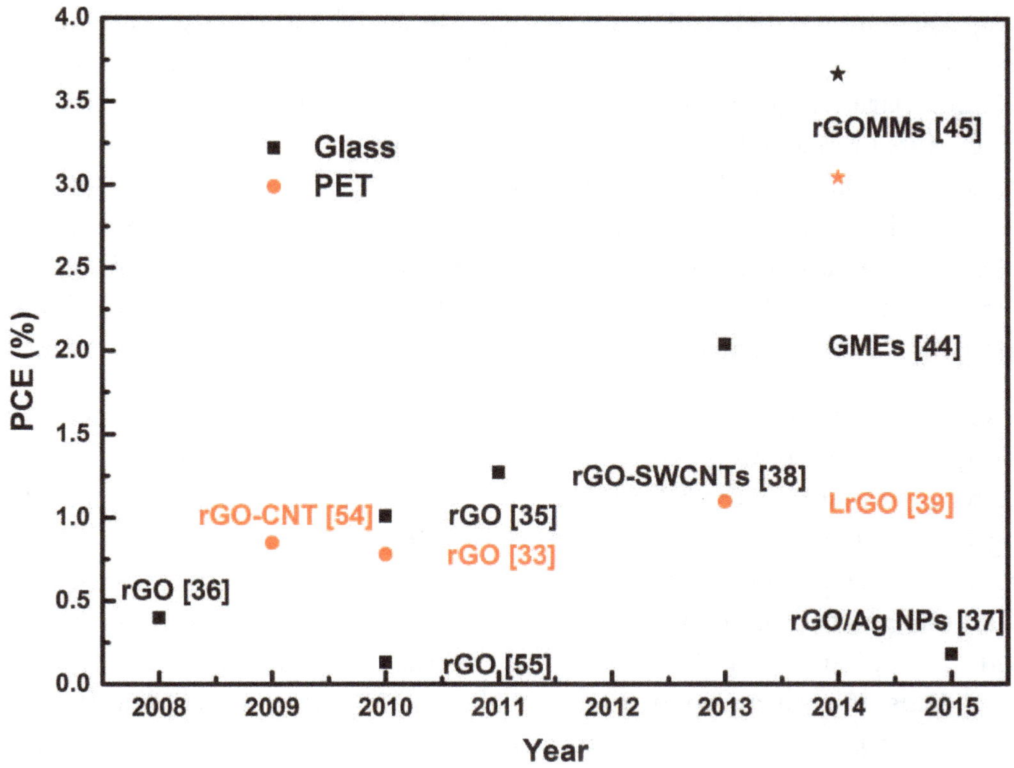

Figure 13. OSCs performance of solution-processed graphene-based TCE with different active layers extracted from literature [44, 48–51, 57, 58, 66, 67]. Squares stand for rigid devices, whereas circles for devices on flexible substrates. Stars illustrate the champion PCE achieved. Reprinted with permission from Konios et al. [58].

Graphene-based electrodes	R_s (kΩ/sq)	Transparency (%)	Reference
Thermal reduced GO	3.2	65	[50]
Thermal reduced GO	1	80	[51]
LrGO	1.6	70	[52]
rGO-SWCNTs	0.331	65.8	[49]
rGO mesh electrodes	0.700	65	[57]
LrGO mesh electrode	0.565	59.1	[58]
rGO/Ag NPs	83	47	[48]

Reprinted with permission from Petridis et al. [26]. Reproduced by permission of The Royal Society of Chemistry.

Table 1. Summary of the optoelectrical properties of graphene-based TCEs in OSC devices.

3. Conclusion

In this chapter, the concept of solution-processed graphene-based films as anode electrodes in OSCs has been presented, with respect to the fabrication techniques used: (a) thermally reduced graphene, (b) chemically and photochemically rGO, (c) composite rGO-CNTs, and (d) rGO mesh films. The application of rGO-based TCEs in OSCs was also demonstrated. The successful transferring of such schemes in large scale poses a significant engineering challenge and ongoing effort is needed for preparing high efficient printed optoelectronic devices such as flexible and stretchable OSCs. Despite graphene-based materials advantages regarding the manufacturing and purchase cost, compatibility with flexible materials, WF tuning (by controlling graphene size, layer and through functionalization), transparency, and solubility, significant improvement is required in order to overcome the corresponding properties of ITO films. Even though the current performance of graphene-based TCE OSCs approaches that of the conventional ITO technology, there is more space for the improvement of OSCs adapting graphene as an anode electrode.

Future research efforts should be focused on (a) inventing cost-effective techniques for large-scale graphene or rGO production, (b) effectiveness improvement of the GO reduction methods, (c) enhancing the electrical conductivity of the graphene-based TCEs without disrupting their transmittance, and (d) improving the graphene-based films morphology. For improving the TCE conductivity, metal nanoparticles (e.g., Au, Ag, and Al) with different sizes and shapes can be incorporated in the graphene-based electrodes, leading to the production of graphene-NPs electrodes with even lower resistance and high mechanical stability; and thus, may improve the overall stability of the devices, contributing to the lifetime enhancement. On top of that, hybrid films consisting of rGO and 1D conductive bridging materials, such as metal nanowires, which exhibit better mechanical properties than ITO films, can significantly decrease the film resistance. The metal nanowires can provide electric pathways among the graphene layers and thus may enhance the conductivity, without sacrificing their optical properties, leading to improved performance of graphene-based TCEs. The incorporation of metal-graphene–based TCEs into optoelectronic devices demonstrates their potential for ITO replacement in a broad range of optoelectronic applications.

Acknowledgements

We acknowledge funding from the European Union's Horizon 2020 research and innovation programme under grant agreement No. 696656 – GrapheneCore1.

Author details

Minas M. Stylianakis[1*], Dimitrios Konios[1], Konstantinos Petridis[1,2] and Emmanuel Kymakis[1]

*Address all correspondence to: stylianakis@staff.teicrete.gr

1 Department of Electrical Engineering, Center of Materials Technology and Photonics, School of Applied Technology, University of Applied Sciences of Crete (Technological Educational Institute (TEI) of Crete), Heraklion, Crete, Greece

2 Department of Electronic Engineering, University of Applied Sciences of Crete (Technological Educational Institute (TEI) of Crete), Chania, Crete, Greece

References

[1] Bonaccorso F, Colombo L, Yu G, Stoller M, Tozzini V, Ferrari AC, Ruoff RS, Pellegrini V. Graphene, related two-dimensional crystals, and hybrid systems for energy conversion and storage. Science. 2015;**347**:6217. DOI: 10.1126/science.1246501

[2] Chang DW, Choi HJ, Filer A, Baek JB. Graphene in photovoltaic applications: Organic photovoltaic cells (OPVs) and dye-sensitized solar cells (DSSCs). Journal of Materials Chemistry A. 2014;**2**:12136-12149. DOI: 10.1039/C4TA01047G

[3] Chowdhury TH, Islam A, Mahmud Hasan AK, Terdi MAM, Arunakumari M, Prakash Singh S, Alam MDK, Bedja IM, Hafidz Ruslan M, Sopian K, Amin N, Akhtaruzzaman MD. Prospects of graphene as a potential carrier-transport material in third-generation solar cells. The Chemical Record. 2016;**16**:614-632. DOI: 10.1002/tcr.201500206

[4] Balis N, Stratakis E, Kymakis E. Graphene and transition metal dichalcogenide nanosheets as charge transport layers for solution processed solar cells. Materials Today. 2016;**19**:580-594. DOI: 10.1016/j.mattod.2016.03.018

[5] Stylianakis MM, Sygletou M, Savva K, Kakavelakis G, Kymakis E, Stratakis E. Photochemical synthesis of solution-processable graphene derivatives with tunable bandgaps for organic solar cells. Advanced Optical Materials. 2015;**5**:658-666. DOI: 10.1002/adom.201400450

[6] Bonaccorso F, Balis N, Stylianakis MM, Savarese M, Adamo C, Gemmi M, Pellegrini V, Stratakis E, Kymakis E. Functionalized graphene as an electron-cascade acceptor for air-processed organic ternary solar cells. Advanced Functional Materials. 2015;**25**:3870-3880. DOI: 10.1002/adfm.201501052

[7] Stylianakis MM, Konios D, Kakavelakis G, Charalambidis G, Stratakis E, Coutsolelos AG, Kymakis E, Anastasiadis SH. Efficient ternary organic photovoltaics incorporating a graphene-based porphyrin molecule as a universal electron cascade material. Nanoscale. 2015;**7**:17827-17835. DOI: 10.1039/C5NR05113D

[8] Stylianakis MM, Mikroyannidis JA, Kymakis E. A facile, covalent modification of single-wall carbon nanotubes by thiophene for use in organic photovoltaic cells. Solar Energy Materials & Solar Cells. 2010;**94**:267-274. DOI: 10.1016/j.solmat.2009.09.013

[9] Stylianakis MM, Spyropoulos GD, Stratakis E, Kymakis E. Solution-processable graphene linked to 3,5-dinitrobenzoyl as an electron acceptor in organic bulk heterojunction photovoltaic devices. Carbon. 2012;**50**:5554-5561. DOI: 10.1016/j.carbon.2012.08.001

[10] Kakavelakis G, Konios D, Stratakis E, Kymakis E. Enhancement of the efficiency and stability of organic photovoltaic devices via the addition of a lithium-neutralized graphene oxide electron-transporting layer. Chemistry of Materials. 2014;**26**:5988-5993. DOI: 10.1021/cm502826f

[11] Stylianakis MM, Stratakis E, Koudoumas E, Kymakis E, Anastasiadis SH. Organic bulk heterojunction photovoltaic devices based on polythiophene-graphene composites. ACS Applied Materials & Interfaces. 2012;**4**:4864-4870. DOI: 10.1021/am301204g

[12] Balis N, Konios D, Stratakis E, Kymakis E. Ternary organic solar cells with reduced graphene oxide-Sb_2S_3 hybrid nanosheets as the cascade material. ChemNanoMat. 2015;**1**:346-352. DOI: 10.1002/cnma.201500044

[13] Hecht DS, Hu L, Irvin G. Emerging transparent electrodes based on thin films of carbon nanotubes, graphene, and metallic nanostructures. Advanced Materials. 2011;**23**:1482-1513. DOI: 10.1002/adma.201003188

[14] Kumar A, Zhou C. The race to replace tin-doped indium oxide: Which material will win?. ACS Nano. 2010;**4**:11-14. DOI: 10.1021/nn901903b

[15] Gomez De Arco L, Zhang Y, Schlenker CW, Ryu K, Thompson ME, Zhou C. Continuous, highly flexible, and transparent graphene films by chemical vapor deposition for organic photovoltaics. ACS Nano. 2010;**4**:2865-2873. DOI: 10.1021/nn901587x

[16] Li G, Shrotriya V, Huang J, Yao Y, Moriarty T, Emery K, Yang Y. High-efficiency solution processable polymer photovoltaic cells by self-organization of polymer blends. Nature Materials. 2005;**4**:864-868. DOI: 10.1038/nmat1500

[17] Kim Y, Cook S, Tuladhar SM, Choulis SA, Nelson J, Durrant JR, Bradley DDC, Giles M, McCulloch I, Ha CS, Ree M. A strong regioregularity effect in self-organizing conjugated polymer films and high-efficiency polythiophene:fullerene solar cells. Nature Materials. 2006;**5**:197-203. DOI: 10.1038/nmat1574

[18] Klaus E. Past achievements and future challenges in the development of optically transparent electrodes. Nature Photonics. 2012;**6**:809-812. DOI: 10.1038/nphoton.2012.282

[19] Cheng YJ, Yang SH, Hsu CS. Synthesis of conjugated polymers for organic solar cell applications. Chemical Reviews. 2009;**109**:5868-5923. DOI: 10.1021/cr900182s

[20] Mishra A, Bäuerle P. Small molecule organic semiconductors on the move: Promises for future solar energy technology. Angewandte Chemie International Edition. 2012;**51**:2020-2067. DOI: 10.1002/anie.201102326

[21] Scharber MC, Sariciftci NS. Efficiency of bulk-heterojunction organic solar cells. Progress in Polymer Science. 2013;**38**:1929-1940. DOI: 10.1016/j.progpolymsci.2013.05.001

[22] Tan YW, Zhang Y, Bolotin K, Zhao Y, Adam S, Hwang EH, Das Sarma S, Stormer HL, Kim P. Measurement of scattering rate and minimum conductivity in graphene. Physical Reviews Letters. 2007;**99**:246803. DOI: 10.1103/PhysRevLett.99.246803

[23] Roy-Mayhew JD, Aksay LA. Graphene materials and their use in dye-sensitized solar cells. Chemical Reviews. 2014;**114**:6323. DOI: 10.1021/cr400412a

[24] Layani M, Kamyshny A, Magdassi S. Transparent conductors composed of nanomaterials. Nanoscale. 2014;**6**:5581-5591. DOI: 10.1039/C4NR00102H

[25] Rana K, Singh J, Ahn JH. A graphene-based transparent electrode for use in flexible optoelectronic devices. Journal of Materials Chemistry C. 2014;**2**:2646-2656. DOI: 10.1039/C3TC32264E

[26] Petridis C, Konios D, Stylianakis MM, Kakavelakis G, Sygletou M, Savva K, Tzourmpakis P, Krassas M, Vaenas N, Stratakis E, Kymakis E. Solution processed reduced graphene oxide electrodes for organic photovoltaics. Nanoscale Horizons. 2016;**1**:375-382, DOI: 10.1039/C5NH00089K

[27] Torrisi F, Hassan T, Wu W, Sun Z, Lombardo A, Kulmala TS, Hsieh G, Jung S, Bonaccorso F, Paul PJ, Chu V, Ferrari AC. Inkjet-printed graphene electronics. ACS Nano. 2012;**6**:2992-3006. DOI: 10.1021/nn2044609

[28] Bae S, Kim H, Lee Y, Xu X, Park JS, Zheng Y, Balakrishnan J, Lei T, Kim HR, Song YI. Roll-to-roll production of 30-inch graphene films for transparent electrodes. Nature Nanotechnology. 2010;**5**:574-578. DOI: 10.1038/nnano.2010.132

[29] Na SI, Kim SS, Jo J, Kim DY. Efficient and flexible ITO-free organic solar cells using highly conductive polymer anodes. Advanced Materials. 2008;**20**:4061-4067. DOI: 10.1002/adma.200800338

[30] Eda G, Chhowalla M. Chemically derived graphene oxide: Towards large-area thin-film electronics and optoelectronics. Advanced Materials. 2010;**22**:2392-2415. DOI: 10.1002/adma.200903689

[31] Nair R, Blake P, Grigorenko A, Novoselov K, Booth T, Stauber T, Peres N, Geim A. Fine structure constant defines visual transparency of graphene. Science. 2008;**320**:1308. DOI: 10.1126/science.1156965

[32] Smith AD, Niklaus F, Paussa A, Vaziri S, Fischer AC, Sterner M, Forsberg F, Delin A, Esseni D, Palestri P, Östling M, Lemme MC. Electromechanical piezoresistive sensing in suspended graphene membranes. Nano Letters. 2013;**13**:3237-3242. DOI: 10.1021/nl401352k

[33] Novoselov KS, Geim AK, Morozov SV, Jiang D, Zhang Y, Dubonos SV, Grigorieva IV, Firsov AA. Electric field effect in atomically thin carbon films. Science. 2004;**306**:666-669. DOI: 10.1126/science.1102896

[34] Behabtu N, Lomeda JR, Green MJ, Higginbotham AL, Sinitskii A, Kosynkin DV, Tsentalovich D, Parra Vasquez ANG, Schmidt J, Kesselman E, Cohen Y, Talmon Y, Tour JM, Pasquali M. Spontaneous high-concentration dispersions and liquid crystals of graphene. Nature Nanotechnology. 2010;**5**:406-411. DOI: 10.1038/nnano.2010.86

[35] Kim KS, Zhao Y, Jang H, Lee SY, Kim JM, Kim KS, Ahn JH, Kim P, Choi JY, Hong BH. Large-scale pattern growth of graphene films for stretchable transparent electrodes. Nature. 2009;**457**:706-710. DOI: 10.1038/nature07719

[36] Choucair M, Thordason P, Stride JA. Gram-scale production of graphene based on solvothermal synthesis and sonication. Nature Nanotechnology. 2009;**4**:30-33. DOI: 10.1038/nnano.2008.365

[37] Wang X, Zhi L, Tsao N, Tomovic Z, Li J, Mullen K. Transparent carbon films as electrodes in organic solar cells. Angewandte Chemie International Edition. 2008;**47**:2990-2992. DOI: 10.1002/anie.200704909

[38] Berger C, Song Z, Li T, Orbazghi AY, Feng R, Dai Z, Marchenkov AN, Conrad EH, First PN, De Heer WA, Ultrathin epitaxial graphite: 2D electron gas properties and a route toward graphene-based nanoelectronics. Journal of Physical Chemistry B. 2004;**108**:19912-19916. DOI: 10.1021/jp040650f

[39] Zhuang N, Liu C, Jia L, Wei L, Cai J, Guo Y, Zhang Y, Hu X, Chen J, Chen X, Tang Y. Clean unzipping by steam etching to synthesize graphene nanoribbons. Nanotechnology. 2013:**24**:325604. DOI: 10.1088/0957-4484/24/32/325604

[40] Konios D, Stylianakis MM, Stratakis E, Kymakis E. Dispersion behaviour of graphene oxide and reduced graphene oxide. Journal of Colloid Interface Science. 2014;**430**:108-112. DOI: 10.1016/j.jcis.2014.05.033

[41] Stankovich S, Dikin DA, Piner RD, Kohlhaas KA, Kleinhammes A, Jia Y, Wu Y, Nguyen ST, Ruoff RS. Synthesis of graphene-based nanosheets via chemical reduction of exfoliated graphite oxide. Carbon. 2007;**45**:1558-1565. DOI: 10.1016/j.carbon.2007.02.034

[42] Mattevi C, Eda G, Agnoli S, Miller S, Mkhoyan KA, Celik O, Mastrogiovanni D, Granozzi G, Garfunkel E, Chhowalla M. Evolution of electrical, chemical, and structural properties of transparent and conducting chemically derived graphene thin films. Advanced Functional Materials. 2009;**19**:2577-2583. DOI: 10.1002/adfm.200900166

[43] Gengler RYN, Badali DS, Zhang D, Dimos K, Spyrou K, Gournis D, Miller RJD. Revealing the ultrafast process behind the photoreduction of graphene oxide. Nature Communication. 2013;**4**:2560. DOI: 10.1038/ncomms3560

[44] Wu J, Becerril H, Bao Z, Liu Z, Chen Y, Peumans P. Organic solar cells with solution-processed graphene transparent electrodes. Applied Physics Letters. 2008;**92**:263302. DOI: 10.1063/1.2924771

[45] Yun JM, Jung CH, Noh YJ, Jeon YJ, Kim SS, Kim DY, Na SI. Morphological, optical, and electrical investigations of solution-processed reduced graphene oxide and its application to transparent electrodes in organic solar cells. Journal of Industrial and Engineering Chemistry. 2015;**21**:877-883. DOI: 10.1016/j.jiec.2014.04.026

[46] Eda G, Fanchini G, Chhowalla M. Large-area ultrathin films of reduced graphene oxide as a transparent and flexible electronic material. Nature Nanotechnology. 2008;3:270-274. DOI: 10.1038/nnano.2008.83

[47] Eda G, Lin YY, Miller S, Chen CW, Su WF, Chhowalla M. Field emission from graphene based composite thin films. Applied Physics Letters. 2008;**92**:23305. DOI: 10.1063/1.3028339

[48] Moaven S, Naji L, Taromi FA, Sharif F. Effect of bending deformation on photovoltaic performance of flexible graphene/Ag electrode-based polymer solar cells. RSC Advances. 2015;**5**:30889-30901. DOI: 10.1039/C5RA00057B

[49] Huang JH, Fang JH, Liu CC, Chu CW. Effective work function modulation of graphene/carbon nanotube composite films as transparent cathodes for organic optoelectronics. ACS Nano. 2011;**5**:6262-6271. DOI: 10.1021/nn201253w

[50] Yin Z, Sun S, Salim T, Wu S, Huang X, He Q, Lam YM, Zhang H. Organic photovoltaic devices using highly flexible reduced graphene oxide films as transparent electrodes. ACS Nano. 2010;**4**:5263-5268. DOI: 10.1021/nn1015874

[51] Geng J, Liu L, Yang SB, Youn SC, Kim DW, Lee JS, Choi JK, Jung HT. A simple approach for preparing transparent conductive graphene films using the controlled chemical reduction of exfoliated graphene oxide in an aqueous suspension. Journal of Physical Chemistry C. 2010;**114**:14433-14440. DOI: 10.1021/jp105029m

[52] Kymakis E, Savva K, Stylianakis MM, Fotakis C, Stratakis E. Flexible organic photovoltaic cells with in situ nonthermal photoreduction of spin-coated graphene oxide electrodes. Advanced Functional Materials. 2013;**23**:2742-2749. DOI: 10.1002/adfm.201202713

[53] Wu H, Hu L, Rowell MW, Kong D, Cha JJ, McDonough JR, Zhu J, Yang Y, McGehee MD, Cui Y. Electrospun metal nanofiber webs as high-performance transparent electrode. Nano Letters. 2010;**10**:4242-4248. DOI: 10.1021/nl102725k

[54] Kang MG, Kim MS, Kim J, Guo LJ. Organic solar cells using nanoimprinted transparent metal electrodes. Advanced Materials. 2008;**20**:4408-4413. DOI: 10.1002/adma.200800750

[55] Yang L, Zhang T, Zhou H, Price S, Wiley BJ, You W. Solution-processed flexible polymer solar cells with silver nanowire electrodes. ACS Applied Materials & Interfaces. 2011;**3**:4075-4084. DOI: 10.1021/am2009585

[56] Zou J, Yip HL, Hau SK, Jen AKY, Metal grid/conducting polymer hybrid transparent electrode for inverted polymer solar cells. Applied Physics Letters. 2010;**96**:203301. DOI: 10.1063/1.3394679

[57] Zhang Q, Wan X, Xing F, Huang L, Long G, Yi N, Ni W, Liu Z, Tian J, Chen Y. Solution-processable graphene mesh transparent electrodes for organic solar cells. Nano Research. 2013;**6**:478-484. DOI: 10.1007/s12274-013-0325-7

[58] Konios D, Petridis C, Kakavelakis G, Sygletou M, Savva K, Stratakis E, Kymakis E. Reduced graphene oxide micromesh electrodes for large area, flexible, organic photovoltaic devices. Advanced Functional Materials. 2015;**25**:2213-2221. DOI: 10.1002/adfm.201404046

[59] Bai J, Zhong X, Jiang S, Huang Y, Duan X. Graphene nanomesh. Nature Nanotechnology. 2010;**5**:190-194. DOI: 10.1038/nnano.2010.8

[60] Sosa NE, Liu J, Chen C, Marks TJ, Hersam MC. Nanoscale writing of transparent conducting oxide features with a focused ion beam. Advanced Materials. 2009;**21**:721-725. DOI: 10.1002/adma.200802129

[61] Zhu Y, Murali S, Stoller MD, Ganesh MJ, Cai W, Ferreira PJ, Pirkle A, Wallace RM, Cychosz KA, Thommes M, Su D, Stach EA, Ruoff RS. Carbon-based supercapacitors produced by activation of graphene. Science. 2011;**332**:1537-1541. DOI: 10.1126/science.1200770

[62] Wang ZL, Xu D, Wang HG, Wu Z, Zhang XB. In situ fabrication of porous graphene electrodes for high-performance energy storage. ACS Nano. 2013;**7**:2422-2430. DOI: 10.1021/nn3057388

[63] Oh JS, Kim SH, Hwang T, Kwon HY, Lee TH, Bae AH, Choi HR, Nam JD. Laser-assisted simultaneous patterning and transferring of graphene. Journal of Physical Chemistry C. 2013;**117**:663-668. DOI: 10.1021/jp309382w

[64] Bie YQ, Zhou YB, Liao ZM, Yan K, Liu S, Zhao Q, Kumar S, Wu HC, Duesberg GS, Cross GLW, Xu J, Peng H, Liu Z, Yu DP. Site-specific transfer-printing of individual graphene microscale patterns to arbitrary surfaces. Advanced Materials. 2011;**23**:3938-3943. DOI: 10.1002/adma.201102122

[65] He Q, Sudibya HG, Yin Z, Wu S, Li H, Boey F, Huang W, Chen P, Zhang H. Centimeter-long and large-scale micropatterns of reduced graphene oxide films: Fabrication and sensing applications. ACS Nano. 2010;**4**:3201-3208. DOI: 10.1021/nn100780v

[66] Tung VC, Chen LM, Allen MJ, Wassei JK, Nelson K, Kaner RB, Yang Y. Low-temperature solution processing of graphene-carbon nanotube hybrid materials for high-performance transparent conductors. Nano Letters. 2009;**9**:1949-1955. DOI: 10.1021/nl9001525

[67] Xu Y, Long G, Huang L, Huang Y, Wan X, Ma Y, Chen Y. Polymer photovoltaic devices with transparent graphene electrodes produced by spin-casting. Carbon. 2010;**48**:3308-3311. DOI: 10.1016/j.carbon.2010.05.017

10

Fluorinated Graphene Dielectric and Functional Layers for Electronic Applications

Irina V. Antonova and Nadezhda A. Nebogatikova

Abstract

Future electronics technology is expected to develop from rigid to flexible devices. This process requires breakthroughs in material properties, especially flexibility, in combination with desirable electrical insulating, semiconducting, or metallic properties. Graphene, being one of the recently developed two-dimensional (2D) materials, presents great promise as an active layer in a wide spectrum of electronics devices and, first of all, in field-effect transistors (FET). The development of optimized dielectrics for the graphene active layer is critical for graphene applications. The carrier transport in graphene films takes place at interfaces with dielectric or semiconductor substrates; therefore, the quality of such interface and the interaction of graphene films with nearby dielectric layers (charge carrier scattering) determine the device performance. Generally, the development of dielectric materials aiming at high performance device operation, proper mechanical properties, and low-temperature fabrication is not progressing well since the graphene thin film is very sensitive to surface conditions of dielectric layers. Solving the problem with dielectric layers in the case of nonorganic printed and flexible electronics is especially acute. As it is demonstrated in the present chapter, dielectric layers fabricated from fluorinated graphene suspension or in its combination with graphene oxide are the most promising for graphene-based flexible, printed, and transparent electronics.

Keywords: flexible electronics, fluorinated graphene, suspension, dielectric films, leakage current, charges, resistive switching, quantum dots, negative differential resistance

1. Introduction

Future technologies evolve toward flexible electronics (bendable, rollable, stretchable, or transparent) developed for a wide spectrum of bio- and medical applications, sensors and gadget displays, textile or clothing electronics, and so on [1–4]. Printing processes have attracted great attention, due to their compatibility with flexible substrates and materials, excellent prospects for applications, technical feasibility for scaling to large-area manufacturing and low cost. The two-dimensional (2D) printing technologies are at present used for creating intelligent components and smart systems of printed electronics, such as memories, displays, electronics, batteries, micromechanical systems, sensors, thin-film transistors, and other devices of modern electronics [1–4]. Flexible device technology does not support high temperatures, required for such materials as SiO_2 or high-k Al_2O_3, HfO_2, and ZrO_2 [5]. So, alternative 2D dielectric materials are highly demanded. Recent developments in technologies of liquid-phase graphene exfoliation lead to a significant progress in methods for creating 2D materials, including graphene, graphene oxide (GO), and other related materials (MoS_2, WSe_2, and so on) [6–10].

Graphene presents great promise as an active layer in a wide spectrum of devices of flexible electronics and, first of all, in field-effect transistors. Recent reports demonstrate successful realization of graphene FETs on flexible or even on stretchable substrates [8–13]. To make such applications possible, the development of optimized dielectrics for the active graphene layer is critical (gate and interlayer dielectrics, or/and supported layer for graphene). Nowadays, traditional high-k materials, such as Al_2O_3, HfO_2, and ZrO_2, are widely used for FETs. Nevertheless, the development of dielectric materials allows achieving high performance of the devices with excellent mechanical properties and low fabrication temperature that is in great demand.

Recently, graphene oxide (GO), one of the most well known graphene derivatives, has been exploited as a gate dielectric for graphene-based FETs [7, 8, 12, 13]. With its good mechanical and optical properties, this material offers a unique advantage for high performance flexible and transparent electronic devices since it can be formed on a graphene channel by solution-based or direct oxidation at room temperatures. Generally, GO has combined an excellent flexibility, a relatively large leakage current, and a strong limitation on enhanced temperature (even under the current flow). This limitation on enhanced temperature is connected with reducing GO: for instance, annealing at 100 °C typically leads to a decrease in GO layer resistivity by 4–5 orders of magnitude [5].

The most stable graphene derivative with dielectric properties is fluorinated graphene (FG), which is a low-k material with k = 1.2 and band gap of 3–8 eV [14–16]. Graphene can be fluorinated by a low-damaged CF_4 plasma treatment [16] or by exposing the graphene to XeF_2 gas to convert it to insulating fluorographene (C_4F) [17]. A new simple approach for graphene fluorination (treatment in aqueous solution of hydrofluoric acid) was recently suggested in Refs. [18, 19]. In the case of graphene suspension, such treatment leads not only to fluorination of flakes but also to additional flake fragmentation and exfoliation [19, 20]. As a result, a considerable decrease in thickness and lateral sizes of graphene flakes is accompanied by simultaneous transition of the flakes from conducting to insulating state. Smooth and uniform insulating films with low roughness can be created from the suspension on different substrates. The films from the fluorinated suspension are cheap, practically feasible, and easy to produce. The electrical and structural properties of the films from such fluorinated suspension with variable fluorination degree are discussed in the present chapter. Excellent characteristics of the dielectric film created from high fluorinated FG may be compared only with well developed SiO_2 layers. In addition to excellent dielectric properties, reversible resistive switching effect and negative differential resistance (NDR) were detected in the films, created from the suspension with relatively low fluorination degree. Possible applications of these films are also considered.

2. Fluorination of graphene suspensions

The starting material was a graphene suspension that can be obtained using any of available approaches. It may be suspensions created by mechanical crushing of natural graphite, dimethylformamide (DMF) intercalation, ultrasonic treatment intended for splitting the intercalated particles, and centrifugation aiming at removal of nonsplit graphite particles [20]. At the stage of intercalation, other organic solvents may be used, for instance, N-methyl-2-7 pyrrolidone (NMP) [21, 22]. Suspensions may be also created by electrochemical exfoliation [23, 24], liquid-phase exfoliation in water or water-based solution [25], or shear mixer [26, 27]. During suspension preparation, the natural graphite typically turned into flakes with characteristic size from 1 to several micrometers and thickness from 0.4 (monolayer) up to tens of nanometres. Parameters of flakes strongly depend on the technology used. After obtaining the suspension, it was subjected to a fluorination procedure [20, 28]. To this end, equal volumes of the suspension and a 5–10 % solution of hydrofluoric acid (HF) in water were mixed together. Fluorination time strongly increased with an increase in size and thickness of pristine graphene flakes and could be accelerated by enhancing temperature [20]. Below, we demonstrate the fluorination process in the time scale when pristine graphene flakes have lateral sizes of 1–2 mm and thickness of 3–20 nm, without temperature acceleration. So, we are considering the admittedly inferior option which took sufficiently long time to clarify fluorination stages and all accompanying processes.

Let us consider here the reason for graphene fluorination during treatment in the aqueous solution of hydrofluoric acid (HF treatment). It is very important to note that practical difficulties associated with F reactions in two-dimensional (2D) carbon nanostructures are avoided if working with curved carbon nanosheets. An example of F reactions with corannulene molecules having a curved surface representing 1/3 of the structure of C_{60} fullerene was realized by dos Santos [29]. Corannulene is a polycyclic aromatic hydrocarbon with chemical formula $C_{20}H_{10}$. Density functional theory (DFT) calculations of the HOMO−LUMO were used to demonstrate that gap variation for fluorinated corannulene, used as a prototype of C-based nanostructures, yields in values of band gap of 0.13–3.46 eV due to fluorination. Our experiments on the fluorination of CVD grown or mechanically exfoliated graphene in the aqueous solution of hydrofluoric acid also show corrugation of these materials during fluorination. In other words, the corrugation of domain boundaries is an initial driving force for subsequent F reaction with carbon atoms in 2D plain and subsequent corrugation [19, 30–32]. This type of reactions stimulated by deformations is also realized for the graphene suspension, and only thin (few layer) flakes are suitable for fluorination during HF treatment.

Liu et al. [33] have employed the first-principles method within the density functional theory to study the structural, electronic, and magnetic properties of the fluorinated graphene with different coverage of fluorine. The authors have found a strong variation of the graphene properties already at the early stage of fluorination (lower than $CF_{0.25}$). Duan [34] has demonstrated that when fluorine bonds to a carbon atom, the latter one is pulled slightly above the graphene plane, creating what is referred to as a CF defect. These CF defects cause the graphene surface buckling. This corrugation was experimentally observed under fluorination in the aqueous solution of the hydrofluoric acid [33]. Duan further showed that the addition of fluorine to graphene in some cases leads to the formation of an energy band gap near the Fermi level of 0.37–0.24 eV [34]. Generally, it was demonstrated that the adsorption of fluorine on graphene surfaces is a promising approach to modify the properties of graphene, which may lead to more flexible electrooptical applications of graphene in the future.

To investigate the properties of FG suspension, some portions of the suspension were periodically used for the study and preparation of films. The substrates for films were silicon (Si) wafers. In depositing the films, the native oxide was removed from the surface of silicon by the hydrofluoric acid available in the solution. The deposited films were dried and rinsed with deionized water for removing the residual hydrofluoric acid and the organic component of the suspension, and then subjected to a second drying treatment for water removal. In other cases (especially for 2D printing), the solution in suspension was substituted for water. Due to hydrophilicity of FG flakes, the water-based suspension was stable.

3. The evolution of graphene suspension structure during fluorination

Fluorination of the DMF-based graphene suspension with a solution of hydrofluoric acid in water is considered in the present part. It has been found that, as a result of the interaction, the suspension particles undergo an additional splitting and size reduction by one-two orders of magnitude. **Figure 1(a)–(d)** shows the surface image of films created of Si substrate from the suspension after different times of fluorinated treatment. A schematic pattern for the action of the HF-containing solution on suspension particles resulting in additional exfoliation of graphene sheets, and formation of thinner and finer fluorinated graphene flakes has been proposed (**Figure 1(e)**). The driving force of these structural changes is found to be mechanical stresses, due to a difference in lattice constants and wettability of fluorographene [20, 28]. The similar fragmentation and exfoliation of graphene flakes are found for different kinds of graphene suspensions during the fluorination process.

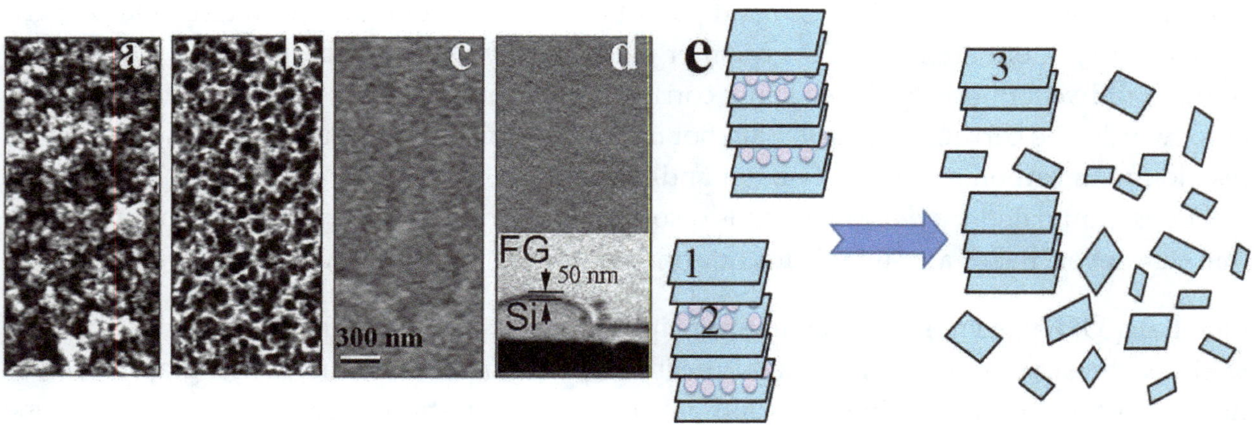

Figure 1. Scanning electron microscopy (SEM) images of the surface for films with different fluorination degree: (a)—pristine (nonfluorinated) film; (b), (c), and (d)—films fluorinated respectively during 2, 10, and 40 days. The inset in (d) shows an image of an edge of the film taken at the angle of 45° to the surface; the film thickness indicated in the figure was evaluated with allowance for measurement geometry. (e) A sketch illustrating initial flakes splitting and their fractionation in finer flakes that occurred during fluorination process (1—initial particle, 2—split of partially fluorinated flakes, 3—intercalated DMF layer). Reprinted with permission from Ref. [20].

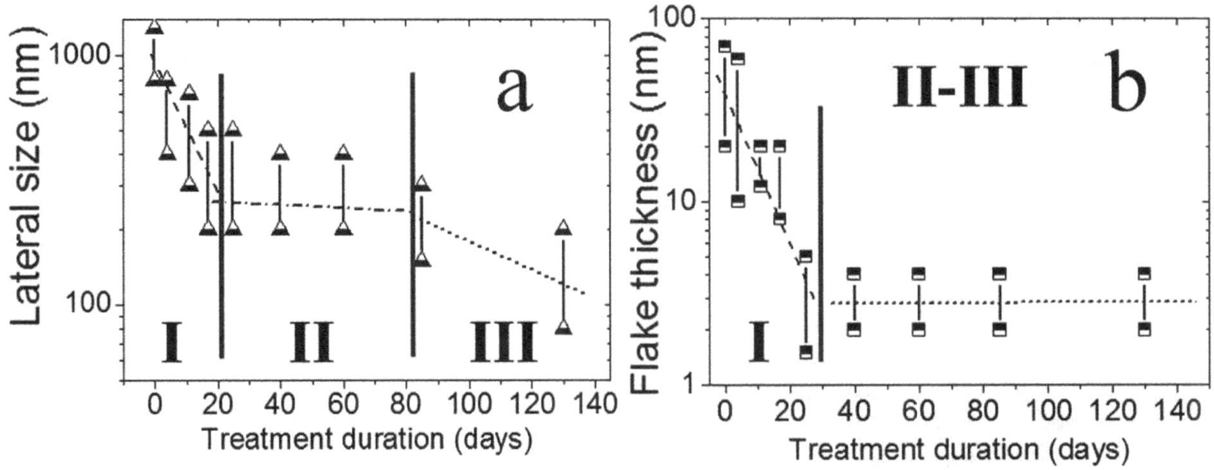

Figure 2. Dependence of the lateral size and thickness of multilayered graphene flakes in suspension with optimum composition of 0.16 mg/ml on fluorination time. Reprinted with permission from Ref. [28].

Dependences of FG flake size on time of HF treatment are given in **Figure 2**. It is seen that the process of flakes fragmentation may be divided into several stages—up to 20 days, from 20 to 80 days, and over 80 days. These intervals correlate with changes in electrical properties of the films obtained from suspensions. About 20-day treatment was required for the film transition from conducting to insulating state. During fluorination for 20 to 80 days, the insulating properties of the films improved, and the current through the films and the charge density in the films decreased. As it will be demonstrated below, the most optimal property for the majority of applications is the fluorination within 60–80 days. The observed stages well agree with the behavior of leakage current through the FG film, which are also given below (**Figure 4(a)**).

Evidences proving the fluorination of suspension particles during the suspension treatment in a solution of HF in water were obtained by means of analysis techniques such as X-ray photoelectron spectroscopy (XPS), Fourier transform infrared (FTIR) spectroscopy, and Raman scattering spectroscopy (RSS) (**Figure 3(a)** and **(b)**) [17]. Strong IR bands with maxima 1107, 1166, and 1230 cm^{-1} are clearly seen in the FTIR spectra. The observed modes supposedly correspond to the fluorine atoms, connected with sp^3-hybridized carbon atoms. The studies of FTIR spectra with an increasing fluorination time have revealed that an adsorption band at 1112 cm^{-1} gradually changes into 1211 cm^{-1}. The modes observed for the studied films are supposed to correspond to C-F bond and change depending on fluorination time. The gradual emergence of fluorinated graphene properties was traced in measured Raman spectra (**Figure 3 (b)**) for the films, which was obtained from suspensions with different fluorination times. It is seen that an increased duration of the treatment results in a decreased intensity of peaks in Raman spectra. This effect is typical for fluorination of graphene and connected with band gap opening. The images of pristine graphene suspension and FG suspension also clarify the fluorination process. XPS spectra show a signal from fluorine F1s and carbon C1s, detected in the energy region of 687.7 and 284–286 eV, respectively (**Figure 3(d)**). The energy position and shape of the peaks are indicative of partial fluorination of suspension particles. Decomposition of C1s demonstrates peaks C-F with position at 288.3 eV and a peak C-CF at 286.5 eV. These XPS spectra correspond to fluorination degree ~ C_4F and are typical for the case when suspension becomes transparent and transfers to a nonconductive state (see details in Section 6). In a

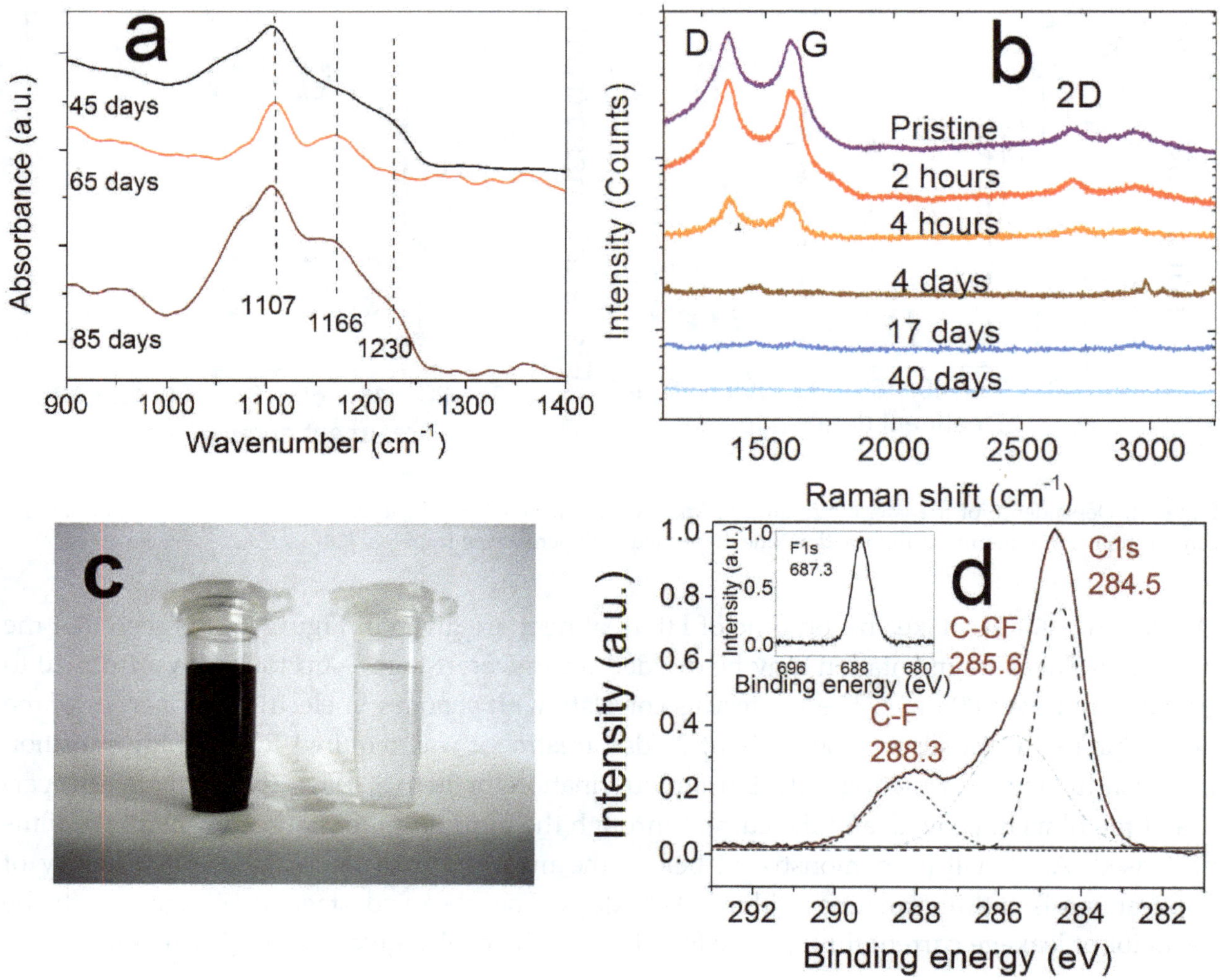

Figure 3. (a, b) Evolution of IR and Raman spectra of pristine fluorinated films with time of HF treatment. (c) Images of pristine graphene suspension and FG suspension after 50 days of HF treatment. (d) XPS study of the fluorinated graphene film: the spectrum near the peak of C1s with decomposition into component lines, an insert shows a part of the spectrum with F1s peak. Fluorination time was 50 days. Reprinted with permission from Ref. [20].

longer HF treatment, it is possible to obtain higher fluorination degrees limited to the value ~ C_2F. Due to this fact, we have a partially fluorinated suspension in all cases, and namely this FG suspension demonstrates many properties that are attractive for applications.

4. Insulating properties of hardly fluorinated graphene suspension

As it has been shown earlier, a considerable decrease in thickness and lateral sizes of graphene flakes (up to 1–5 monolayers in thickness and 20–30 nm in diameter) during fluorination is found to accompany the transition of the grapheme flakes from conducting to insulating state. The change in leakage current through the FG films as a function of the fluorinated time is given in **Figure 4(a)**. One can compare these currents with leakage current of 17 mA/cm^2 for 100 nm thick GO films [12] and ~20 A/cm^2 for 4 nm GO films [8]. So, the electrical and structural properties of the films suggest their use as insulating elements in thin-film nano- and microelectronics devices/structures. We have performed an analysis of the capacitance-

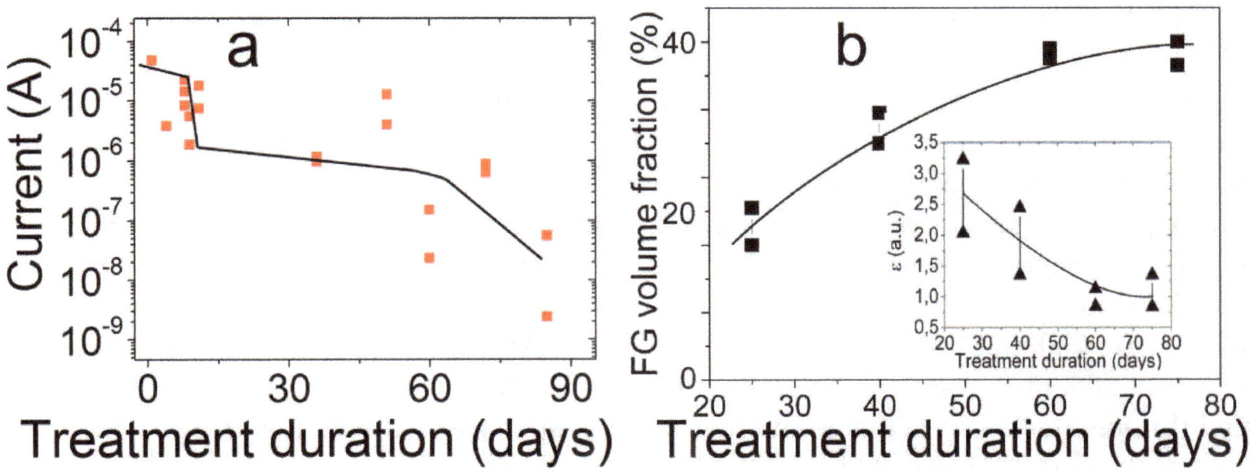

Figure 4. (a) The electric current density across the vertical Me/FG film/Si structures biased with voltage U = ± 0.1 V versus the duration of the suspension fluorination procedure. (b) Film fluorination degree versus the duration of fluorination treatment in the solution of HF in water. The inset in (b) shows the value of dielectric constant ε versus the fluorination treatment duration. (b) is reprinted with permission from Ref. [20].

voltage characteristics of metal (Ag, Au)/insulator (FG)/semiconductor (Si) (MIS) structures on different substrates. Dependence of relative dielectric constant ε on fluorinated time is shown in **Figure 4(b)**. The dielectric constant of the films is determined to range from 1.1 to 3.2, depending on the fluorination degree of the material. It is important to mention that value of ε = 1.2 known for fluorographene [16, 20], appears in our films for relatively low fluorinated degree (between C_4F and C_2F).

The density of the fixed charge in the film Q_f extracted from the flat band voltage of the capacitance-voltage characteristics and the density of interface states at the interface with silicon D_{it} are given in **Table 1**. Recall that the fluorinated graphene film was applied onto a silicon surface free of native oxide. The densities Q_f and D_{it} proved to range from 6×10^{11} cm^{-2} to $(0.3–0.5) \times 10^{11}$ cm^{-2}. With increasing duration of fluorination treatment, the densities Q_f and D_{it} were found to decrease in value. The quality of obtained dielectric films was improved by increasing the treatment duration. On the whole, the obtained values proved to be much lower than the values of Q_f and D_{it}, typical of many traditional dielectric coatings used in nano- and microelectronics such as Al_2O_3, HfO_2, and ZrO_2 [5]. The typical values of built-in or interface charge density in the widely used Al_2O_3 films range within 10^{12}–10^{13} cm^{-2} [35]; in HfO_2, they vary within 10^{11}–10^{12} cm^{-2} [36, 37], and in the Si/SiO_2graphene/ZrO_2 structures, they range within $(1–15) \times 10^{11}$ cm^{-2} eV^{-1} [38].

Substrate	Time of HF treatment	20 days	40 days	60 days	80 days	150 days
Si	Q_f, cm^{-2}	$(5–6) \times 10^{11}$	1×10^{11}	5×10^{10}	$(2–6) \times 10^{10}$	$(5–7) \times 10^{10}$
Si	D_{it}, cm^{-2}	$(3–5) \times 10^{11}$	5×10^{10}	$(1–3) \times 10^{10}$	2×10^{10}	$(3–5) \times 10^{10}$

Table 1. Specific values of fixed charge density Q_f, calculated on the flat band voltage, determined from the capacitance-voltage characteristics, and the state density D_{it} at the interface with silicon, obtained from the difference between the middle-gap and flat band voltages.

5. Relation between structural and electronic properties of FG films

Figure 5 shows SEM and AFM images of partially fluorinated films, created from the graphene suspension, treated in an aqueous solution of HF for 7 days (~$CF_{0.10}$) and 20 days ($CF_{0.23}$) [31]. In both cases, the continuous films consisting of separate flakes may be observed in **Figure 5**. The bright corrugated areas of the film correspond to fluorinated regions, and the dark round regions correspond to nonfluorinated or weakly fluorinated graphene islands. The estimated flake sizes in suspension are 100–300 nm with a thickness of 0.5–5 nm for less fluorinated films (see **Figure 2(a)** and **(b)**), and 20–100 nm with a thickness of 0.5–2 nm for more fluorinated ones. The film thickness varies from 20 to 150 nm depending on the drop volume. **Figure 5(a)** and **(b)** presents the AFM image and the current map, measured with the probe for the same part of the weakly fluorinated FG film. Conductivity was found for graphene islands (dark areas), and bright areas (fluorinated part) were characterized by insulating properties. A comparison of AFM images clearly demonstrates that an increase in the HF treatment time leads to a decrease in the size of graphene islands and an increase in the size of the fluorinated areas between the graphene islands. So, generally, the films contain graphene islands and their properties were examined for pertinence to quantum dots (QDs).

The charge deep level transient spectroscopy (Q-DLTS) and transport measurements were used to characterize partially fluorinated films [31]. It has been found that at the temperature range from 330 to 250 K, the current is described by equation $I = I_o \exp(-E/kT)$, where E = 0.48 eV is the activation energy of the current flow through the film, and k is the Boltzmann constant. At lower temperatures, the current is close to be constant. The activationless current flow through the film is, most likely, associated with carrier migration (tunneling) through the traps in the FG band gap or graphene inclusions. Temperature dependence of the current is given in **Figure 7**, curve 1. Q-DLTS measurements demonstrate that only one type of activated process with energy 0.50 eV and density of localized states of ~ 10^{11} was observed. This activated process most likely corresponds to carriers passing over 0.5 eV potential barriers between graphene islands and fluorinated part of the film. It means that QDs with quantized electronic properties are not seen in these films. It may be connected with varying sizes of graphene islands and existing edge-related defects which provide the possibility of activationless transport in the films. It is worth mentioning that

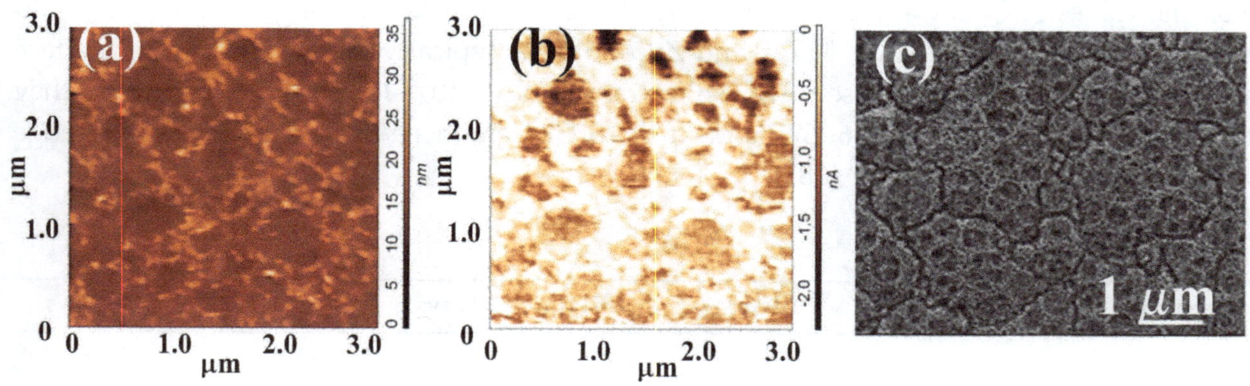

Figure 5. (a) AFM image of the FG film created by drops from a partially fluorinated suspension (7 days, $CF_{0.10}$) and (b) current map for the same place on the film. The bright regions in both images correspond to fluorinated graphene. (c) SEM image of the films created from a graphene suspension and treated in an aqueous solution of HF for 20 days ($CF_{0.23}$), respectively. Reprinted with permission from Ref. [31].

the arrays of QDs formed after similar fluorination of graphene or few layer graphene demonstrate the size quantization levels in Q-DLTS and transport measurements [31, 39–41]

6. Resistance switching effects in fluorinated films

Currently, the resistive memory approach is considered the most promising, as it allows obtaining a significantly lower time of memory operation (data overwriting). The most important parameters for the memory devices are a low overwriting voltage (lower than 3 V) and a high number of rewriting cycles and stability [42, 43]. The main problems of traditional metal oxide materials in the case of flexible electronics are relatively high fabrication temperature and very high cost. So, recent interest in the graphene-based materials results from its possible use for flexible resistive memory elements. Recently, some efforts have been made to create the resistive memory from graphene oxide (GO) or MoS_2/graphene oxide composite [5, 10]. We have investigated fluorographene that is much more stable than GO graphene derivatives. The resistive switching effect for partially fluorinated graphene films (a two-phase system of graphene islands embedded in FG matrix) was observed for the first time [32].

The reversible resistive switching effect was found for the films, created from the suspension of partially fluorinated graphene for fluorination level of about ~ C_4F (fluorination time of 40–60 days). **Figure 6(a)** and **(b)** shows current-voltage (I-V) characteristics for vertical Au/FG/Si/InGa structures and lateral Au/FG/Au structures, measured at two voltage sweeps. It is seen that the transition to the lower resistance of the layer occurs at voltage of about −1.5 V in vertical configuration and at about −4 V in lateral configuration. The reverse transition occurs by changing the polarity of the applied voltage. Repeated measurements prove the repeatability of such switching. The variation of temperature measurements from 80 to 350 K shows that these transitions may be observed in the entire studied temperature range. Daylight illumination is very important for carrier transport in fluorinated films (compare **Figure 6(b)** and **(c)**). The reset loops observed in dark conditions degrade under daylight illumination. The reason for this effect is connected with a strong decrease in the time of nonequilibrium carrier relaxation, directly observed in similar structures [34].

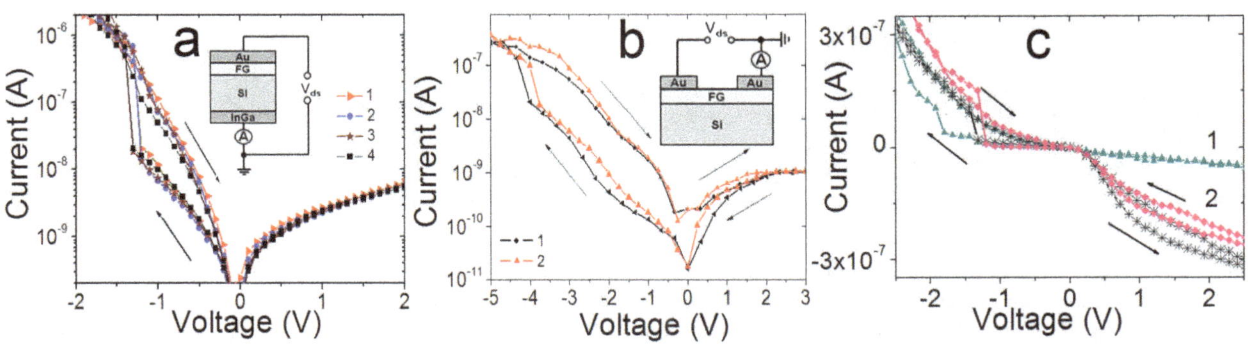

Figure 6. Current-voltage characteristics measured at daylight conditions of (a) vertical Au/FG/Si/InGa structures and (b) lateral Au/FG/Au structures (contact size was 0.5 mm, and period between contacts was 1.5 mm) measured several times at two voltage sweeps at the temperature of 300 K. (c) vertical Au/FG/Si/InGa structures measured in daylight (1) and dark (2) conditions. Fluorinated graphene film had thickness of 80 nm. Reprinted with permission from Ref. [32].

For a lower fluorination degree, the relation of resistances decreased, and for a higher fluorination degree, films became insulated.

One of the most important parameters of the material for resistive memory is the time of switching from high to low resistance states [5, 42, 43]. Q-DLTS measurements allow direct determination of the time of carrier emission (relaxation of nonequilibrium charge) from the specified traps [32]. Time is one of the switching process parameters. The emission (relaxation) time of nonequilibrium carriers is determined from the position of the corresponding peak on Q-DLTS spectrum. It was found that after applying relatively high voltage (~4 V) we can fix the low resistive state of films for Q-DLTS measurements, and after 160°C annealing, the high resistive state can be restored.

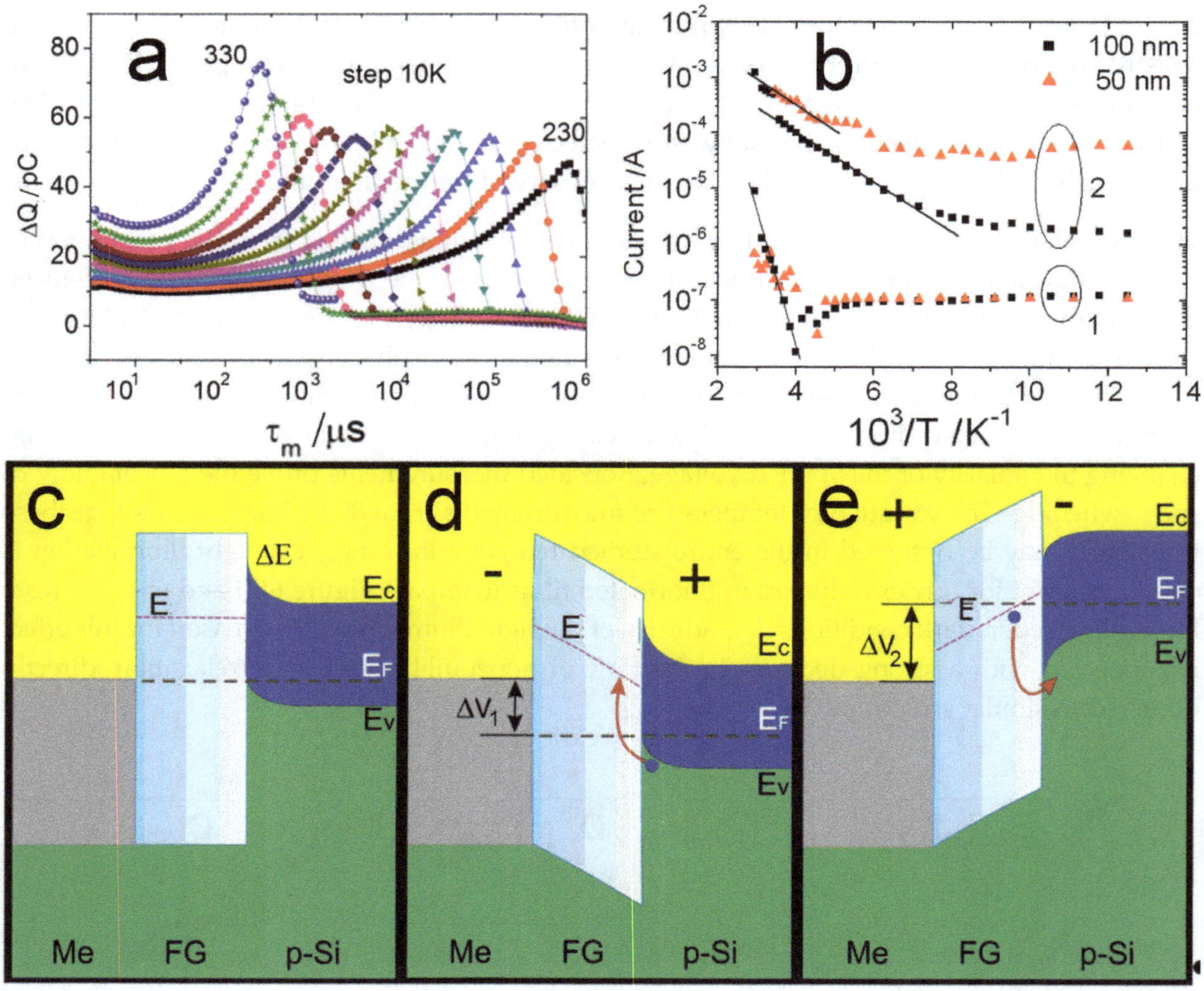

Figure 7. (a) Q-DLTS spectra for FG/p-Si structure in low resistance state, associated with the holes, which are consistent with the activation energy of the holes 0.34 eV in FG/p-Si structures. (b) Temperature dependences of the current through the fluorinated graphene film on p-Si substrate in the high (curve 1) and low (curve 2) resistance states, measured in vertical configuration. Lines correspond to the activation energy 0.48 eV (1) and 0.09 eV (2). (c–e) Expected band diagram of (c) Au/FG/p-Si structures at no voltage, (d) for the filling pulse ΔV_1, (e) during the carriers emission from the traps for holes at constant voltage, applied to the sample ΔV_2. Only one trap is considered in this diagram for simplicity. Reprinted with permission from Ref. [32].

The charge spectroscopy Q-DLTS and transport measurements were used to study traps in the films from partially fluorinated suspension in the states of both low and high resistance. The activation energy of traps E_{01} (0.50 eV), obtained from Q-DLTS method for the films in high resistive states, as mentioned above, may be interpreted as a potential barrier between graphene and fluorinated graphene areas, or between FG film and valence bands of silicon substrate. Transport and nonequilibrium recharging processes in the high resistance state were found to occur above all, due to carrier tunnelling through potential barriers in the films. Several types of traps for electrons and holes (one trap is demonstrated in **Figure 7**) with the density of 10^{10}–10^{12} cm^{-2} were formed in the low resistive state of FG films. Among them are electron traps with activation energies of 0.15, 0.12, and 0.08 eV, and hole traps with energy of 0.34 eV. The minimum relaxation time of nonequilibrium carriers from different traps was found to be about 700 ns. The energy level position of corresponding traps from the conduction band of a silicon substrate equals 0.08 eV. The origin of the observed traps is supposed to correspond to traces of organic components, which are used during graphene suspension creation.

Figure 7(b) demonstrates temperature dependences of the current through the films, with thickness of 50 nm and 100 nm, at two electric field intensities. For the low resistance state (curve 2), the activation energy E for both films with different thickness is 0.09 eV, at the temperature range from 330 to 150–170 K [32]. At lower temperatures, the current again is close to constant. The activationless current flow through the film is, most likely, associated with carrier tunneling on the traps in the FG band gap and graphene inclusions. Comparison of transport and Q-DLTS data demonstrates that the carrier transport in the low-resistance state is determined by the same traps (traps with activation energy 0.08 – 0.09 eV), and they form conductive channels in the films.

7. Films with negative differential resistance

Negative differential resistance (NDR) devices with nonohmic current-voltage characteristics are used in a wide range of applications, including frequency multipliers, memories, fast switches, high-frequency oscillators operating up to the THz range, etc., [44–46]. Currently, the theoretical predictions of NDR pertaining to graphene appear to prevail [47–49], and only a few experimental observations of NDR are presented in the literature (in fact, NDR is experimentally observed in lateral structures only in Ref. [50]). In the case of a two-barrier structure, it is possible to observe different regimes, from oscillation behavior in conductivity to only positive differential conductance [47]. It is this situation that was observed in our experiments for FG films with a relatively low fluorination degree (F/C ratio) [30].

In our case for films created from suspensions, NDR was observed for relatively weakly fluorinated layers CF_x with fluorination degree ranged from $0.10 < x < 0.23$ (**Figure 8(a)** and **(b)**) [30]. Two types of the film with slightly different flake size demonstrate similar I-V curves, but with different numbers of peaks. These films demonstrate one or few peaks in I-V curves, depending on the flake sizes, and that is explained by formation of barrier or multi-barrier systems. The current steps also seen in I-V curves are most likely connected with the presence of electrically active traps considered above. It's worth noting that the obtained films were multibarrier systems due to the presence of a fluorinated network and non-fluorinated graphene islands. An increase in the fluorination degree first results in the increased number of the FG barriers. The origin of NDR for I-V curves shown in **Figure 8(a)** and **(b)** is mostly likely associated with the theoretically predicted gap in the transmission coefficient for carriers in the barrier between fluorinated graphene and graphene areas. It is caused by the competition of hole-to-electron transport and Klein tunnelling with resonant tunnelling in structures with potential barrier(s) [48].

The origin of the NDR was demonstrated to vary with an increase in the F/C ratio. Thick FG films (150 nm) with F/G ratio $CF_{0.23}$ showed two peaks in the I-V curves, observed for voltages swept from negative to positive values (**Figure 8(c)** and **(d)**). The peaks were observed for the vertical configuration of measurements. A reverse voltage sweep produced I-V curves without any peaks (inset of **Figure 8(d)**). It is suggested that electrochemical oxidation-reduction reactions involving organic contaminants (traces of DMF, used to produce the graphene suspension) are located near defects, fluorinated carbons, or other special areas in our films. The temperature dependence of the conductivity at peaks exhibited activation behavior $\sigma = \sigma_0 \exp(-E_i/kT)$, where E_i is the activation energy, k is the Boltzmann constant, and T is the temperature. E_i takes values 0.04 and 0.09 for peak 1, and 0.16 for peak 2.

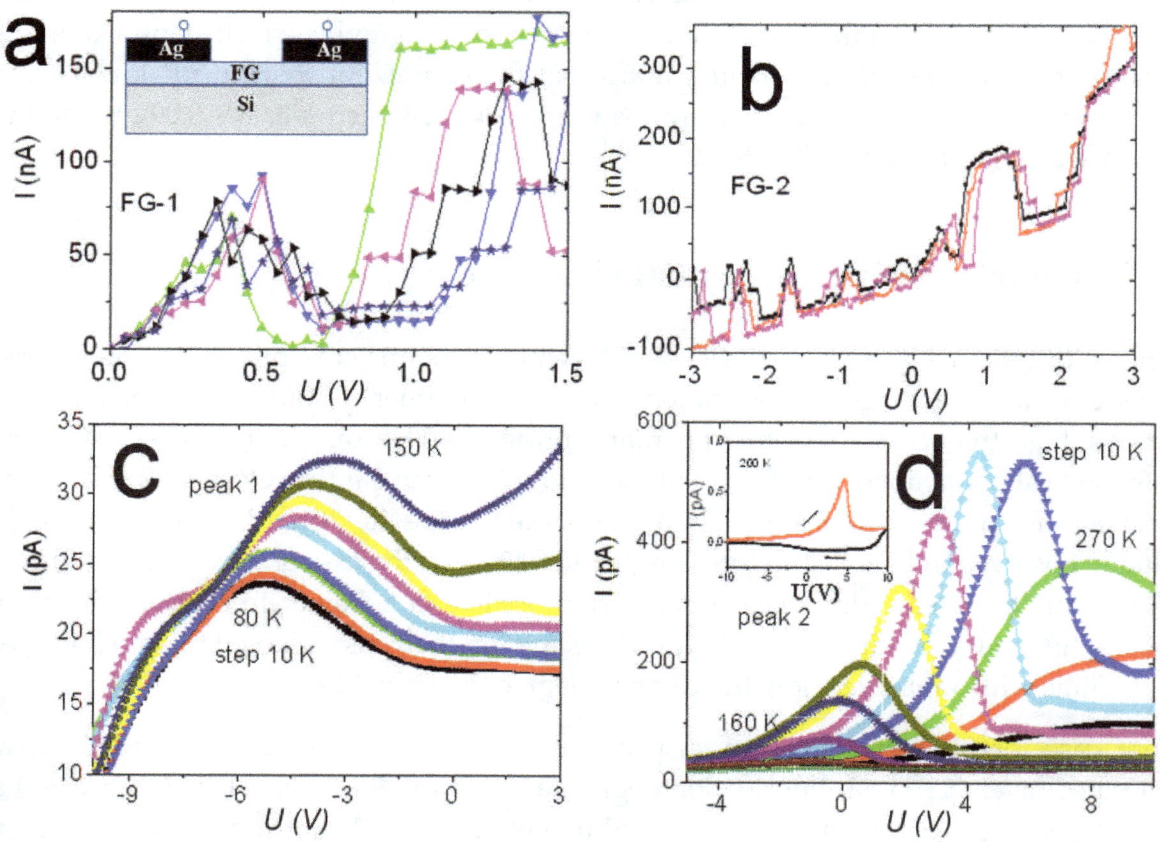

Figure 8. (a) and (b) I-V characteristics measured in lateral configurations for FG films ~$CF_{0.10}$ for films created from suspension with smaller flake size in the case (b). Different curves correspond to repeated measurements on the same structures. (c, d) I-V characteristics measured in vertical configuration for FG film ($CF_{0.23}$) over the temperature range of 80–300 K. Peak 1 (c) and peak 2 (d) were observed in thick 150 nm FG film. Reprinted with permission from Ref. [30].

The effect of NDR in fluorinated films widens the range of application including active device layers fabricated using 2D printed technologies on rigid and flexible substrates.

8. FG suspension for 2D inkjet printing technologies

The discovered control of the size of flakes in the graphene suspension during its fluorination serves to create inks for 2D inkjet printing [28, 51–53]. Suspensions of fluorinated graphene with nanometer size flakes are of interest for the development of 2D inkjet printing technologies, and production of thermally and chemically stable dielectric films for nanoelectronics [28]. The printed fluorinated graphene films on silicon and flexible substrates have been demonstrated first time, and the charges in MIS structures have been estimated as ultra low values of $(0.5–2) \times 10^{10}$ cm^{-2}.

The properties of the graphene oxide films may be greatly improved by adding the FG top layer. The thin FG and GO films are printed on silicon and flexible polyethylene terephthalate (PET) substrates (**Figure 9 (a)** and **(b)**). FG flakes have lateral sizes ranging within 20–100 nm and thickness of 0.4–1.5.nm. **Figure 9 (a)** contains the photo of printed graphene oxide lines with different number of layers. Layers in the left part of the substrate are coated by a layer of printed FG. The boundary between areas with and without FG is marked with arrows. The FG

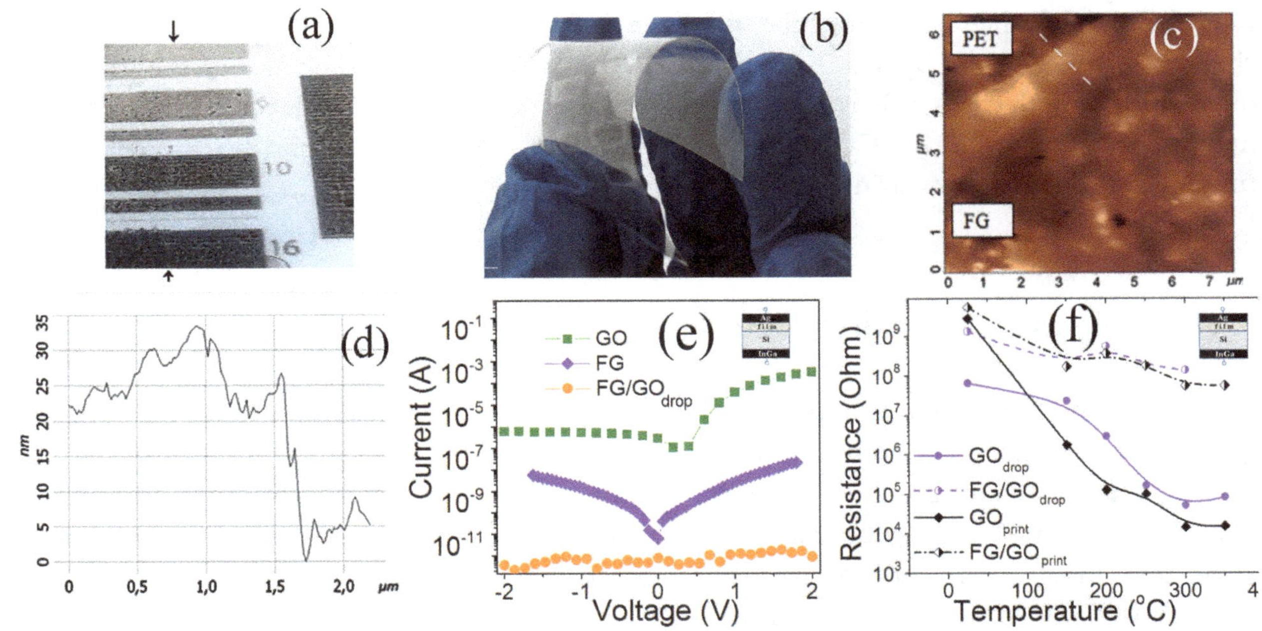

Figure 9. (a) FG printed 10-layer films on the PET substrate. (b) Images of graphene oxide printed 4, 6, 10, and 16-layer films with an additional FG 8-layered coating printed on top of the left half of the wafer. The boundary of the FG layer is slightly seen in the image, and for better visualization, it is marked with arrows. (c) An AFM image and height profile (d) near the edge of the FG film from (b). FG layer thickness can be estimated at ~17 nm. (e) Comparison of the current values through the GO, FG, and FG/GO films deposited by droplets. (f) Dependence of the current through the two-layer FG/GO and GO films on the isochronous (20 min) annealing temperature; the considered films were obtained by printing and applying droplets. Annealings were carried out in an inert atmosphere (Ar with addition of 10% H_2). Reprinted with permission from Ref. [28].

layers on the PET substrate are presented in **Figure 9 (b)**. An AFM image of the FG layer on PET and the profile near the edge of FG film are demonstrated in **Figure 9 (c)** and **(d)**.

The properties of two-layer films of FG on GO were the most extensively studied due to their revealed stability. **Figure 9 (e)** shows the current-voltage characteristics of the two-layer films of FG/GO, created by printing and dropping, measured in lateral configuration. It has been found out that the magnitude of the current through the two-layer structure is significantly lower than that through the separate films of GO or FG. The increase in the two-layer film thickness was shown to be not that significant, compared with the effect of reducing the current through the film by several orders of magnitude, especially in the case of printed layers. Applying the FG layer causes primarily the "healing" of structural defects in the GO film. The origin of this effect lies in good affinity between GO and FG, and in the electrostatic interaction between GO structural defects and FG flakes. Surface roughness of different films on rigid substrates was extracted from AFM images: the surface roughness for GO film was 8.1 nm, for FG/GO film, it was 5.7 nm, and for FG film, it was 1.5 nm. This effect is supposed to result from blocking of graphene oxide conductivity with small FG flakes. Significant decrease in FG/GO film roughness suggests formation of FG few layers on structural defects of GO films (local insulating island with thickness exceeding the average FG film thickness).

Assessment of thermal stability of different structures required investigation and comparison of the conductivity of GO and FG/GO films, obtained by applying drops and printing. **Figure 9** (f) presents dependences of film resistance on temperature of isochronous annealing. The graphs show that the resistance of the two-layer structures, both printed and created by drops, changes not more than by 1–1.5 orders, while the change in the resistance of the graphene oxide films in both cases is 3–5 orders of magnitude. Therefore, the two-layer FG/GO films are much more stable than the GO films, since the application of the second layer greatly suppresses the GO recovery during heating.

9. Outlook and conclusion remarks

Creation of the fluorinated graphene suspension and inks are shown to extend the range of graphene-based materials from conductive to insulating and functional layers (**Figure 10**). The suggested approach for graphene or graphene suspension fluorination in the aqueous solution of hydrofluoric acid allows obtaining the partially fluorinated graphene with fluorine content lower than in $CF_{0.5}$. This approach is cheap, practically feasible, and easy to produce. Moreover, films obtained from the partially fluorinated graphene may be conducting or insulating, depending on the suspension fluorination degree.

Fluorinated graphene cannot be structured using traditional technological approaches such as plasma treatment. This problem of the fluorinated graphene nanostructuring is successfully overcome by means of spin-coating or printed technologies. As a result, the use of FG suspension is the most convenient approach for a wide spectrum of applications, especially for flexible technologies.

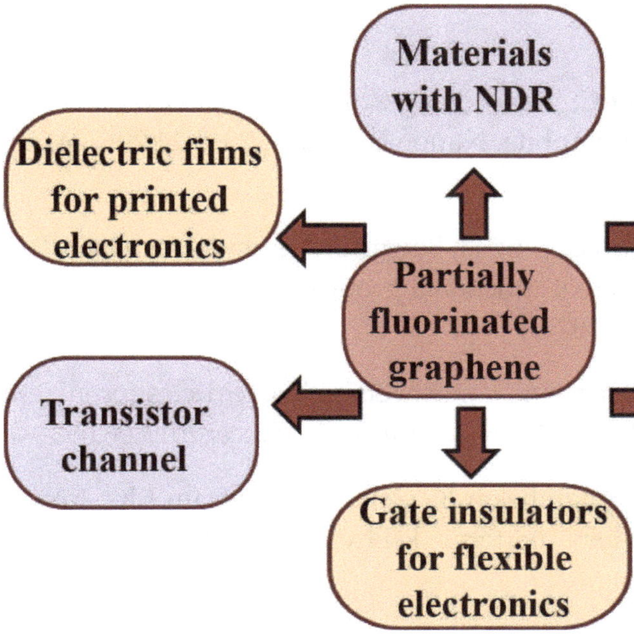

Figure 10. Schematic illustration of possible applications for partially fluorinated graphene films created from suspensions.

The fluorination process in the case of graphene suspension causes additional flakes splitting and fragmentation. The small size of the fluorinated flakes makes them an excellent base for graphene-based inks. Dropped and printed films, obtained from the inks both on rigid and flexible substrates, demonstrate a great potential for a wide spectrum of electronic devices, especially for flexible electronics. Partial fluorination provides for FG films such effects as NDR and resistive switching that are promising for applications. The development of optimized dielectrics for the graphene active layer (active layer gate and interlayer dielectrics or/and substrate for graphene) has been described. Dielectric layers fabricated from fluorinated graphene or in combination with graphene oxide are the most promising graphene-based flexible and transparent electronics.

Acknowledgements

The study was financially supported by the Russian Science Foundation, grant 15-12-00008. **Figures 1**, **3**, and **4** are reproduced from Ref. [20] with permission from the PCCP Owner Societies.

Author details

Irina V. Antonova* and Nadezhda A. Nebogatikova

*Address all correspondence to: nadonebo@gmail.com, antonova@isp.nsc.ru

A. V. Rzhanov Institute of Semiconductor Physics, SB RAS, Novosibirsk State Technical University, Novosibirsk, Russia

References

[1] Han X, Chen Y, Zhu H, Preston C, Wan J, Fang Z, Hu L: Scalable, printable, surfactant-free graphene ink directly from graphite. Nanotechnology. 2013;**24**:205304. DOI: 10.1088/0957-4484/24/20/205304

[2] Wei D, Li H, Han D, Zhang Q, Niu L, Yang H, Bower C, Andrew P, Ryhänen T: Properties of graphene inks stabilized by different functional groups. Nanotechnology. 2011;**22**:245702. DOI: 10.1088/0957-4484/22/24/245702

[3] Lee CL, Chen CH, Chen CW: Graphene nanosheets as ink particles for inkjet printing on flexible board. Chem. Eng. J. 2013;**230**:296–2. DOI: 10.1016/j.cej.2013.06.093

[4] Secor EB, Prabhumirashi PL, Puntambekar K, Geier ML, Hersam MC: Inkjet printing of high conductivity flexible graphene patterns. J. Phys. Chem. Lett. 2013;**4**:1347–51. DOI: 10.1021/jz400644c

[5] Antonova IV: Non-organic dielectric layers for graphene and flexible electronics. Int. J. Nanomater. Nanotechnol. Nanomed. 2016;**2**:018–24. DOI: 10.17352/2455-3492.000010

[6] Song L, Ci L, Lu H, Sorokin PB, Jin C, Ni J, Kvashnin AG, Kvashnin DG, Lou J, Yakobson BI, Ajayan PM: Large scale growth and characterization of atomic hexagonal boron nitride layers. Nano Lett. 2010;**10**:3209–15. DOI: 10.1021/nl1022139

[7] Eda G, Fanchini G, Chhowalla M: Large-area ultrathin films of reduced graphene oxide as a transparent and flexible electronic material. Nat. Nanotechnol. 2008;**3** 270−4. DOI: 10.1038/nnano.2008.83

[8] Standley B, Mendez A, Schmidgall E, Bockrath M: Graphene−graphite oxide field-effect transistors. Nano Lett. 2012;**12**:1165−9. DOI: 10.1021/nl2028415

[9] Petrone N, Chari T, Meric I, Wang L, Shepard KL, Hone J: Flexible graphene field-effect transistors encapsulated in hexagonal boron nitride. ACS Nano. 2015;**9**:8953–9. DOI: 10.1021/acsnano.5b02816

[10] Lee GH, Yu YJ, Cui X, Petrone N, Lee CH, Choi MS, Lee DY, Lee C, Yoo WJ, Watanabe K, Taniguchi T, Nuckolls C, Kim P, Hone J: Flexible and transparent MoS_2 field-effect transistors on hexagonal boron nitride graphene heterostructures. ACS Nano. 2013;**7**:7931–6. DOI: 10.1021/nn402954e

[11] Lee SK, Kim BJ, Jang H, Yoon SC, Lee C, Hong BH, Rogers JA, Cho JH, Ahn JH: Stretchable graphene transistors with printed dielectrics and gate electrodes. Nano Lett. 2011;**11**:4642–6. DOI: 10.1021/nl202134z

[12] Lee SK, Jang HY, Jang S, Choi E, Hong BH: All graphene-based thin film transistors on flexible plastic substrates. Nano Lett. 2012;**12**:3472−6. DOI: 10.1021/nl300948c

[13] Jewel MU, Siddiquee TA, Islam MR. Flexible graphene field effect transistor with graphene oxide dielectric on polyimide substrate. In: Proceeding of the International

Conference on Electrical Information and Communication Technology (EICT); 13–15 February 2013; Khulna, Bangladesh: IEEE; 2013. pp. 1–5. DOI: 10.1109/EICT.2014.6777834

[14] Nair RR, Ren W, Jalil R, Riaz I, Kravets VG, Britnell L, Blake P, Schedin F, Mayorov AS, Yuan S, Katsnelson MI, Cheng HM, Strupinski W, Bulusheva LG, Okotrub AV, Grigorieva IV, Grigorenko AN, Novoselov KS, Geim AK: Fluorographene: a two dimensional counterpart of Teflon. Small. 2010;**6**:2877–84. DOI: 10.1002/smll.201001555

[15] Robinson JT, Burgess JS, Junkermeier CE, Badescu SC, Reinecke TL, Perkins FK, Zalalutdniov MK, Baldwin JW, Culbertson JC, Sheehan PE, Snow ES: Properties of fluorinated graphene films. Nano Lett. 2010;**10**:3001–5. DOI: 10.1021/nl101437p

[16] Ho KI, Huang CH, Liao JH, Zhang W, Li LJ, Lai CS, Su CY: Fluorinated graphene as high performance dielectric materials and the applications for graphene nanoelectronics. Sci. Rep. 2014;**4**:5893. DOI: 10.1038/srep05893

[17] Lee W-K, Robinson JT, Gunlycke D, Stine RR, Tamanaha CR, King WP, Sheehan PE: Chemically isolated graphene nanoribbons reversibly formed in fluorographene using polymer nanowire masks. Nano Lett. 2011;**11**:5461–4. DOI: 10.1021/nl203225w

[18] Nebogatikova NA, Antonova IV, Komonov AI, Prinz VY: Producing arrays of graphene and few-layer graphene quantum dots in a fluorographene matrix. Optoelectronics, Instrumentation and Data Processing. 2014:**50**:298–303. DOI: 10.3103/S8756699014030145

[19] Nebogatikova NA, Antonova IV, Volodin VA, Prinz VY: Functionalization of graphene and few-layer graphene with aqueous solution of hydrofluoric acid. Phys. E. 2013;**52**:106–11. DOI: 10.1016/j.physe.2013.03.028

[20] Nebogatikova NA, Antonova IV, Prinz VY, Kurkina II, Alexandrov GN, Timofeev VB, Smagulova SA, Zakirov ER, Kesler VG: Fluorinated graphene dielectric films obtained from functionalized graphene suspension: preparation and properties. Phys. Chem. Chem. Phys. 2015;**17**:13257–66. DOI: 10.1039/C4CP04646C

[21] O'Neill A, Khan U, Nirmalraj PN, Boland J, Coleman JN: Graphene dispersion and exfoliation in low boiling point solvents. J. Phys. Chem. C. 2011;**115**:5422-8. DOI: 10.1021/jp110942e

[22] Wang J, Manga KK, Bao Q, Loh KP: High-yield synthesis of few-layer graphene flakes through electrochemical expansion of graphite in propylene carbonate electrolyte. J. Am. Chem. Soc. 2011;**133**:8888–91. DOI: 10.1021/ja203725d

[23] Su CY, Lu AY, Xu Y, Chen FR, Khlobystov AN, Li LJ: High-quality thin graphene films from fast electrochemical exfoliation. ACS Nano. 2011;**5**:2332–9. DOI: 10.1021/nn200025p

[24] Zhou M, Tang J, Cheng Q, Xu G, Cui P, Qin LC: Few-layer graphene obtained by electrochemical exfoliation of graphite cathode. Chem. Phys. Lett. 2013;**572**:61–5. DOI: 10.1016/j.cplett.2013.04.013

[25] Lotya M, Hernandez Y, King PJ, Smith RJ, Nicolosi V, Karlsson LS, Blighe FM, De S, Wang Z, McGovern IT, Duesberg GS, Coleman JN: Liquid phase production of graphene by

exfoliation of graphite in surfactant/water solutions. J. Am. Chem. Soc. 2009;**131**:3611–20. DOI: 10.1021/ja807449u

[26] Shinde DB, Brenker J, Easton CD, Tabor RF, Neild A, Majumder M: Shear assisted electrochemical exfoliation of graphite to graphene. Langmuir. 2016;**32**:3552–9. DOI: 10.1021/acs.langmuir.5b04209

[27] Paton KR, Varrla E, Backes C, Smith RJ, Khan U, Neill AO, Boland C, Lotya M, Istrate OM, King P, Higgins T, Barwich S, May P, Puczkarski P, Ahmed I, Moebius M, Pettersson H, Long E, Coelho J, O'Brien SE, McGuire EK, Sanchez BM, Duesberg GS, McEvoy N, Pennycook TJ, Downing C, Crossley A, Nicolosi V, Coleman JN: Scalable production of large quantities of defect-free few-layer graphene by shear exfoliation in liquids. Nat. Mater. 2014;**13**:624–30. DOI: 10.1038/nmat3944

[28] Nebogatikova NA, Antonova IV, Kurkina II, Soots RA, Vdovin VI, Timofeev VB, Smagulova SA, Prinz VY: Fluorinated graphene suspension for inkjet printed technologies. Nanotechnology. 2016;**27**:205601. DOI: 10.1088/0957-4484/27/20/205601

[29] dos Santos RB, Rivelino R, Mota FB, Gueorguiev GK: Exploring hydrogenation and fluorination in curved 2D carbon systems: a density functional theory study on Corannulene. J. Phys. Chem. A. 2012;**116**:9080–7. DOI: 10.1021/jp3049636

[30] Nebogatikova NA, Antonova IV, Prinz VY, Timofeev VB, Smagulova SA: Graphene quantum dots in fluorographene matrix formed by means of chemical functionalization. Carbon. 2014;**77**: 1095–103. DOI: 10.1016/j.carbon.2014.06.026

[31] Antonova IV, Kurkina II, Nebogatikova NA, Smagulova SA. Films fabricated from partially fluorinated graphene suspension: structural, electronic properties and negative differential resistance. Nanotechnology. 2017;**28**:074001. DOI: 10.1088/1361-6528/28/7/074001

[32] Kurkina II, Antonova IV, Nebogatikova NA, Kapitonov AN, Smagulova SA: Resistive switching effect and traps in partially fluorinated graphene films. Journal of Physics D: Applied Physics. 2016;**49**:095303. DOI: 10.1063/1.4953239

[33] Liu HY, Hou ZF, Hu CH, Yang Y, Zhu ZZ. Electronic and magnetic properties of fluorinated graphene with different coverage of fluorine. J. Phys. Chem. C. 2012;**116**:18193–201. DOI: 10.1021/jp303279r

[34] Duan Y, Stinespring CD, Chorpening B: Electronic structures, bonding configurations, and band-gap-opening properties of graphene binding with low-concentration fluorine. Chem. Open. 2015;**4**:642–50. DOI: 10.1002/open.201500074

[35] Hoex B, Gielis JJH, Van de Sanden MCM, Kessels WMM: On the *c*-Si surface passivation mechanism by the negative-charge-dielectric Al_2O_3. J. Appl. Phys. 2008;**104**:113703. DOI: 10.1063/1.3021091

[36] Garg R, Chowdhury NA, Bhaskaran M, Swain PK, Misra D. Electrical characteristics of thermally evaporated HfO_2. J. Electrochem. Soc. 2004;**151**:F215–9. DOI: 10.1149/1.1784212

[37] Terlinden NM, Dingemans G, Vandalon V, Bosch RHEC, Kessels WMM. Influence of the SiO_2 interlayer thickness on the density and polarity of charges in $Si/SiO_2/Al_2O_3$ stacks as studied by optical second-harmonic generation. J. Appl. Phys. 2014;**115**:033708. DOI: 10.1063/1.4857075

[38] Hasan M, Jang M, Kim D-H, Nguyen MC, Yang H, Jeong JK, Choi R. Improved electrical properties of solution-processed ZrO_2 gate dielectric for large-area flexible electronics. Jap. J. Appl. Phys. 2013;**52**:100206. DOI: 10.7567/JJAP.52.100206

[39] Antonova IV, Nebogatikova NA, Prinz VY: Self-organized arrays of graphene and few-layer graphene quantum dots in fluorographene matrix: charge transient spectroscopy. Appl. Phys. Lett. 2014;**104:**193108. DOI: 10.1063/1.4878262

[40] Antonova IV, Nebogatikova NA, Prinz VY, Popov VI, Smagulova SA: Light-assisted recharging of graphene quantum dots in fluorographene matrix. J. Appl. Phys. 2014;**116:**134310. DOI: 10.1063/1.4897231

[41] Antonova IV, Nebogatikova NA, Prinz VY: Fluorinated graphene films with graphene quantum dots for electronic applications. J. Appl. Phys. 2016;**119;**224302. DOI: 10.1063/1.4953239

[42] Yang JJ, Zhang MX, Strachan JP, Miao F, Pickett MD, Kelley RD, Medeiros-Ribeiro G, Williams RS: High switching endurance in TaOx memristive devices. Appl. Phys. Lett. 2010;**97**:232102. DOI: 10.1063/1.3524521

[43] Jeong DS, Thomas R, Katiyar RS, Scott JF, Kohlstedt H, Petraru A, Hwang CS: Emerging memories: resistive switching mechanisms and current status. Reports on Progress in Physics. 2012;**75**:076502. DOI: 10.1088/0034-4885/75/7/076502

[44] Lyo IW, Avouris P: Negative differential resistance on the atomic scale: implications for atomic scale devices. Science. 1989;**245**:1369–71. DOI: 10.1126/science.245.4924.1369

[45] Chen J, Reed MA, Rawlett AM, Tour JM: Large on-off ratios and negative differential resistance in a molecular electronic device. Science. 1999;**286**:1550–2. DOI: 10.1126/science.286.5444.1550

[46] Léonard F, Tersoff J: Negative differential resistance in nanotube devices. J. Phys. Rev. Lett. 2000;**85**:4767–70. DOI: 10.1103/PhysRevLett.85.4767

[47] Song Y, Wu HC, Guo Y: Negative differential resistance in graphene double barrier resonant tunneling diode. Appl. Phys. Lett. 2013;**102**:093118. DOI: 10.1063/1.4794952

[48] Nguyen HC,Nguyen VL: Tunneling of dirac electrons through one-dimensional potentials in graphene: a T-matrix approach. J. Phys.: Condens. Matter. 2009;**21**:045305. DOI: 10.1088/0953-8984/21/4/045305

[49] Ren H, Li QX, Luo Y, Yang JL: Graphene nanoribbon as a negative differential resistance device. Appl. Phys. Lett. 2009;**94**:173110. DOI: 10.1063/1.3126451

[50] Wu Y, Farmer DB, Zhu W, Han SJ, Dimitrakopoulos CD, Bol AA, Avouris P, Lin YM. Three-terminal graphene negative differential resistance devices. ACS Nano. 2012;**6**:2610–2. DOI: 10.1021/nn205106z

[51] Kamyshny A, Magdassi S: Conductive nanomaterials for printed electronics. Small. 2014;**10**:3515–35. DOI: 10.1002/smll.201303000

[52] Li J, Ye F, Vaziri S, Muhammed M, Lemme MC, Östling M: Efficient inkjet printing of graphene. Adv. Mater. 2013;**25**:3985-92. DOI: 10.1002/adma.201300361

[53] Xu Y, Hennig I, Freyberg D, Strudwick AJ, Schwab MG, Weitz T, Cha KCP: Inkjet-printed energy storage device using graphene/polyaniline inks. J. Power Sources. 2014;**248**:483–8. DOI: 10.1016/j.jpowsour.2013.09.096

11

Adsorption of Metal Clusters on Graphene and their Effect on the Electrical Conductivity

Roxana M. Del Castillo and Luis E. Sansores Cuevas

Abstract

When adsorbates are introduced in graphene, the electric conductivity is highly modified. This chapter discusses how to estimate the electrical conductivity of graphene sheets with adsorbates, using electronic structure calculations and some theoretical approaches. Also, we discussed how the clustering of adsorbates attached to the graphene can impact electrical conductivity. We will focus in using metallic clusters as adsorbates (M_n; M = Ag, Au, Pt, and Pd; $n = 1, 2, 3$, and 4). The electrical conductivity is found using theoretical approaches, which are summarized in this chapter. We compare these approaches between each other to determine which is the most appropriate for each system.

Keywords: graphene, metallic clusters adsorption, conductivity

1. Introduction

It is well known that the experimental measurements of electrical conductivity are not so complicated; this is done with a four-point method to measure the sheet resistivity or conductivity [1, 2]. However, making a theoretical estimate of the electrical conductivity requires complicated mathematical elements, e.g., Lippmann-Schwinger equation [3, 4]. But there are some analytical methods based on Green's function to estimate the theoretical conductivity of pure graphene [5–8]. These methods can be extrapolated to estimate the conductivity of graphene with adsorbates or vacancies. The main advantage of these methods consists in that the conductivity can be calculated with structural parameters (deformation radius and heights). We propose to use these approximations to estimate the conductivity of graphene with the adsorption of metallic clusters (transition metal clusters).

By studying the way in which adsorption occurs, we can obtain strong elements to predict the behavior of the electrons into the graphene, wherein the scattering centers will be modeled as potential steps in a two-dimensional waveguide. We consider that, when introducing the adsorbate in the graphene, the size of the scattering centers increases. If there is physisorption, the scattering centers are small, i.e., the potential step is small, and a good number of electrons can pass without scattering. If there is chemisorption, the scattering centers grow, so the potential step in the waveguide will increase in size and hinder the passage of electrons, causing more electrons to be scattered. This phenomenon of the creation of scattering centers is directly seen in the deformation of the graphene sheet, since the adsorption sites break the two-dimensionality of the graphene sheet, creating obstacles for the electrons and near the adsorption zone the sp_2 hybridization is broken, creating a clearly deformation of the graphene network.

The novelty of this work is to propose a direct methodology to estimate the electrical conductivity directly from calculations of electronic structure, in particular, optimized geometries. This methodology will allow a significant simplification of the theoretical calculations of conductivity. Recently, it has been reported that pure graphene has theoretical conductance and conductivity, but so far it has done anything on systems with dopants, and even more, on systems doped with metal clusters. Also, it consists of having direct applications in the field of nanoelectronic devices, catalytic materials, molecular electronics, developed sensors, and solar cells.

1.1. Review overview

Since the experimental production of graphene in 2004 by Novoselov et al. [9], graphene has been much studied by the scientific community. From the first moment, graphene presented an outstanding electronic behavior, since its electrons behave like Dirac Fermions with an effective speed around 10^6 m/s [10], reaching the regimes of electronic ballistic transport. Graphene has a minimum conductivity with a theoretical value of $e^2/\pi h$ (per valley per spin) [11], but the minimum conductivity measured experimentally is e^2/h (per valley per spin); this is known as the *missed-π-problem*. In 2007, Miao et al. [12] solved the problem by measuring the conductivity of small pieces of graphene as a function of an applied voltage, and to these pieces of graphene were changed the length and depth. They concluded that the theory is true, but only for small sheets. In addition to this, it is known that the mobility of graphene depends slightly on temperature and has a characteristic V-behavior [13]. The fact that electronic mobility depends slightly on temperature caused the theoretical conductivity of graphene to be investigated. In 2010, Barbier et al. [14] published a review where they calculated the theoretical conductivity of graphene in the limit of δ-function barriers. Recently, Kolasiński et al. calculated the conductivity with the Lippmann-Schwinger equation [15], and Garcia et al. published a theoretical approximation of graphene conductivity using the Kubo-Bastin approach [16]. In 2015, Kirczenow [17] calculated the scattering state wave functions of electrons on graphene nanostructures with the Lippmann-Schwinger equation, and he observed that when the Fermi energy is around the Dirac point (tunneling regimen), there is an inversion symmetry breaking to take into account. There is an alternative picture to describe the conductivity of graphene, Green function formalism with a different type of potential. In 2006, Ando [18] described graphene as a two-dimensional waveguide, where the impurities are treated as Coulomb centers. The main advantage of treating impurities as centers of Coulomb is that it allows to propose in the equations a potential with long-range terms. However, this description was not enough, and there are other more sophisticated descriptions, e.g., graphene is developed as different regions of interest, connecting with self-energy terms in the Green function [19]; or three-terminal equilibrium Green's functions are used to determine the transmission phenomenon [20], using the structural parameters and boundary condition of graphene. In this chapter, will address three alternative approaches mention before, based on structural parameters and a Green function with a long-range potential. The three approximations present a behavior very similar to experiments being able to describe numerically the behavior at several temperatures.

However, for electronic devices, it is important to find the materials with good electrical conductivity and with an energy gap. When doping occurs to graphene, it is introduced as an energy gap and in some cases, continues to maintain a high electrical conductivity [21]. Mainly, the doping by the adsorption of atoms or clusters has been a well-supported way, since by means of the adsorbates the electronic and structural properties of the graphene can be modified. For example, introducing nitrogen atoms to the graphene surface induces a N-type doping, while by introducing oxygen atoms to the graphene, the Fermi energy remains at the point of Dirac (i.e. undoped) [22]. Depending on the adsorbate on the graphene, will be the type of interaction present in the system. Graphene with Cu, Ag atoms are adsorbed weakly on the surface of the graphene [23–25], whereas graphene with Pd are strongly adsorbed [25, 26]. It is of interest, to see that when doping the graphene with different adsorbate, we modify the electronic transport in the system, in particular, the conductivity [27]. In 2015, Liu et al. [28] distributed randomly weighted atoms on a sheet of graphene, this in order to modify the electronic transport, and they estimated the conductivity with the Kubo-Bastin formula. On the other hand, the studies made with graphene and adsorbates have been extrapolated to surface of high-quality boron and nitrogen co-doped graphene on silicon carbide substrate [29], indicating that these studies are important for the field of electronic devices.

2. Some theoretical conductivity approaches

In this chapter, we estimate conductivity with three approaches: frozen ripples approach (FRA), charged impurities scattering approach (CISA), and resonant scattering approach (RSA).

2.1. Frozen ripples approach (FRA)

In the frozen ripples approach (FRA), the main idea consists in using the deformation as a scattering center and then calculated the conductivity. Katsnelson and Geim [5] probed that in pure graphene, the scattering of the electrons behaves like as a controlled ripple introduced into the graphene surface (microscopic corrugations of a graphene sheet). They used the fact that certain types of ripples create a long-range scattering potential, as if they were Coulomb scatters, and provide charge-carriers almost independent of carrier concentration. Specifically, electrons are dispersed by a potential proportional to the square of the local curvature of graphene. If electrons are scattered, then the resistivity increases, diminishing the ballistic behavior. Katsnelson and Geim approximated the excess of conductivity as follows:

$$\delta\sigma \approx \frac{4e^2}{h}\frac{a^2R^2}{z^4} \tag{1}$$

where z and R are the characteristic height and radius of ripples, respectively. In graphene with adsorbates, the scattering centers would be defined in the function of the adsorption properties (see Section 4).

2.2. Charged impurities scattering approach (CISA)

Once again, Katsnelson with Guinea and Geim analyzed how electrons in graphene, behaving like Dirac fermions, are scattered by clusters of charged impurity clusters [6]. Likewise, they showed that for graphene with a very low doping level, in some cases, the disorder increase the conductivity [6], and for the graphene with impurities, the disorder occurs in the form of large circular clusters, losing the scattering cross section, in comparison with isolated single atoms [7], and conductivity could be in ballistic regimen. The experimental manifestation of this phenomenon is observed as an increasing in the mobility [9–11, 13] at different temperatures. In these theories, there are two regimes:

- $k_FR << 1$ (the cluster is small compared to the Fermi wavelength), thus weakly perturbing the electronic wave function, and the adiabatic approximation can be used to obtain a constant value for $\sigma \propto k_F R^2\left(\frac{V}{\hbar v_F R^{-1}}\right)^2$,
- $k_FR >> 1$ in this point is the cross section that is a function of the incident angle θ and the total cross section. The resonances are associated with the quasibound states inside the cluster. The conductivity (σ) is a function of Fermi wavelength (k_F), the potential (V), the Fermi velocity (v_F), the cluster concentration density (n_c), and the distortion radius (R) to the fourth power.

$$\sigma = \frac{e^2}{h}k_F l \sim \begin{cases} \frac{e^2}{h}\frac{1}{n_c R^2}\left(\frac{\hbar v_F R^{-1}}{V}\right)^2, & k_F R \ll 1 \\ \frac{e^2}{h}\left(\frac{k_F^2}{n_c}\right)^2, & k_F R \gg 1 \end{cases} \tag{2}$$

2.3. Resonant scattering approach (RSA)

Wehling et al. [8] used the semiclassical Boltzmann theory to calculate conductivity with a linear dependence of the carrier density. The principal scattering mechanism used is the midgap states [30] in the scattering mechanism. This mechanism is designed to describe the material defects as boundaries, cracks, adsorbates, or vacancies because these materials' defects induced a high potential difference with respect to the graphene sheet. Therefore, the phase shift tends to zero for wave vectors near to the Dirac point. If we considered a short-range contact potential, the behavior of the phase shift is logarithmic. In some instances, when the Fermi energy of the graphene-adsorbates is close to the Dirac point, it appears as a resonance [31]. It is possible to compute the conductivity, using Boltzmann equation and the T matrix $\sigma \approx \left(\frac{2e^2}{h}\right)\left(2\pi n_i\left|\frac{T(E_F)}{D}\right|^2\right)^{-1}$, with $D = \sqrt{\sqrt{3}\pi t}$, and t is the nearest-neighbor hopping parameter. In the limit of resonant impurities, conductivity is,

$$\sigma \approx \left(\frac{4e^2}{\pi h}\right)\frac{n_e}{n_i} ln^2\left|\frac{E_F}{D}\right|, \tag{3}$$

$n_e = E_F^2/D^2$ is the number of charge carriers per carbon atom. The deformation radius is $D = hv_F/R$; the R parameter is seen directly to the geometry optimization; and $v_F = \sqrt{2E_F/m_e}$. For pure graphene, in a ballistic regimen, the Boltzmann approach becomes questionable because the quantum correction is predominant. However, small clusters has been proved to be sufficient to estimate conductivity of the graphene sheets with the adsorption of the metallic clusters [25].

3. Principal aspects of adsorption of metallic clusters on graphene

In this section, we examine the adsorption process of metallic clusters on graphene. We give some general trends, utilizing some specific systems to exemplify the adsorption phenomenon. The main features that will be studied in this section are ground-state properties, adsorption sites, adsorption energies, and density of states.

3.1. Ground states, adsorption sites, and adsorption energies

The first characteristic to study is the adsorption energy, which indicates if physisorption or chemisorption is occurring. In the next step, the adsorption site and how the deformation occurs in terms of the adsorption sites are observed. It is important to see under what conditions the deformation occurs, and the study of the adsorption sites gives us the radius and height of the scattering centers. The radius and heights are introduced into the three approaches to estimate the electrical conductivity. For physisorbed systems, the deformation is very low (small radius and height). For chemisorption systems, the deformation increases because the graphene feels strongly the metallic clusters.

To find the ground-state structures, adsorption sites, and adsorption energies, we optimized all our systems using a graphene sheet as a 6 × 6 supercell with a periodic condition in z-axis of 30 Å. All initial positions of the graphene systems with the clusters are listed below:

1. For one metal atom, the atom was placed in a top-site, right in the middle of the graphene sheet, over a distance of 3 Å (**Figure 1**).
2. For the dimer case, there are two possible positions: two metallic atoms at the same height forming a horizontal line and two metallic atoms aligned over a carbon atom forming a vertical line (**Figure 2**).

3. For the trimer case, there exists four possible positions: three metallic atoms at the same height conforming a horizontal line; three metallic atoms aligned over a carbon atom conforming a vertical line; three metallic atoms are at the same height, outlining a horizontal triangle over the graphene sheet; and three metallic atoms modeling a vertical triangle, and the two metallic atoms are at the same height of two adjacent carbon atoms (**Figure 3**).

4. For the tetramer case, there are eight initial configurations: four metallic atoms at the same height in a horizontal line; four metallic atoms aligned with one carbon atom of the graphene sheet forming a vertical line; four metallic atoms at the same height forming a rhombohedral shape or a diamond-shape in a horizontal plane; four metallic atoms aligned at one carbon atom of the graphene sheet forming a rhombohedral shape or a diamond-shape in a vertical plane; four metallic atoms at the same height forming a planar triangle plus 1 shape or a Y-shape in a horizontal plane; four metallic atoms aligned at one carbon atom of the graphene sheet forming a planar triangle plus 1 shape or a Y-shape in a vertical plane; four metallic atoms aligned at two carbon atoms of the graphene sheet forming a planar triangle plus 1 shape or a Y-shape in a vertical plane; and a tetrahedral isomer (the base of the tetrahedron is close to the graphene sheet) (**Figure 4**).

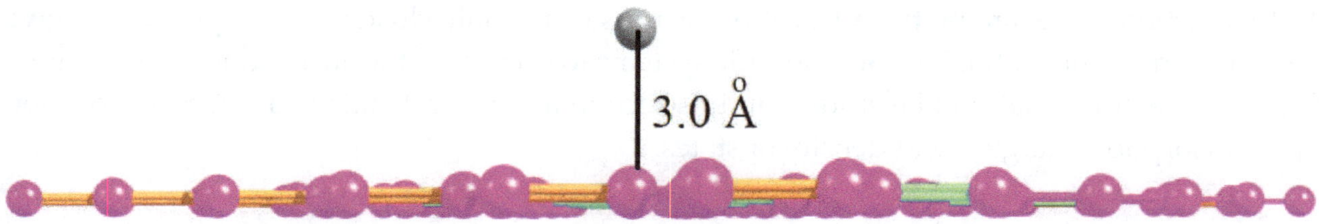

Figure 1. Initial position for one metal atom over a graphene sheet.

Figure 2. Initial position for a metallic dimer over a graphene sheet.

Figure 3. Initial position for a metallic trimer over a graphene sheet.

The computational calculations were performed with QUANTUM ESPRESSO computational package [32], and specifically performed DFT calculations using a plane wave basis set and pseudopotentials. The exchange and correlation interactions were carried out with the Perdew-Burke-Ernzerhof parameterization (GGA-PBE) [33]. The kinetic energy cutoff was taken as 40 Ry with a convergence criterion of 10^{-8} Ry. The Brillouin-zone integrations have been achieved with Methfessel–Paxton smearing special-point technique [34] centering de Γ point with a $4 \times 4 \times 1$

k-point mesh. The pseudopotentials of Ag and Pd were a Kr-like core; meanwhile, the pseudopotential of Au and Pt were Xe-like core. All the metals were approximate with smooth pseudopotentials generated with the Rappe-Rabe-Kaxiras-Joannopoulos (RRKJ) [35]. The adsorption energies have been computed with $E_a = E_{graphene + Mn} - (E_{graphene} + E_{Mn})$. The adsorption energies, the bond length from carbon to metallic atoms, and metallic atom to metallic atoms are presented in **Table 1** [25]. In this table, we indicate, in bold, the ground states. These ground states are taken to calculate the conductivity.

The ground-state system is shown in **Figure 5**. The systems with Ag and Au present that the Ag-Ag and Au-Au interactions are stronger than Ag-C and Au-C, the clusters almost do not

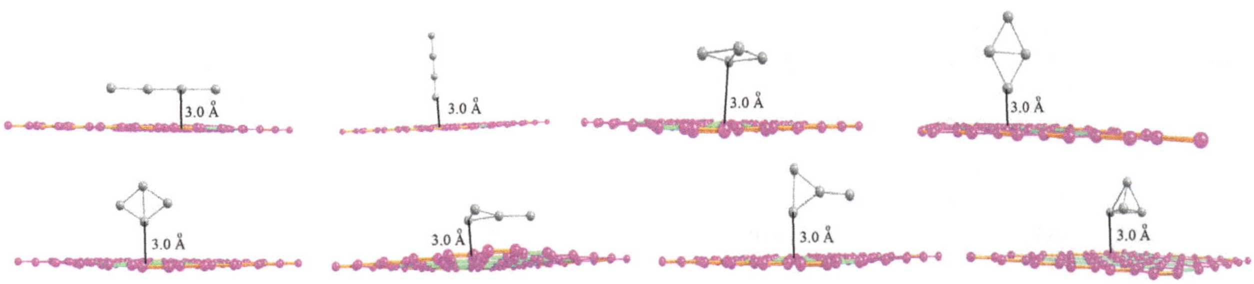

Figure 4. Initial position for a metallic tetramer over a graphene sheet.

		Graphene-Ag_n			Graphene-Au_n			Graphene-Pt_n			Graphene-Pd_n		
Structure	**n**	d_{Ag-C}	d_{Ag-Ag}	E_a	d_{Au-C}	d_{Au-Au}	E_a	d_{Pt-C}	d_{Pr-Pt}	E_a	d_{Pd-C}	d_{Pd-Pd}	E_a
Single atom	1	**3.26**		**−0.03**	**2.61**		**−0.31**	**2.09**		**−1.20**	**2.15**		**−1.20**
Horizontal line	2	Horizontal line convert in vertical line			Horizontal line convert in vertical line			3.87, 3.98	2.38	−0.14	2.13	2.70	−1.35
Vertical line	2	**2.63**	**2.58**	**−0.13**	**2.32**	**2.53**	**−0.64**	**2.23**	**2.41**	**−0.94**	**2.21**	**2.54**	**−0.65**
Horizontal line	3	Horizontal line convert in vertical line			Horizontal line convert in vertical line			3.94(2) 4.07	2.37(2)	−0.34	Horizontal line convert in horizontal triangle		
Vertical line	3	2.91	2.66(2)	−0.08	2.35	2.58, 2.56	−0.49	2.18	2.40, 2.42	−0.94	Vertical line convert in vertical triangle		
Horizontal triangle	3	3.34 3.33(2)	2.72(3)	−0.13	4.01 4.00(2)	2.69(3)	−0.34	3.89, 3.92, 3.97	2.51(3)	−0.54	3.90. 3.93. 4.08	2.51(3)	−1.80
Vertical triangle	3	**2.49 2.52**	**2.78, 2.70, 2.66**	**−0.36**	**2.34, 2.35**	**2.65 (2), 2.75**	**−0.75**	**2.17, 2.23, 4.55**	**2.53(2), 2.59**	**−1.82**	**2.13, 4.25, 4.32**	**2.46, 2.55(2)**	**−1.98**
Vertical line	4	2.63	2.62(2), 2.72	−0.053	2.36	2.54 (2), 2.60	−0.51	3.2	2.3(4)	−0.78	Vertical line convert in tetrahedral isomer		
Diamond shape horizontal	4	3.47, 3.67, 3.63, 3.62	2.75(2), 2.76(2), 2.64	−0.02	4.06, 4.07 (2), 4.11	2.66 2.69(4) 2.67	−0.17	3.98, 4.1(2), 3.98	2.53, 2.54(4)	−0.13	D-shape horizontal convert in tetrahedral isomer		

		Graphene-Ag_n			Graphene-Au_n			Graphene-Pt_n			Graphene-Pd_n		
Structure	n	$d_{Ag\text{-}C}$	$d_{Ag\text{-}Ag}$	E_a	$d_{Au\text{-}C}$	$d_{Au\text{-}Au}$	E_a	$d_{Pt\text{-}C}$	$d_{Pr\text{-}Pt}$	E_a	$d_{Pd\text{-}C}$	$d_{Pd\text{-}Pd}$	E_a
Diamond shape vertical	4	5.1(2), 2.74, 3.64	2.66, 2.72, 2.74(2), 2.76	−0.002	2.72, 5.00 (2), 7.34	2.67, 2.69 2.70(3)	−0.14	2.22, 4.27, 4.31, 6.17	2.52 (2),2.55 (2),2.58	−0.89	D-shape vertical convert in tetrahedral isomer		
Diamond Shape Vertical 1	4	2.56, 4.07, 4.2 5.2	2.70(2), 2,81(2), 2.65	−0.240	2.33, 4.00, 4.20, 4.97	2.63 2.65 (2), 2.78(2)	−0.44	2.49(2) 2.61(2) 2.53	2.23, 3.79(2) 4.7	−0.86	D-shape vertical convert in tetrahedral isomer		
Y-shape horizontal	4	3.59, 3.44, 3.75, 3.87	2.62, 2.63, 2.76, 2.62	−0.05	3.89, 3.92(3)	2.62, 2.69(2) 2.54	−2.47	4.08 (3), 4.13	2.37, 2.47, 2.52(2)	−0.25	Y-shape horizontal convert in tetrahedral isomer		
Y-shape vertical	4	3.33 (2), 5.83, 8.45	2.63(2), 2.75, 2.77	−0.05	Y-shape vertical convert in Y-shape vertical 1.			2.22 (2), 4.50, 6.90	2.52(2) 2.58, 2.41	−0.84	Y-shape vertical convert in tetrahedral isomer		
Y-shape vertical 1	4	**2.48, 4.27 (2), 5.03**	**2.61, 2.63, 2.76(2)**	**−0.30**	**2.32, 4.18, 4.57, 4.93**	**2.55, 2.61, 2.67, 2.70**	**−0.66**	2.25, 3.93, 4.10, 4.72	2.37, 2.47, 2.51, 2.58	−1.14	Y-shape vertical convert in tetrahedral isomer		
Tetrahedral	4	Tetrahedral isomer convert in D-shape horizontal			Tetrahedral isomer convert in D-shape horizontal			**2.17, 2.20, 3.80, 4.64**	**2.53, 2.65(2) 2.66(3)**	**−1.27**	**2.19, 3.83, 4.44, 4.77**	**2.55, 2.57(2) 2.62, 2.66(2)**	**−0.61**

All the distances are in Å, the adsorption energies (E_a) are in eV, and n is the number of atoms of the clusters. The ground states are presented in bold letter.

Table 1. Optimized geometries and adsorption energies of Ag, Au, Pt, and Pd clusters are supported on graphene.

distort the graphene surface. Meanwhile, the systems with Pd and Pt show that the interactions Pt-C and Pd-C is stronger than Pt-Pt and Pd-Pd, the clusters distorts very much the graphene sheet, and they settle into a valley created by them.

From the adsorption energies, adsorption sites, and the bond lengths, the most stable system is the trimer metallic cluster. The stability depends, mostly, in the covalent atomic radius. When the covalent atomic radius is smaller, the metallic atom can be accommodated between a carbon-carbon bond, distorting the graphene network. It is energetically favorable that the metallic atoms stick together and form clusters than to be separate between them onto the surface.

3.2. Electronic structure

The density of states (DOS) indicates how the adsorbates modified the type of doping in the systems. There are two charge transfer mechanisms observed in the density of states and in the population analysis [36], and these mechanisms are given in terms of the highest occupied orbital (HOMO) and the lowest unoccupied orbital (LUMO) of the adsorbates:

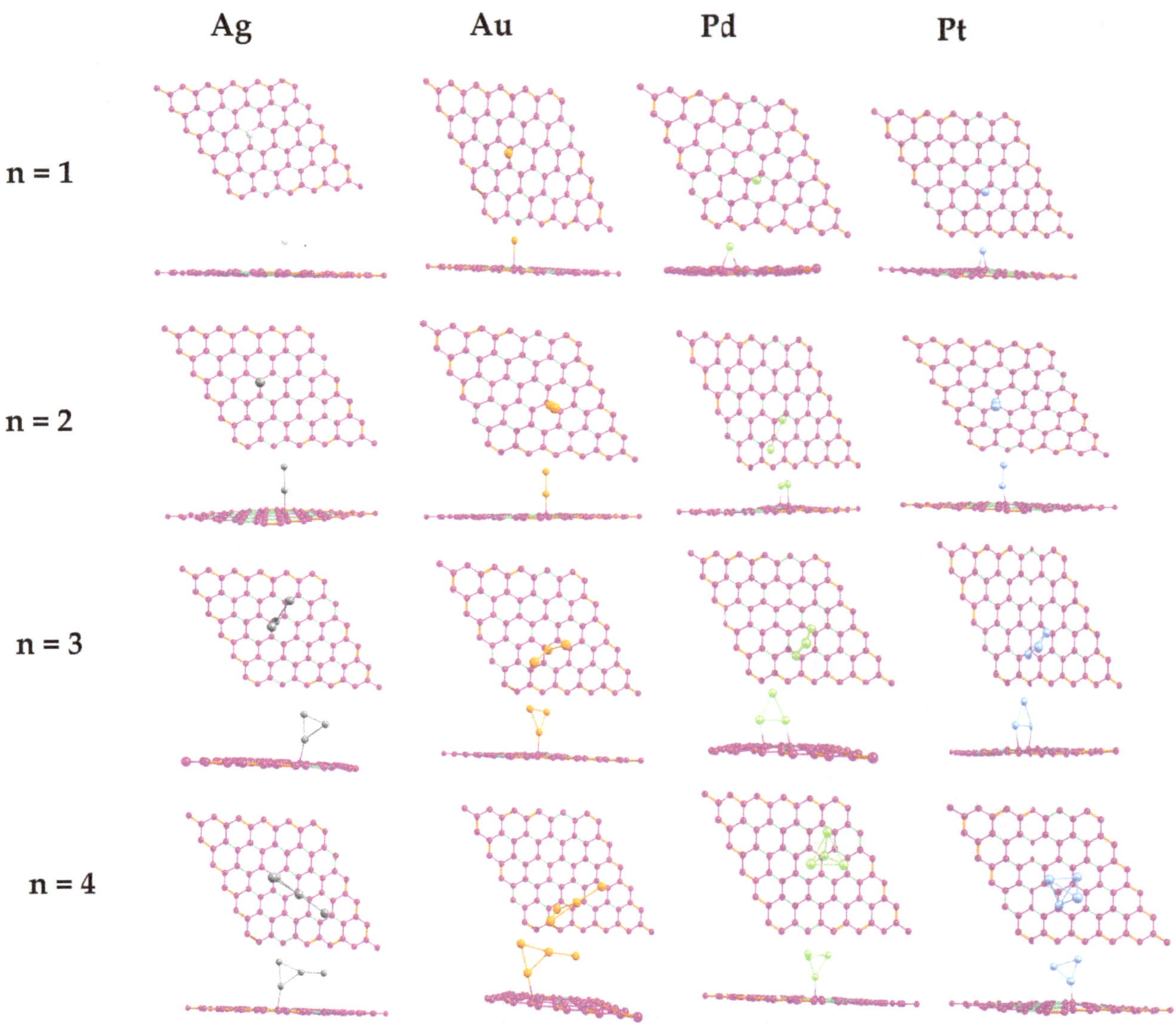

Figure 5. Ground-state structures of graphene-M_n (M = Ag, Au, Pt, and Pd; n = 1, 2, 3, and 4) [25].

- When the HOMO exceeds the Dirac point, the Fermi level of pure graphene occurs that the charge transfer goes from the cluster to the graphene. When the LUMO is below the Dirac point, the charge will transfer from the graphene to the cluster.
- When the charge transfer occurs between the cluster and the graphene sheet, this charge transfer is partially determined by the mixing of HOMO and LUMO orbitals (due to the hybridization).

The density of states (DOS) indicates how the charge transfer occurs and gives us a good supposition of how many free states are available to contribute to the conductivity. Also, indicates to us how the structure has been modified in terms of electronic behavior. For other side, the DOS is not conclusive to determine the conductivity, and we need other mechanism to complete the description.

The DOS of pure graphene has two basic properties: the DOS has a linear dependence of the energy around the Dirac point; and there are two Van Hove singularities around the Γ point, in the Brillouin zone. When we introduced adsorbates, the Van Hove singularities displace away from the Γ point, and the DOS is modified and the Fermi level is shifted away from the Dirac point. The graphene-Ag_n (**Figure 6**) and graphene-Pt_n (**Figure 8**) are P-type doping, in which the charge is transferred from the graphene to the metallic clusters. Meanwhile, the graphene-Au_n (**Figure 7**) and graphene-Pd_n (**Figure 9**) are N-type doping, and the charge is transferred from the metallic cluster to the graphene surface.

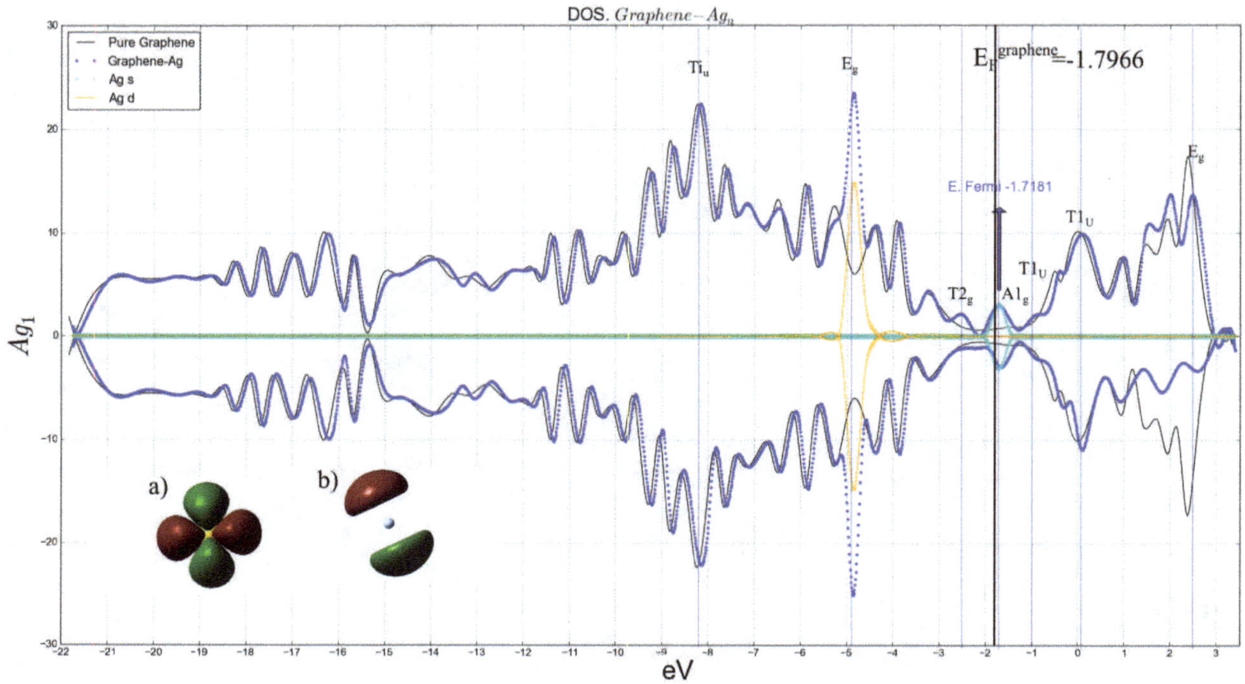

Figure 6. DOS of graphene-Ag_1 systems. Pure graphene is presented in a black line. The Dirac point is presented in −1.79 eV. It is presented the contributions of the MOs of Ag_1 are indicated with dotted lines [25], (a) the HOMO of Ag_1 and (b) the LUMO of Ag_1.

With the density of states, we can see the metallic cluster orbitals that distort the density of states of pure graphene. By distorting the density of states, the electronic transport properties are modified. For graphene-Ag_n systems, the silver cluster has that its 5s states are very close to the Fermi energy, contributing directly to the electronic transport. In the particular case of a single silver atom on the surface of graphene (Graphene-Ag), the 6s state of the silver enters directly into the Fermi energy, which causes states to fill up and improves electronic transport. For graphene-Pt_n systems, the 6s states of platinum enter the border orbitals, mostly in the LUMO, whereas the 5d orbitals enter the HOMO orbital. Both graphene-Ag_n and graphene-Pt_n present P-type doping and their orbital p is not present in the transport of charge. For systems with N-type doping, it is seen that the orbitals involved in the transfer of charge remain the orbital s and d of metallic clusters. Although in these cases, the contribution of the p orbitals of the metallic clusters in the virtual orbitals of the final structure is already perceived. To see more detail, we recommend the article [25]

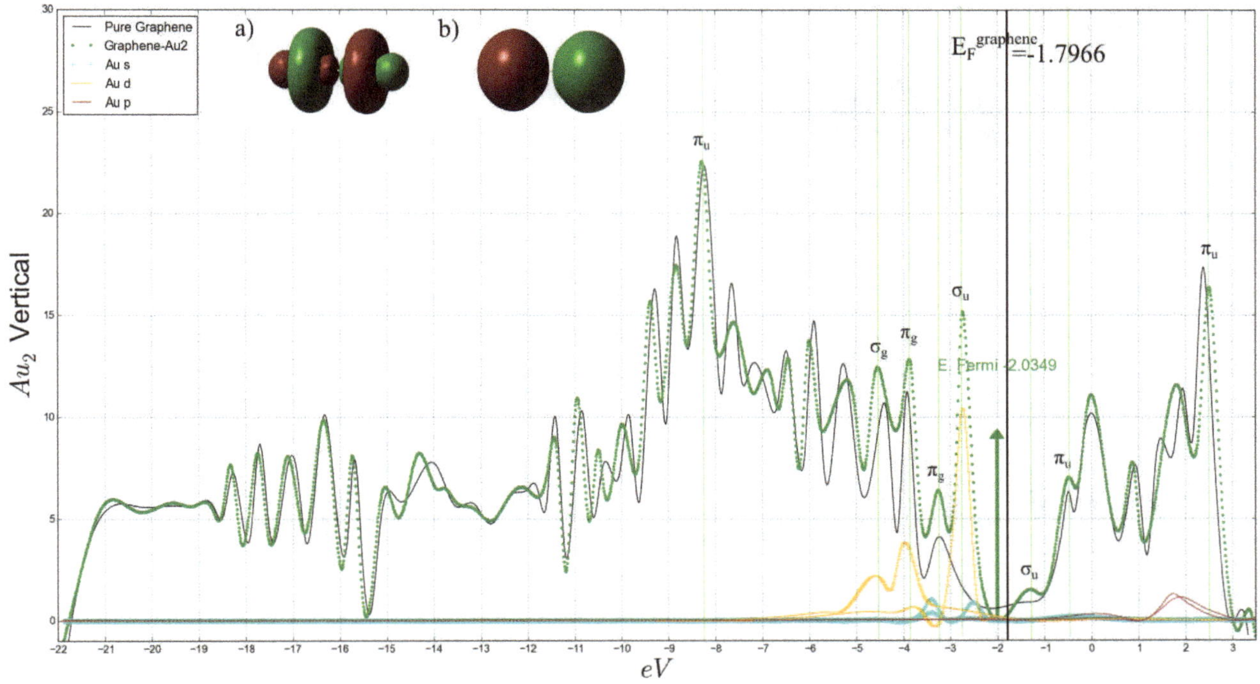

Figure 7. DOS of graphene-Au2 systems. Pure graphene is presented in a black line. The Dirac point is presented in −1.79 eV. It is presented the contributions of the MOs of Au_2 are indicated with dotted lines [25], (a) the HOMO of Au_2 and (b) the LUMO of Au_2.

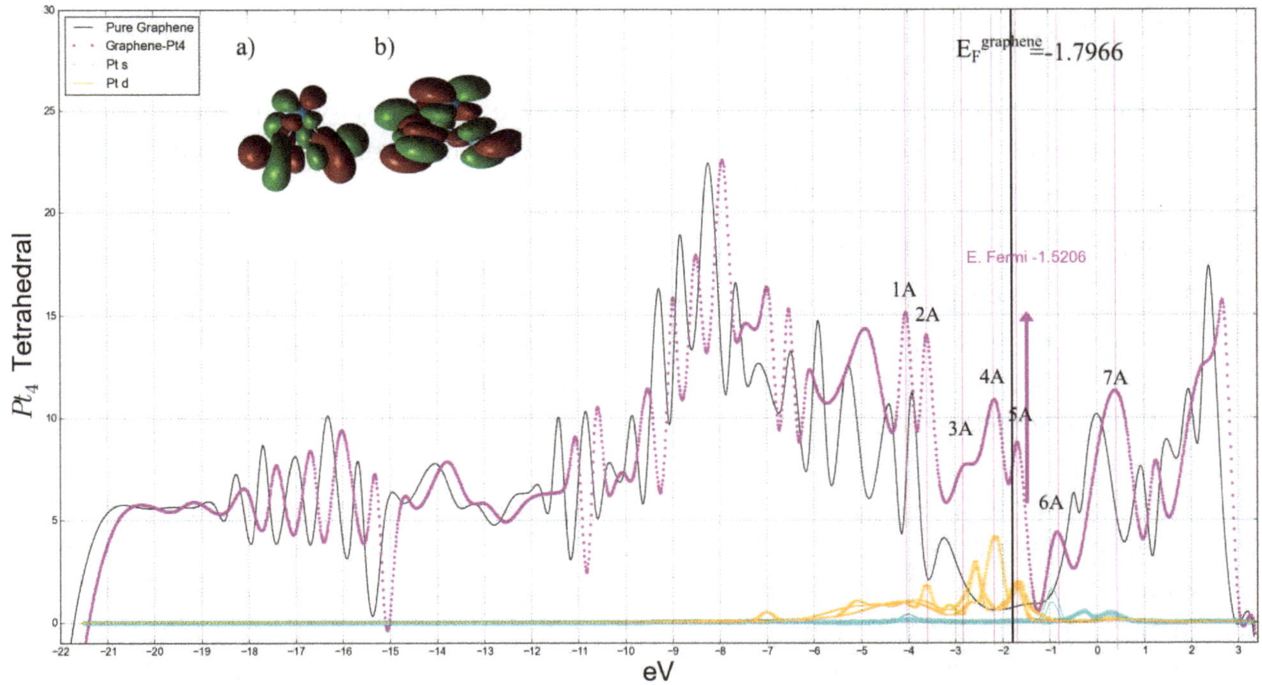

Figure 8. DOS of graphene-Pt_4 systems. Pure graphene is presented in a black line. The Dirac point is presented in −1.79 eV. It is presented the contributions of the MOs of Pt_4 are indicated with dotted lines [25], (a) the HOMO of Pt_4 and (b) the LUMO of Pt_4.

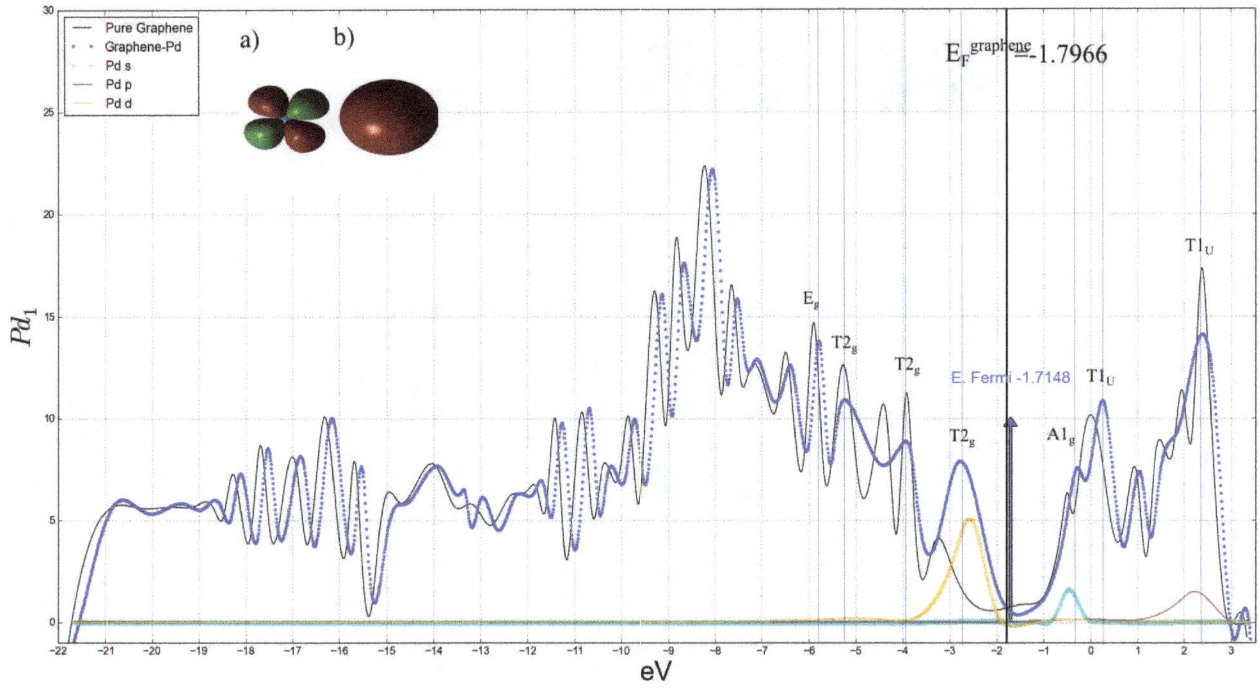

Figure 9. DOS of graphene-Pd_1 systems. Pure graphene is presented in a black line. The Dirac point is presented in −1.79 eV. It is presented the contributions of the MOs of Pd_1 are indicated with dotted lines [25], (a) the HOMO of Pd_1 and (b) the LUMO of Pd_1.

4. Conductivity

We used the approaches from Section 2 to estimate the conductivity of the graphene with metallic adsorbates. To calculate the conductivity or conductance of the graphene with adsorbates, it is necessary to follow the next steps:

1. Optimized the geometry. When we optimized the geometry, in most computational package, we obtained the Fermi Energy E_F and the deformation. With the Fermi energy, it is possible to estimate v_F ($v_F = \sqrt{2E_F/m_e}$, m_e is the electron mass). The deformation is directly obtained with the coordinates of the final structure. Specifically, the deformation is obtained with the adsorption radius and height of the sheet of graphene produced by the adsorption, i.e., the graphene surface is placed in the $z = 0$ plane, the metallic cluster will pull to the surface, this distance is to what we call height. The deformation radius takes us to the distance between the minimum height and the maximum height around the adsorbate. Other method to obtain the deformation is with the charge density surface (CDS); the CDS displays the scattering section due to the adsorption of the metal cluster on the graphene sheet. With the CDS, it is possible to obtain the deformation radius (see **Figure 10**) [37].

2. Determine which approach will be used:

   a. Evaluate the gap between the Dirac point and the Fermi Energy of the graphene-M_n systems $|\varepsilon - E_F|$ (see **Table 2**). If E_F of the adsorbed system is closed to Dirac point $|\varepsilon - E_F| = 0$, the more convenient approach to used is RSA.

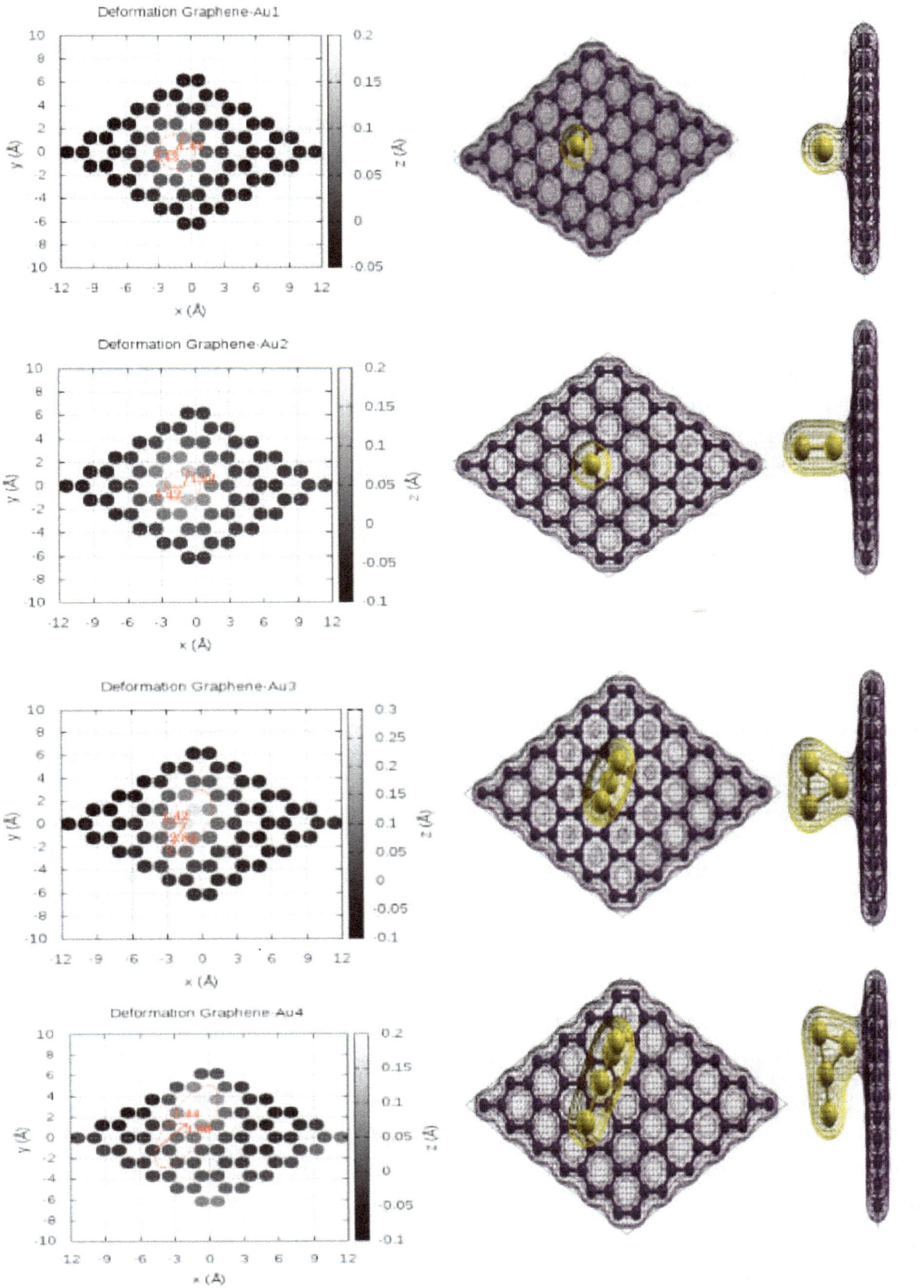

Figure 10. Optimized geometry of graphene-Au_n. It shows the deformation produced by the adsorption of the golden cluster with the coordinates and with the charge density surface (CDS) [25, 37].

b. If we are going to use CISA, it is important to determine which is the regimen that we are going to use. This is done by calculating the $k_F R$ parameter. For our systems, we have $k_F R \approx (1.6 - 3.05) \times 10^{-19} \ll 1$ for $1.4Å \leq R \leq 2.7Å$.

3. Obtain the adsorption energy, which can be used as of the V potential in CISA (suppose the graphene sheet as a two-dimensional wave guide with a step potential as the scattering center).

4. For RSA, it is imperative to determine the cluster concentration n_c. We calculate n_c with the unit's supercell parameters (see **Figure 10**), for our systems, we have $n_c = 2.9 \times 10^{13}$ cm^2 for cluster. Likewise, it is necessary to calculate the parameter D, which is calculated in terms of v_F and the deformation radius (R), as $D = hv_F/R$,

Following these steps, it is possible to estimate the conductivity with one of the three available approaches (FRA, CISA, or RSA), see **Table 3**. It is important to notice, that in some cases, the deformation would be created in an ellipse center and not a circle center. In this case, the radius of deformation was taken as the average of the major and minor radii. The ellipse center is previously considered in bilayer graphene [38, 39].

$\lvert \varepsilon - E_F \rvert$ (eV)	1	2	3	4
Ag	0.0785	0.0987	0.7409	0.5436
Au	0.2434	0.2383	0.2830	0.0060
Pt	0.0413	0.2355	0.0171	0.2760
Pd	0.0798	0.0191	0.0457	0.2811

Table 2. Gap between the Dirac point (ε) and the Fermi Energy of the graphene-M_n systems $\lvert \varepsilon - E_F \rvert$.

			FRA		CISA			RSA		
System	**E_F [eV]**	**R [Å]**	**z [Å]**	**g [S/cm]**	**v_F [m/s]**	**V [eV]**	**g [S/cm]**	**D(R) [eV]**	**n_e/C-atom**	**g [S/cm]**
Graphene-Ag	−1.72	1.43	0.025	**148.78**	7.77×10^5	0.03	**128.53**	1.89	0.82	**153.27**
Graphene-Ag_2	−1.70	2.46	0.03	**23.34**	7.73×10^5	0.13	**60.73**	2.06	0.67	**47.94**
Graphene-Ag_3	−1.06	2.66	0.163	**6.03**	6.09×10^5	0.36	**3.60**	1.51	0.49	**14.18**
Graphene-Ag_4	−1.25	3.36	0.073	**1.24**	6.64×10^5	0.3	**2.49**	1.30	0.93	**21.32**
Graphene-Au	−2.04	1.43	0.020	**34.11**	8.47×10^5	0.31	**40.77**	3.90	0.27	**33.24**
Graphene-Au_2	−2.04	1.42	0.12	**12.16**	8.46×10^5	0.64	**27.05**	3.92	0.27	**15.63**
Graphene-Au_3	−1.51	2.13	0.25	**0.36**	7.30×10^5	0.75	**2.89**	2.25	0.45	**8.83**
Graphene-Au_4	−1.80	2.65	0.061	**1.63**	7.96×10^5	0.66	**1.86**	1.98	0.83	**13.00**
Graphene-Pt	−1.84	1.94	0.33	**0.15**	8.04×10^5	1.2	**1.99**	2.73	0.45	**6.09**
Graphene-Pt_2	−2.03	1.98	0.12	**6.78**	8.45×10^5	0.94	**3.32**	2.81	0.52	**9.67**
Graphene-Pt_3	−1.78	2.39	0.21	**1.27**	7.91×10^5	1.82	**0.37**	2.18	0.67	**3.23**
Graphene-Pt_4	−1.52	2.65	0.33	**0.19**	7.31×10^5	1.27	**0.42**	1.82	0.70	**4.32**
Graphene-Pd	−1.72	1.94	0.25	**1.04**	7.77×10^5	1.2	**1.86**	2.64	0.42	**5.54**
Graphene-Pd_2	−1.53	2.52	0.12	**9.93**	7.35×10^5	1.35	**4.62**	1.92	0.64	**4.01**

			FRA		CISA			RSA		
System	E_F [eV]	R [Å]	z [Å]	g [S/cm]	v_F [m/s]	V [eV]	g [S/cm]	D(R) [eV]	n_e/C-atom	g [S/cm]
Graphene-Pd_3	−1.76	2.27	0.21	**1.27**	7.85×10^5	1.98	**3.75**	2.28	0.59	**2.79**
Graphene-Pd_4	−1.52	2.66	0.145	**3.21**	7.30×10^5	0.61	**1.88**	1.81	0.70	**1.55**

The conductivity is present in bold letter [25].

Table 3. Parameters used to obtain the conductivity in the three approximations (FRA, CISA, and RSA).

We compare our results with some other experiments previously reported. For graphene-Ag system, the experimental conductivity value is of 155 Scm^{-1} [40], whereas our theoretical value is 148.78 Scm^{-1} with FRA, 123.53 Scm^{-1} with CISA, and 153.27 Scm^{-1} with RSA. The best approach for this system is RSA, consistently with a gap between the Dirac point and the

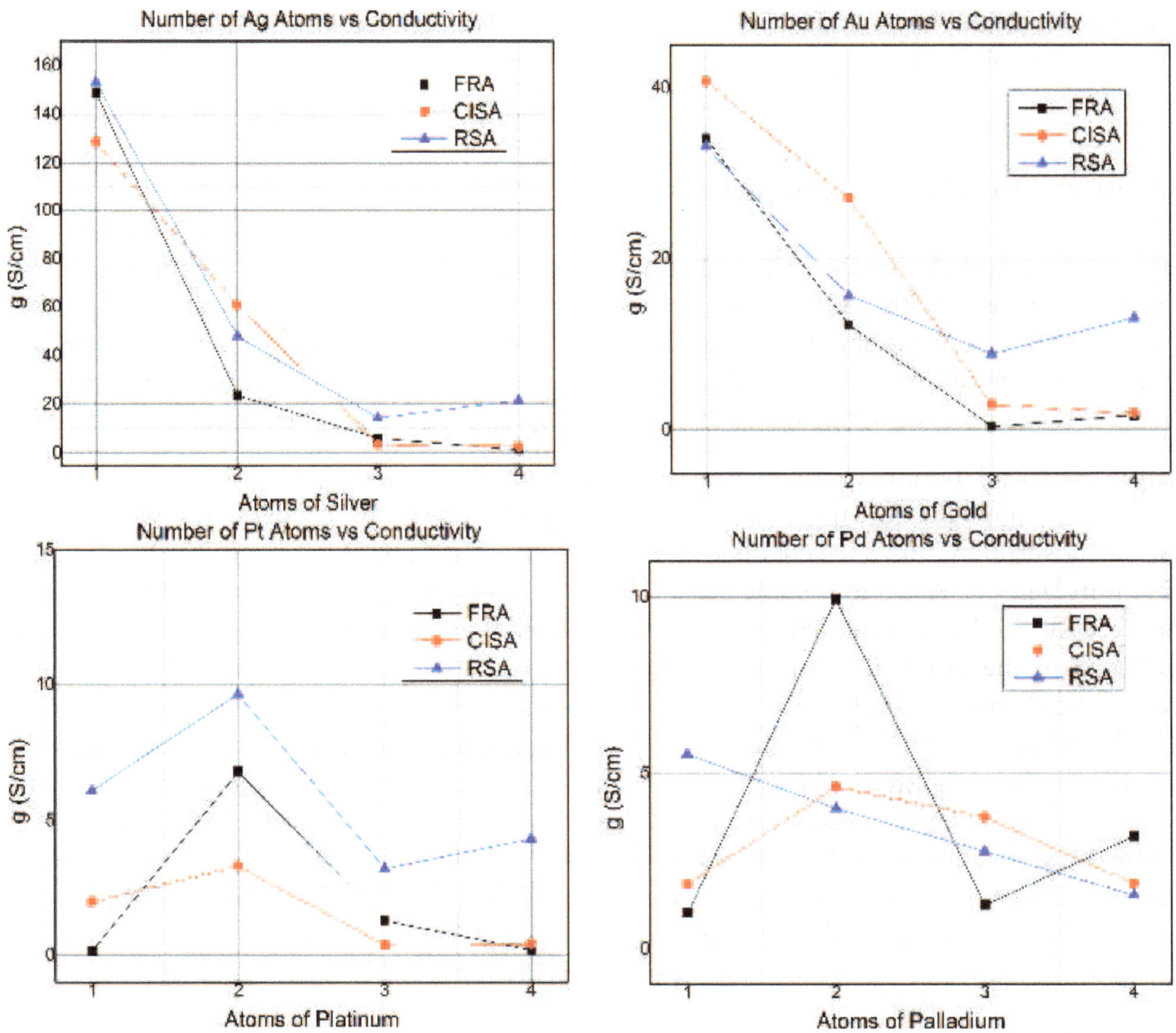

Figure 11. Conductivity of the graphene-Mn (M = Ag, Au, Pt, and Pd; n = 1, 2, 3, and 4) as a function of the number atoms. The conductivity is calculated with the FRA, CISA, and RSA approach [25]© The European Physical Journal.

Fermi Energy of 0.078 eV. In **Figure 11**, it is possible to observe the conductivity value for each system. We present the conductivity in the function of a number of atoms. It is evident that the conductivity decreases as the number increases for graphene-Ag_n and graphene-Au_n systems. Till the graphene-Pt_n systems have an even-odd behavior, i.e., for an even number of metallic atoms in the cluster, the conductivity increases. For the graphene-Pd_n systems, the behavior of the conductivity depends on the approach. It is important to notice that the graphene-Pd_1 system, graphene-Pd_2, and graphene-Pd_3 meet the condition of $|\varepsilon - E_F| = 0$, making the RSA approach more convenient than the others. Also for graphene-Pd_n, the RSA approach is the only approach with a linear curve giving us a little more confidence since it coincides with the behavior of the other metals.

In general terms, the FRA approximation is not exact but allows to estimate the conductivity in orders of magnitude, only with the ground-state geometry. The CISA is a good approach; it is designed for graphene with charged impurities, which would be our case. The main advantage is that it uses few parameters obtained from an optimization of the geometry. But nevertheless, do not consider midgap states. The RSA approach needs some little extra calculations, in addition to geometry optimization, but it seems to be the one that most closely approximates the experimental results.

5. Future trends and conclusions

It has been proposed that a simple method to make a qualitative estimation of the conductivity of graphene surfaces with dopants, leaving the conductivity as a function of the structural deformation of the graphene surface. The proposed method is based on three theoretical approaches (FRA, CISA, and RSA). The main advantage of this method is that the calculation of the conductivity is made from the optimized geometry and three approaches are used to calculate it.

From our calculations of conductivity, it is observed that the phenomenon of adsorption affects much the conductivity. The weak adsorption slightly affects the conductivity, whereas the strong adsorption causes the conductivity to decrease drastically. For weak adsorption, involving pure van der Waals forces, the FRA, CISA, and RSA approximations are quite acceptable since they predict conductivity in the same order of magnitude. However, for the strong adsorption, the situation changes, since these approximations are designed for graphene and when the chemisorption breaks the two-dimensionality of the graphene. The approximations are based on the fact that graphene is a two-dimensional waveguide, where the scattering centers are potential steps. These potential steps are of small value so that most of the electrons pass without being dispersed, and the ballistic conductivity is preserved. As the adsorbate enter the system, the size of dispersing centers increases. This phenomenon is presented in the method when we take the energy of adsorption as the value of potential step. It is recommended to be careful when applying approaches because if the Fermi energy is very close to the Dirac point, the only valid approach is RSA.

With respect to the adsorption of metallic clusters on graphene surfaces, it was observed that there are metallic clusters adsorbed in graphene that induces a band gap in the Fermi energies and conserved the high conductivity (graphene-Ag).

It is observed that the physisorbed systems on the surface of graphene have a lower deformation of the network and they have a high conductivity, as is the case of graphene-Ag_n, whereas the chemisorbed systems have a high deformation of the network and drastically reduced the conductivity, such as the graphene-Pt_n system. It was observed that it is important to consider the midgap states, since if they occur in our systems, we cannot use the simplest approximations to estimate the conductivity and we need to use RSA, which is the approximation that does consider these states (observe the case of graphene-Pd_n). It is also observed that there is no correlation between the type of doping and the conductivity of the material. The partial state density of the systems gives us the orbital of the metallic cluster that interact with the graphene and that directly modify the energy gap, necessary for the creation of transistors.

For future research, we propose conducting the conductivity measurement with more direct methods, such as conductivity measurement with Landauer-Büttiker. And compare it with our calculations presented here. It will also serve to understand, as fully as possible, the behavior of electronic transport. This can be done by looking at the energy bands and the transport coefficient, calculated with Landauer-Büttiker. It is also proposed that these results would be compared with experimental conductivity measurements. To carry out this, it is essential to know the process of production of the graphene samples, to understand what the type of doping would be and then to model the graphene sheet computationally to estimate the conductivity.

In an article made by Wehling et al., it was proposed to expand the RSA approximation into defective Graphene/h-BN hetero-structures [41]. Since in these structures, it is observed that there are states of midgap and band gap modification. We believe that these studies can be extrapolated to any two-dimensional surface in which impurities are joined by Van der Waals forces. This assumption will have to be verified posteriori.

Author details

Roxana M. Del Castillo* and Luis E. Sansores Cuevas

*Address all correspondence to: roxanadelcastillo@ciencias.unam.mx

Materials Research Institute, National Autonomous University of Mexico, Circuito Exterior, Campus Universitario, Mexico City, Mexico

References

[1] Lotz M.R., Boll M., Østerberg F.W., Hansen O., Petersen D.H. Mesoscopic current transport in two-dimensional materials with grain boundaries: Four-point probe resistance and Hall effect. Journal of Applied Physics. 2016;**120**(12), 134303:10.1063. DOI: http://dx.doi.org/10.1063/1.4963719

[2] Wang L., Meric I., Huang P.Y., Gao Q., Gao Y., Tran H., Taniguchi T., Watanabe K., Campos L. M., Muller D.A., Guo J., Kim P., Hone J., Shepard K.L., Dean C.R. One-dimensional electrical

contact to a two-dimensional material. Science. 2013;**342**(6158):614–617. DOI: 10.1126/science.1244358

[3] François Fillion-Gourdeau F., MacLean S. Time-dependent pair creation and the Schwinger mechanism in graphene. Physical Review B. 2015;**92**:035401. DOI: https://doi.org/10.1103/PhysRevB.92.035401

[4] Schneider M., Faria D., Viola Kusminskiy S., Sandler N. Local sublattice symmetry breaking for graphene with a centrosymmetric deformation. Physical Review B. 2015;**91**:161407 (R). DOI: https://doi.org/10.1103/PhysRevB.91.161407

[5] Katsnelson M.I., Geim A.K. Electron scattering on microscopic corrugations in graphene. Philosophical Transactions of the Royal Society A. 2008;**366**(1863). DOI: 10.1098/rsta.2007.2157

[6] Katsnelson M.I., Guinea F., and Geim A.K. Scattering of electrons in graphene by clusters of impurities. Physical Review B. 2009;**79**:195426. DOI: https://doi.org/10.1103/PhysRevB.79.195426

[7] Guinea F. Models of electron transport in single layer graphene. Journal of Low Temperature Physics. 2008;**153**(5):359–373. DOI: 10.1007/s10909-008-9835-1

[8] Wehling T., Yuan S., Lichtenstein A.I., Geim A.K., Katsnelson M.I. Resonant scattering by realistic impurities in graphene. Physical Review Letters. 2010;**105**(5):056802. DOI: https://doi.org/10.1103/PhysRevLett.105.056802

[9] Novoselov K.S., Geim A.K., Morozov S.V., Jiang D., Zhang Y., Dubonos S.V., Grigorieva I. V., Firsov A.A. Electric field effect in atomically thin carbon films. Science. 2004;**306** (5696):609–666. DOI: 10.1126/science.1102896

[10] Partoens B. and Peeters F.M. From graphene to graphite: Electronic structure around the K point. Physical Review B. 2006;**74**:075404. DOI: https://doi.org/10.1103/PhysRevB.74.075404

[11] Katsnelson M.I., Novoselov K.S. Graphene: New bridge between condensed matter physics and quantum electrodynamics. Solid State Communications. 2007;**143**(1–2):3–13. DOI: http://dx.doi.org/10.1016/j.ssc.2007.02.043

[12] Miao F., Wijeratne S., Zhang Y., Coskun U.C., Bao W., Lau C.N. Phase-coherent transport in graphene quantum billiards. Science. 2007;**317**(1530):1530–1535. DOI: DOI: 10.1126/science.1144359

[13] Morozov S.V., Novoselov K.S., Katsnelson M.I., Schedin F., Elias D.C., Jaszczak J.A., and Gei A.K. Giant intrinsic carrier mobilities in graphene and its bilayer. Physical Review Letters. 2008;**100**:016602. DOI: https://doi.org/10.1103/PhysRevLett.100.016602

[14] Barbier M., Vasilopoulos P., Peeters F.M. Single-layer and bilayer graphene superlattices. Philosophical Transactions of the Royal Society A. 2010;**368**:5499–5524. DOI: doi:10.1098/rsta.2010.0218

[15] Kolasiński K., Mreńca-Kolasińska A., and Szafran B. Theory of ballistic quantum transport in the presence of localized defects. Physical Review B. 2016;**94**:115406. DOI: https://doi.org/10.1103/PhysRevB.94.115406

[16] Garcia J.H, Rappoport T.G. Kubo–Bastin approach for the spin Hall conductivity of decorated graphene. 2D Materials. 2016;**3**(2). DOI: http://dx.doi.org/10.1088/2053-1583/3/2/024007

[17] Kirczenow G. Valley currents and nonlocal resistances of graphene nanostructures with broken inversion symmetry from the perspective of scattering theory. Physical Review B. 2015;**95**:125425. DOI: https://doi.org/10.1103/PhysRevB.92.125425

[18] Ando T. Screening effect and impurity scattering in monolayer graphene. Journal of Physical Society of Japan. 2006;**75**:074716. DOI: http://dx.doi.org/10.1143/JPSJ.75.074716

[19] Settnes M., Power S.R., Lin J., Petersen D.H., Jauho A. Patched Green's function techniques for two-dimensional systems: Electronic behavior of bubbles and perforations in graphene. Physical Review B. 2015;**91**:125408. DOI: https://doi.org/10.1103/PhysRevB.91.125408

[20] Jacobsena K.W., Falkenberga J.T., Papiora N., Bøggilda P., Jauhoa A.P., Mads Brandbyge M. All-graphene edge contacts: Electrical resistance of graphene T-junctions. Carbon. 2016;**101**:101–106. DOI: http://dx.doi.org/10.1016/j.carbon.2016.01.084

[21] McCreary K.M., Pi K., Swartz A.G., Han W., Bao W., Lau C.N., Guinea F., Katsnelson M. I., Kawakami R.K. Effect of cluster formation on graphene mobility. Physical Review B. 2010;**81**:115453. DOI: https://doi.org/10.1103/PhysRevB.81.115453

[22] Marsden A.J., Brommer P., Mudd J.J., Dyson M.A., Cook R., Asensio M. Effect of oxygen and nitrogen functionalization on the physical and electronic structure of graphene. Nano Research. 2015;**8**(8):2620–2635. DOI: 10.1007/s12274-015-0768-0

[23] Vanin M., Mortensen J.J., Kelkkanen A.K., Garcia-Lastra J.M., Thygesen K.S., and Jacobsen K.W. Graphene on metals: A van der Waals density functional study. Physical Review B. 2010;**81**:081408(R). DOI: https://doi.org/10.1103/PhysRevB.81.081408

[24] Amft M., Lebègue S., Eriksson O. and Skorodumova N.V. Adsorption of Cu, Ag, and Au atoms on graphene including van der Waals interactions. Journal of Physics: Condensed Matter. 2011;**23**(39). DOI: http://dx.doi.org/10.1088/0953-8984/23/39/395001

[25] Del Castillo R.M., Sansores L.E. Study of the electronic structure of Ag, Au, Pt and Pd clusters adsorption on graphene and their effect on conductivity. European Physical Journal B. 2015;**88**(248). DOI: http://dx.doi.org/10.1140/epjb/e2015-60001-2

[26] Lopez M.J., Cabria I., Alonso J.A. Palladium clusters anchored on graphene vacancies and their effect on the reversible adsorption of hydrogen. Journal of Physical Chemistry C 2014;**118**(10):5081–5090. DOI: 10.1021/jp410262t

[27] Hofmann M., Hsieh Y.P., Chang K.W., Tsai H.G, Chen T.T. Dopant morphology as the factor limiting graphene conductivity. Scientific Reports. 2015;**5**:17393. DOI: 10.1038/srep17393

[28] Liu Z., Zhu M., and Zheng Y. Quantum transport properties of graphene in the presence of randomly distributed. Physical Review B 2015;**92**:245438. DOI: 10.1103/PhysRevB.92.245438

[29] Telychko M., Mutombo P., Merino P., Hapala P., Ondráček M., Bocquet F.C, et al. Electronic and chemical properties of donor, acceptor centers in graphene. ACS Nano. 2015;**9** (9):9180–9187. DOI: 10.1021/acsnano.5b03690

[30] Stauber T., Peres N.M.R., and Guinea F. Electronic transport in graphene: A semiclassical approach including midgap states. Physical Review B. 2007;**76**:205423. DOI: https://doi.org/10.1103/PhysRevB.76.205423

[31] Guinea F., Katsnelson M.I. Many-body renormalization of the minimal conductivity in graphene. Physical Review Letters. 2014;**112**:116604. DOI: https://doi.org/10.1103/PhysRevLett.112.116604

[32] Giannozzi P., Baroni S., Bonini N., Calandra M., Car R., Cavazzoni C., et al. QUANTUM ESPRESSO: A modular and open-source software project for quantum simulations of materials. Journal of Physics: Condensed Matter. 2009;**21**(39). DOI: http://dx.doi.org/10.1088/0953-8984/21/39/395502

[33] Perdew J.P, Burke K., and Ernzerhof M. Generalized gradient approximation made simple. Physical Review Letters. 1996;**77**:3865. DOI: https://doi.org/10.1103/PhysRevLett.77.3865

[34] Methfessel M. and Paxton A.T. High-precision sampling for Brillouin-zone integration in metals. Physical Review B. 1989;**40**(3616)DOI: https://doi.org/10.1103/PhysRevB.40.3616

[35] Rappe A.M., Rabe K.M., Kaxiras E., and Joannopoulos J.D. Optimized pseudopotentials. Physical Review B. 1991;**41**:1227. DOI: https://doi.org/10.1103/PhysRevB.41.1227

[36] Masir, M.R., Leenaerts, O., Partoens, B., and Peeters, F.M. Theory of the Structural, Electronic and Transport Properties of Graphene. In: Houssa M., Dimoulas A., Molle A., editors. 2D Materials for Nanoelectronics. Boca Raton, FL, CRC Press; 2016. pp. 3–36. DOI: 10.1201/b19623

[37] Del Castillo R.M. Ballistic Conductivity in Graphene [thesis]. Mexico: UNAM; 2016. 134 p. Available from: http://132.248.9.195/ptd2016/enero/099083380/Index.html

[38] Katsnelson M.I., Scattering of charge carriers by point defects in bilayer graphene. Physical Review B. 2007;**76**:073411. DOI: https://doi.org/10.1103/PhysRevB.76.073411

[39] Lotfi E., Rezania H., Arghavaninia B., Yarmohammadi M. Impurity effects on electrical conductivity of doped bilayer graphene. Chinese Physics B. 2016;**25**(7):076102. DOI: 10.1088/1674-1056/25/7/076102

[40] Pasricha R., Gupta S., and Srivastava A.S. A facile and novel synthesis of Ag–graphene-based. Small. 2009;**5**(20):2253–2259. DOI: 10.1002/smll.200900726

[41] Sachs B., Wehling T.O., Katsnelson M.I., and Lichtenstein A.I. Midgap states and band gap modification in defective graphene/h-BN heterostructures. Physical Review B. 2016;**94**:224105. DOI: https://doi.org/10.1103/PhysRevB.94.224105

Permissions

All chapters in this book were first published in GMAA, by InTech Open; hereby published with permission under the Creative Commons Attribution License or equivalent. Every chapter published in this book has been scrutinized by our experts. Their significance has been extensively debated. The topics covered herein carry significant findings which will fuel the growth of the discipline. They may even be implemented as practical applications or may be referred to as a beginning point for another development.

The contributors of this book come from diverse backgrounds, making this book a truly international effort. This book will bring forth new frontiers with its revolutionizing research information and detailed analysis of the nascent developments around the world.

We would like to thank all the contributing authors for lending their expertise to make the book truly unique. They have played a crucial role in the development of this book. Without their invaluable contributions this book wouldn't have been possible. They have made vital efforts to compile up to date information on the varied aspects of this subject to make this book a valuable addition to the collection of many professionals and students.

This book was conceptualized with the vision of imparting up-to-date information and advanced data in this field. To ensure the same, a matchless editorial board was set up. Every individual on the board went through rigorous rounds of assessment to prove their worth. After which they invested a large part of their time researching and compiling the most relevant data for our readers.

The editorial board has been involved in producing this book since its inception. They have spent rigorous hours researching and exploring the diverse topics which have resulted in the successful publishing of this book. They have passed on their knowledge of decades through this book. To expedite this challenging task, the publisher supported the team at every step. A small team of assistant editors was also appointed to further simplify the editing procedure and attain best results for the readers.

Apart from the editorial board, the designing team has also invested a significant amount of their time in understanding the subject and creating the most relevant covers. They scrutinized every image to scout for the most suitable representation of the subject and create an appropriate cover for the book.

The publishing team has been an ardent support to the editorial, designing and production team. Their endless efforts to recruit the best for this project, has resulted in the accomplishment of this book. They are a veteran in the field of academics and their pool of knowledge is as vast as their experience in printing. Their expertise and guidance has proved useful at every step. Their uncompromising quality standards have made this book an exceptional effort. Their encouragement from time to time has been an inspiration for everyone.

The publisher and the editorial board hope that this book will prove to be a valuable piece of knowledge for researchers, students, practitioners and scholars across the globe.

List of Contributors

Zhi Jiang, Yang Wang and Cheng Yang
Division of Energy & Environment, Graduate School at Shenzhen, Tsinghua University, Shenzhen City, Guangdong Province, China

Mohammad Razaul Karim
Department of Chemistry, Shahjalal University of Science and Technology, Sylhet, Bangladesh

Shinya Hayami
Department of Chemistry, Graduate School of Science and Technology, Kumamoto University, Kumamoto, Japan

Liu Shuan, Zhao Xia, Jiang Xin, Pu Jibin, Wang Liping
Key Laboratory of Marine Materials and Related Technologies, Zhejiang Key Laboratory of Marine Materials and Protective Technologies, Ningbo Institute of Materials Technologies and Engineering, Chinese Academy of Sciences, Ningbo, China

Zhao Xiaorong
College of Hydraulic and Environmental Engineering, China Three Gorges University, Yichang, China

Zhou Kaihe
State Grid Zhejiang Electric Power Corporation Ningbo Power Supply Company, Ningbo, China

Bongiwe Silwana
SensorLab, Department of Chemistry, University of the Western Cape, Bellville, South Africa
Department of Chemistry, Durham University, Durham, United Kingdom

Charlton van der Horst and Emmanuel Iwuoha
Department of Chemistry, Faculty of Applied Sciences, Cape Peninsula University of Technology, Bellville, South Africa

Vernon Somerset
Department of Chemistry, Durham University, Durham, United Kingdom

Rafael Vargas-Bernal
Department of Materials Engineering, Higher Technological Institute of Irapuato, Irapuato, Guanajuato, Mexico

Golap Kalita and Masaki Tanemura
Department of Applied Physics and Engineering, Nagoya Institute of Technology, Japan

Jian Wang and Xiao Hu
Wuhan National Laboratory for Optoelectronics, School of Optical and Electronic Information, Huazhong University of Science and Technology, Wuhan, Hubei, China

Mehrdad Siahsar
OIC Research Group, University of Tabriz, Tabriz, Iran

Mahboubeh Dolatyari and Ghasem Rostami
SP-EPT Labs, ASEPE Company, Industrial Park of Advanced Technologies, Tabriz, Iran

Ali Rostami
OIC Research Group, University of Tabriz, Tabriz, Iran
SP-EPT Labs, ASEPE Company, Industrial Park of Advanced Technologies, Tabriz, Iran

Minas M. Stylianakis, Dimitrios Konios and Emmanuel Kymakis
Department of Electrical Engineering, Center of Materials Technology and Photonics, School of Applied Technology, University of Applied Sciences of Crete (Technological Educational Institute (TEI) of Crete), Heraklion, Crete, Greece

Konstantinos Petridis
Department of Electrical Engineering, Center of Materials Technology and Photonics, School of Applied Technology, University of Applied Sciences of Crete (Technological Educational Institute (TEI) of Crete), Heraklion, Crete, Greece
Department of Electronic Engineering, University of Applied Sciences of Crete (Technological Educational Institute (TEI) of Crete), Chania, Crete, Greece

Irina V. Antonova and Nadezhda A. Nebogatikova
A. V. Rzhanov Institute of Semiconductor Physics, SB RAS, Novosibirsk State Technical University, Novosibirsk, Russia

Roxana M. Del Castillo and Luis E. Sansores Cuevas
Materials Research Institute, National Autonomous University of Mexico, Circuito Exterior, Campus Universitario, Mexico City, Mexico

Index

P

Q

R

S

T

W

www.ingramcontent.com/pod-product-compliance
Lightning Source LLC
LaVergne TN
LVHW060557210926
848485LV00005B/1325